Metamaterials Modeling and Design

Metamaterials Modeling and Design

edited by

Didier Felbacq
Guy Bouchitté

Pan Stanford Publishing

Published by

Pan Stanford Publishing Pte. Ltd.
Penthouse Level, Suntec Tower 3
8 Temasek Boulevard
Singapore 038988

Email: editorial@panstanford.com
Web: www.panstanford.com

British Library Cataloguing-in-Publication Data
A catalogue record for this book is available from the British Library.

Metamaterials Modeling and Design

ISBN 978-981-4316-12-5 (Hardcover)
ISBN 978-1-315-36500-8 (eBook)

Printed in Canada

Contents

Section II

General methods: waves in periodic media

Preface

The domain of metamaterials now covers many areas of physics: electromagnetics, acoustics, mechanics, thermics, and even seismology. Huge literature is now available on the subject but the results are scattered. Although many ideas and possible applications have been proposed, which of these will emerge as a viable technology will only unfold with time. This book is concerned with electromagnetic waves only and deals essentially with the hard science, mathematical and numerical, behind the often spectacular, but somewhat oversold, possible applications of metamaterials. In a rapidly evolving field, with lots of would-be revolutions, spending too much pages on the zoology of metamaterials would certainly condemn this book to a rapid obsolescence. By contrast, the theoretical and numerical methods presented here are the basis upon which future trends will be built.

The first chapter is a survey of Maxwell's equations and their main properties. After a short historical introduction, potentials and conservation laws are addressed. Then comes a brief presentation of the formulation of Maxwell equations using differential forms. Finally, causality and its consequences are addressed.

Chapter 2 provides the elements of the physics of materials required to bridge semiconductor and metal sciences with electromagnetism, and Chapter 3 is a general reflection upon the notion of averaging and the definition of effective properties.

Chapter 4 is a crash course on basic principles of transformation optics. Simple examples in cylindrical geometry are given using radial transformations that show the unifying power of the concept: mapping an open domain on a bounded domain, perfectly matched layers, invisibility cloaks, and superlenses. Some numerical simulations are presented as an illustration including cloaks of arbitrary shapes and mimetism.

Chapter 5 is concerned with wave propagation in periodic media. The theory of Bloch waves is described in detail. The situation where the medium does not cover the entire space is addressed, because in that situation the boundary of the periodic medium is decorated with evanescent modes. Evanescent waves are then investigated. They are shown to be a complexification (in the mathematical meaning) of the Bloch spectrum.

Chapter 6 tackles the problem of diffraction of electromagnetic waves by a bidimensional grating. A new formulation based on finite element method is proposed. A lot of academic cases and more challenging cases are given for highlighting both the versatility and the powerfulness of the method described in this chapter. The second part of the chapter is devoted to the method of multiple scattering, which is presented for a collection of parallel cylinders.

Chapter 7 is the first chapter devoted to effective properties of metamaterials. Periodic structures are considered and the period of the materials are small with regard to the wavelength of the incoming wave. Besides, the materials are supposed to be of low contrast: this is the framework of soft problems. Closed formulae are given in some academic cases such as small spherical and small circular cylindrical inclusions. A special attention is drawn on spherical inclusions and the mixing formulae (Rayleigh, Maxwell Garnett, Bruggeman) are compared to the two-scale theory.

Chapter 8 addresses the homogenization of highly contrasted objects. The first situation investigated is that of a periodic collection of thin metallic wires. It is shown that the effective medium obtained is dispersive and has a plasmonic resonance. In the second part, the theory is extended to deal with finite-length rods. It is proven that the effective medium becomes spatially dispersive. The chapter closes with numerical investigations of the properties of the effective medium.

The final chapter is also devoted to homogenization theory. It deals with the possibility of homogenizing metamaterials for frequency above the first band and taking into account the Mie resonances. Bidimensional resonant dielectric metamaterials are addressed and the onset of an effective magnetic activity is proven.

Didier Felbacq
Guy Bouchitté
Spring 2017

Section I

Elements of electromagnetic fields in media

Chapter 1

General Introduction

Didier Felbacq,[a] André Nicolet,[b] and Frédéric Zolla[b]

[a]*Laboratory Charles Coulomb UMR CNRS-UM 5221, University of Montpellier, Place Bataillon, 34095 Montpellier Cedex 05, France*
[b]*Institut FRESNEL, University of Aix-Marseille, Avenue Escadrille Normandie Niemen, 13013 Marseille, France*
didier.felbacq@umontpellier.fr

1.1 Maxwell Equations

An amazing characteristic of Maxwell equations is their robustness: Discovered several decades before the rise of Einsteinian relativity and quantum mechanics, they nevertheless remain true in almost all contexts.

They were initially designed by J. C. Maxwell in around 1865 (Maxwell, 1873) to describe the behavior of electromagnetic fields in free space in the classical context, the only known at that time. Maxwell added the current displacement term in the Ampère equation to restore the mathematical coherence of the electromagnetic theory. This term was the ultimate key of a treasure chest: The new system thus appeared as a wave equation, and the predicted celerity of these waves was the speed of light. Therefore, Maxwell unified not only electricity and magnetism the right way but

Metamaterials Modeling and Design
Edited by Didier Felbacq and Guy Bouchitté

ISBN 978-981-4316-12-5 (Hardcover), 978-1-315-36500-8 (eBook)
www.panstanford.com

also optics in one stroke. The experimental proof of electromagnetic waves was performed by H. Hertz in 1886. Unfortunately, Maxwell died several years before, in 1879, the very year in which D. E. Hughes was experimenting radio transmission with sparks but in such a way that it was not considered a conclusive proof for Maxwell equations.

Nevertheless, there was an apparent flaw in the beautiful building: The equations were not compatible with the Galilean relativity principle, prescribing the invariance of physical laws in inertial reference frames, a fundamental pillar of Newtonian physics. But it turned out that the crack was in the pillar! Physicists had started to design new space–time transformations that leave Maxwell equations invariant: the Lorentz transformations, named by Poincaré, who noticed they form a mathematical group (Poincaré, 1905).

In the same year, 1905, Einstein published his theory of special relativity (Einstein, 1905) stating that the physical theories have to be invariant under the set of transformations corresponding to the Poincaré group: space translations + time translations + space rotations + Lorentz transformations that are in fact "space–time" rotations involving both space and time coordinates while preserving the speed of light. They are all compatible with the fundamental postulate that the velocity of light is the same for all observers, and this is the natural framework not only for Maxwell equations but also for entire physics.

Einstein's efforts to find a gravitation theory led him to "general relativity" where the physical theories have to be invariant under the group of diffeomorphisms (roughly speaking arbitrary differentiable coordinate transformations) and where gravitation is associated to space–time intrinsic curvature. Amazingly, in some sense Maxwell equations are still resilient in this curved space.

Even if relativity has been a deep scientific revolution, the most overwhelming change in physics has been the quantum theory. These upheavals occurred at the end of the twentieth century, and Maxwell equations were also a protagonist of the rise of the quantum world, notably via the photoelectric effect (yet again Einstein!). In this case, light seems to have a particle behavior since the energy may be exchanged only by "packets" of value $E = h\nu$ where ν is the frequency of the light and h is the Planck constant.

Is this a farewell to Maxwell equations? Not really since they are still good equations to describe the behavior of the electromagnetic field at the quantum level! The celebrated Schrödinger equation is, in fact, the quantum non-relativistic description of a massive spinless particle. The accurate description of an electron is that it is a massive spin $\frac{1}{2}$ relativistic particle, and the corresponding quantum equation is the Dirac equation. In 1939, Wigner proposed a systematic description of the relativistic quantum equations of all possible particles (massive or massless and with a given spin) based on the representations of the Poincaré group (Wigner, 1939). In this picture, the "light particle", i.e., a photon, is a massless spin 1 particle and the corresponding relativistic quantum equations are Maxwell equations! Even if it means that the electromagnetic field can be considered in some sense as the "wave function" of photons (Bialynicki-Birula, 1996), this must be considered extremely carefully since it cannot be interpreted in terms of the probability of presence of a photon in a region of space, as it is the case with all other quantum relativistic equations. If fact, a full quantum theory of electromagnetic waves requires a "second quantization" where fields are "operator valued distributions"[a] over space–time, but still satisfying Maxwell equations in a generalized meaning. Coupling all these ingredients led Tomonaga, Schwinger, and Feynman to quantum electrodynamics (jointly awarded the Nobel Prize in physics in 1965 for this theory).

Another step could have undermined the range of validity of Maxwell equations: the transition from the microscopic to the macroscopic level. For instance, the movement of individual molecules in a gas can be described satisfactorily using Newton's laws of mechanics, but their collective behavior at the macroscopic level is more efficiently expressed in terms of the laws of thermodynamics. Maxwell equations work well in a void for an electromagnetic field possibly interacting with given charges (the charges create electromagnetic fields, and the electromagnetic fields act on the charges via the Lorentz force). Consider now the interaction of an incident electromagnetic field with matter from a macroscopic point of view: Even if the considered piece of matter is globally neutral, it is made of a huge number of

---

[a] In the sense of L. Schwartz.

charged electrons and protons interacting with the electromagnetic field at the microscopic level. How can we describe the resulting electromagnetic field since it will suffer hectic fluctuations in space and time due, for instance, to the influence of microscopic charges shaken by thermal agitation? Fortunately, macroscopic equations can be found for an averaged (in space and time) electromagnetic field that takes into account only the large-scale behavior of the field. This smoothed electromagnetic field is named macroscopic electromagnetic field and it is described by Maxwell equations! The reason is that the averaging process is a low-pass filtering performed by convoluting the field with a smooth bounded support function, the mollifier, and this operation commutes with the partial derivations with respect to space coordinates and time. Therefore, the equations for the macroscopic electromagnetic field retain their initial form. Of course, something has to change to take into account the interaction with matter: The free space constitutive relations involving the permittivity ε_0 and the permeability μ_0 are replaced by *ad hoc* constitutive relations involving for instance homogeneous linear isotropic permittivity, permeability, and conductivity or possibly much more unwieldy models. The generic term for this process is homogenization.

Consider now a piece of matter made of two media (or more) in such a way that it is a regular repetition of the same cell (e.g., a cube of the first medium with a spherical inclusion of the second medium) looking locally just like a periodic structure. Can we find another homogenization process at this level to smooth out the fluctuation at the cell level and preserve again the form of the Maxwell equations? The positive answer is just the topic of the present book! Before entering this topic, let us recall a few important facts about Maxwell equations in vacuum.

Given the densities of charges $\varrho(\mathbf{x}, t)$ and current $\mathscr{J}(\mathbf{x}, t)$, Maxwell equations read as:

$$\mathbf{curl}\,\mathscr{E} = -\partial_t \mathscr{B} \tag{1.1}$$

$$\operatorname{div}\mathscr{B} = 0 \tag{1.2}$$

$$\operatorname{div}\mathscr{E} = \frac{\varrho}{\varepsilon_0} \tag{1.3}$$

$$\mathbf{curl}\,\mathscr{B} = \mu_0 \mathscr{J} + \frac{1}{c^2}\partial_t \mathscr{E} \tag{1.4}$$

Together with the Lorentz force

$$\mathbf{F} = q(\mathscr{E} + v \times \mathscr{B}),$$

they give a complete theory of classical electrodynamics.

The sources of the field ϱ and $\mathscr{J}$ cannot be defined by Maxwell equations alone, although they imply the continuity relation:

$$\text{div}\,\mathscr{J} + \partial_t \varrho = 0. \tag{1.5}$$

Charges are complicated quantum objects and a model has to be made: A theory of charged particles is needed to know what the densities of charges and current are. Classically, a charged particle (with charge q) is described as a punctual object associated with a density of charge $\varrho = q\delta(\mathbf{x} - \mathbf{x}(t))$ and a density of current $\mathscr{J} = q\mathbf{v}\delta(\mathbf{x} - \mathbf{x}(t))$. These definitions are compatible with the continuity equation, understood in the distribution meaning. For a continuous repartition of charges, the density of charges can be described by a measure (in the mathematical meaning) of the form: $\rho = f(\mathbf{x}, t)\mu(dV)$, where $\mu(dV)$ is a measure that can be supported by a curve (linear density), a surface (surface density), or a volume (volume density). Because the density of charges only gives the infinitesimal quantity of charges at a point r and a time t, a full description requires that the velocity field of the charges be given, in the form: $\mathbf{V} = \mathbf{v}(\mathbf{x}, t)\mu(dV)$, in such a way that the density of current is given by: $\mathbf{j} = f(\mathbf{x}, t)\mathbf{v}(\mathbf{x}, t)\mu(dV)$.

For an incompressible electron fluid, i.e., div $\mathbf{V} = 0$, the continuity equation, which states the conservation of charges, then reads as:

$$\frac{D\rho}{dt} = 0,$$

where $\dfrac{D}{dt} = \dfrac{\partial}{\partial t} + \mathbf{V} \cdot \nabla$

1.1.1 Potential and Gauge Invariance

Both the magnetic and electric fields can be written in terms of a vector field, the "vector potential" $\mathscr{A}$, and a scalar field, the "scalar potential" ϕ. This is due to the fact that $\mathscr{B}$ is divergence free (or "a closed form" in the language of differential forms explained below). This implies the existence of a vector field A such that:

$$\mathscr{B}(r, t) = \textbf{curl}\,\mathscr{A}(r, t), \;\; \mathscr{E} = -\frac{\partial \mathscr{A}}{\partial t} - \textbf{grad}\,\phi.$$

These fields are not uniquely defined. Adding the gradient of a scalar field to $\mathscr{A}$ does not change its **curl**, and adding a constant to ϕ does not change its **grad**. This results in the following so-called gauge invariance relations: the electromagnetic fields $\mathscr{E}$ and $\mathscr{B}$ remain unchanged, when $(\mathscr{A}, \phi)$ are replaced with $(\mathscr{A}', \phi')$ defined by:

$$\mathscr{A}' = \mathscr{A} + \mathbf{grad}\, f,\ \phi' = \phi - \frac{\partial f}{\partial t}.$$

Using the potentials, Maxwell equations now read as:

$$-\Delta\mathscr{A} + \mathbf{grad}\,\mathrm{div}\,\mathscr{A} = \mu_0\mathscr{J} + \frac{1}{c^2}\left(-\partial_{t^2}^2\mathscr{A} - \mathbf{grad}\,\partial_t\phi\right),$$

$$-\partial_t\,\mathrm{div}\,\mathscr{A} - \Delta\phi = \rho/\varepsilon_0.$$

The non-uniqueness of the potential fields can be used to simplify the equations that they satisfy. This is done by imposing a relation between $\mathscr{A}$ and ϕ, which is called "fixing the gauge." The two most commonly used gauges are the Coulomb gauge and the Lorentz gauge.

- The Coulomb gauge
  One imposes that $\mathrm{div}\,\mathscr{A} = 0$. It is possible because if $\mathscr{A}$ does not satisfy this relation, then, for a smooth f, another vector potential $\mathscr{A}'$ satisfies: $\mathrm{div}\,\mathscr{A}' = \mathrm{div}\,\mathscr{A} + \Delta f$. Then, choosing f in such a way that: $\Delta f = -\,\mathrm{div}\,\mathscr{A}$[b] ensures the condition $\mathrm{div}\,\mathscr{A}' = 0$. With that gauge choice, one has:

  $$-\Delta\mathscr{A} = \mu_0\mathscr{J} + \frac{1}{c^2}\left(-\partial_{t^2}^2\mathscr{A} - \mathbf{grad}\,\partial_t\phi\right),$$

  and the scalar potential satisfies the electrostatic Poisson equation:

  $$-\Delta\phi = \rho/\varepsilon_0.$$

- The Lorentz gauge
  One imposes that $\mathrm{div}\,\mathscr{A} + \frac{1}{c^2}\partial_t\phi = 0$.
  In that case, both potential fields satisfy the wave equation, with their respective sources:

  $$-\Delta\mathscr{A} + \frac{1}{c^2}\partial_{t^2}^2\mathscr{A} = \mu_0\mathscr{J},$$

  $$-\Delta\phi + \frac{1}{c^2}\partial_{t^2}^2\phi = \rho/\varepsilon_0.$$

[b] This is possible for any reasonable $\mathscr{A}$.

1.2 Maxwell Equations in the Fourier Domain

As far as sources are given, Maxwell equations are best discussed by performing a Fourier transform on them. The definition of the Fourier transform is given by:

$$\hat{f}(\mathbf{k}) = \int f(x)e^{i\mathbf{k}\cdot x}dx^N, \quad f(x) = (2\pi)^{-N}\int \hat{f}(\mathbf{k})e^{-i\mathbf{k}\cdot x}d\mathbf{k}^N.$$

After Fourier transforming the space variables, Maxwell equations read as:

$$\mathbf{k}\times\mathcal{E} = -\partial_t\mathcal{B}, \tag{1.6}$$

$$\mathbf{k}\cdot\mathcal{E} = R/\mathbf{e}_0, \tag{1.7}$$

$$\mathbf{k}\cdot\mathcal{B} = 0, \tag{1.8}$$

$$\mathbf{k}\times\mathcal{B} = \mu_0\mathcal{J} - \partial_t\mathcal{E}. \tag{1.9}$$

The third equation implies that $\mathcal{B}$ is perpendicular to $\mathbf{k}$. This suggests to decompose the fields along $\mathbf{k}$ and along a plane perpendicular to $\mathbf{k}$. The decomposition of the fields in components parallel and orthogonal to the vector $\mathbf{k}$ is a Fourier space version of Helmholtz (Hodge) decomposition theorem of vector fields in a sum of a gradient (i.e., a curl-free vector field) and a curl (i.e., a divergence-free vector field). The projection operators over the parallel and orthogonal directions are particularly simple in Fourier space. Let us denote $\kappa = \mathbf{k}/|\mathbf{k}|$ the unit vector in the direction of $\mathbf{k}$. Given a vector field in Fourier space, $\mathbf{F}(\mathbf{k})$, we can define the transverse and orthogonal parts of the field $\mathbf{F}$ by:

$$\mathbf{F}_{||} = (\kappa\cdot\mathbf{F})\ \mathbf{k}, \ \mathbf{F}_{\perp} = \mathbf{F} - \mathbf{F}_{||}.$$

For the electromagnetic fields, we immediately have $\mathcal{B} = \mathcal{B}_{\perp}$ and we obtain the following equations:

$$\mathbf{k}\times\mathcal{E}_{\perp} = -\partial_t\mathcal{B}_{\perp}, \tag{1.10}$$

$$\mathbf{k}\times\mathcal{B}_{\perp} = \mu_0\mathcal{J}_{\perp} - \partial_t\mathcal{E}_{\perp}, \tag{1.11}$$

$$\mathbf{k}\cdot\mathcal{E}_{||} = R/\mathbf{e}_0, \tag{1.12}$$

$$\partial_t\mathcal{E}_{||} = \mu_0\mathcal{J}_{||}. \tag{1.13}$$

1.3 Field Created by Sources

In this section, we investigate the field created by charged particles, described by a current density j and a density of charges ρ. In the Lorentz gauge, the vector potential satisfies the equation:

$$\Delta \mathscr{A} - \frac{1}{c^2}\partial_t^2 \mathscr{A} = -\mu_0 \mathscr{J},$$

which is solved uniquely by using the fundamental solution G (i.e., a Green's function) for the d'Alembertian operator. It satisfies:

$$\frac{1}{c^2}\frac{\partial^2 G}{\partial t^2} - \Delta G = \delta(\mathbf{x}, t),$$

and initial conditions. It is equal to:

$$G(\mathbf{x}, t) = \frac{1}{4\pi c t} H(t)\delta\left(t - \frac{|\mathbf{x}|}{c}\right),$$

where $H(t)$ is the Heaviside function. The expressions for the potential fields are then obtained straightforwardly:

$$\mathscr{A}(\mathbf{x}, t) = \frac{1}{4\pi\varepsilon_0}\int d^3\mathbf{x}' \frac{\mathscr{J}(\mathbf{x}', t - \frac{|\mathbf{x}-\mathbf{x}'|}{c})}{|\mathbf{x}-\mathbf{x}'|}.$$

1.4 Conservation Laws

In this section, we introduce the fundamental quantities that are invariants of the field and matter system. Consider a field created by a set of particles with charges and current given by:

$$\rho(\mathbf{x}, t) = \sum_n q_n \delta(\mathbf{x} - \mathbf{x}_n(t)),$$

$$\mathscr{J} = \sum_n q_n \mathbf{v}_n \delta(\mathbf{x} - \mathbf{x}_n(t)).$$

From Maxwell equations, the Poynting's identity can be derived as:

$$\mathrm{div}(\mathscr{E} \times \mathscr{H}) + \mathscr{E} \cdot \mathscr{J} + \partial_t w = 0,$$

where $w = \frac{1}{2}\varepsilon_0^2 \mathscr{E}^2 + \frac{1}{2}\mu_0 \mathscr{H}^2$ is the density of electromagnetic energy. In order to interpret this relation in terms of physical quantities, let us comment on each term:

- $\mathscr{E} \cdot \mathscr{J}$ represents the coupling of field to matter.
- $\mathscr{P} = \mathscr{E} \times \mathscr{H}$ is a current. Indeed, by integrating the Poynting's identity over the volume V containing the charges, we obtain:

$$\iint \mathscr{P} + \int \mathscr{E} \cdot \mathscr{J} = -\partial_t \int w.$$

The meaning of the identity is thus that the rate of change of the electromagnetic energy in a region is equal to the energy flux through the boundary of the region plus the work done by the field forces.

The expression for the work done, i.e., the transfer of energy to matter, $\mathscr{E} \cdot \mathscr{J}$, can be written by means of the mechanical energy and the associated flux:

$$\frac{\partial u_m}{\partial t} + \operatorname{div} \boldsymbol{S}_m = \mathscr{E} \cdot \mathscr{J},$$

where:

$$u_m = \sum_n \frac{m_n}{2} \dot{\mathbf{x}}_n^2 \delta(\mathbf{x} - \mathbf{x}_n(t)),\ \boldsymbol{S}_m = \sum_n \frac{m_n}{2} \dot{\mathbf{x}}_n \, \dot{\mathbf{x}}_n^2 \delta(\mathbf{x} - \mathbf{x}_n(t)).$$

From Poynting's identity, we obtain the energy conservation law:

$$\frac{\partial (u_m + w)}{\partial t} + \operatorname{div}(\boldsymbol{S}_m + \mathscr{P}) = 0.$$

The Hamiltonian of the system of field and particles is given by:

$$H = \sum_n \frac{1}{2} m \mathbf{v}_n^2 + \frac{1}{2} \int d^3x [\varepsilon_0 \mathscr{E}^2 + \mu_0 \mathscr{H}^2]. \tag{1.14}$$

The Poynting vector is the momentum of the field; its angular momentum is given by:

$$\mathbf{G} = \int \mathbf{x} \times (\mathscr{E} \times \mathscr{H}) d^3x = \int \mathbf{x} \times \mathscr{P} \, d^3x.$$

The total angular momentum of the system of field and particles is:

$$\mathcal{P} = \mathbf{G} + \sum m_n \mathbf{x} \times \mathbf{v}_n. \tag{1.15}$$

Both the Hamiltonian (1.14) and the total angular momentum (1.15) are constant of the motion:

$$\frac{dH}{dt} = 0,\ \frac{d\mathcal{P}}{dt} = 0. \tag{1.16}$$

This is proved by a direct calculation:

$$\frac{dH}{dt} = \sum_n m_n \mathbf{v}_n \cdot \frac{d\mathbf{v}_n}{dt} + \int d^3x \left[\mathbf{e}_0 \mathscr{E} \cdot \frac{\partial \mathscr{E}}{\partial t} + \mu_0 \mathscr{H} \cdot \frac{\partial \mathscr{H}}{\partial t} \right].$$

Let $\mathscr{A}$ be the vector potential, with the radiation gauge $\text{div}(\mathscr{A}) = 0$. Then ones has:

$$\mathbf{G} = \int \mathbf{x} \times \mathscr{P}\, d^3x = \int \sum_k \mathscr{E}_k (\mathbf{x} \times \nabla \mathscr{A}_k) + \int \mathscr{E} \wedge \mathscr{A}\, d^3x.$$

This shows that the total angular momentum of the electromagnetic field is the sum of two terms: the first one involves the spatial variable, but the second one only involves $\mathscr{E}$ and $\mathscr{A}$. This term corresponds to the purely intrinsic part of the angular momentum: It is the (classical) spin of the electromagnetic field.

1.5 A Framework with Differential Forms

The description of fields in terms of functions of space–time coordinates with vector values is probably not the best description, from the point of view of neither mathematics nor physics. Indeed, it is not clear to understand why the magnetic field should depend on the orientation of space, why the circulation of the electric fields is an energy, and so on. Besides, the invariance of the equations with respect to a change in space coordinates is far from obvious without speaking of the Lorentz invariance. For all these reasons (and many others), a much better framework is that of differential forms. The reader being not acquainted with differential forms should consult (do Carmo, 1994) for a very clear and not technical introduction; a very good reference as well is (Bossavit, 1993). As it was said earlier, the circulation of the electric field along a curve produces a scalar, or infinitesimally: $dW = q\mathscr{E} \cdot d\mathbf{x}$. This shows that the electric field acts on vectors to produce a scalar: This is the definition of 1-form. We shall write: $\mathscr{E} = \sum_k E_k dx^k$. Next, we differentiate $\mathscr{E}$ with the exterior derivative $d\mathscr{E} = -\partial_t \mathscr{B}$, which shows that $\mathscr{B}$ is 2-form: $\mathscr{B} = \sum_{n'm} B_{nm} dx^n \wedge dx^m$. Consider now the density of charge, ρ; for a volume repartition, the integral of ρ over a volume is the total charge, a scalar: $Q = \int \rho$. This shows that ρ is necessarily a volume form, or 3-form: $\rho = \sigma(\mathbf{x}) dx^1 \wedge dx^2 \wedge dx^3$. Considering now the density of current, electromagnetism indicates that the flux of $\mathscr{J}$ is the relevant quantity, as it gives the variation of charge through a surface. In order to provide a scalar when integrated over a surface, $\mathscr{J}$ should be 2-form: $\mathscr{J} = \sum_{n\leq m} j_{nm} dx^n \wedge dx^m$. Now the density of charge results from the exterior derivative of a 2-form, showing that $\mathscr{D} = \sum_{n'm} D_{nm} dx^n \wedge dx^m$ and that $\mathscr{H}$ is a 1-form: $\mathscr{H} = \sum_k H_k dx^k$. At that point, it appears that the connection between $\mathscr{E}$ and $\mathscr{D}$, $\mathscr{H}$ and $\mathscr{B}$ transforms 1-form to 2-form. This is the so-called Hodge operator, denoted by a star: $\varepsilon_0 * \mathscr{E} = \mathscr{D}$, $\mu_0 * \mathscr{H} = \mathscr{B}$. This operator allows to transform a 2-form, i.e., a completely antisymmetric tensor of rank 2, to a 1-form, to which a vector that depends on the orientation of space can be associated. The properties of the physical space enter through a metric, i.e., a way of measuring distances. In coordinates, we have (ϵ_{lnm} is the Levi–Civita symbol):

$$(*\mathbf{F})_{nm} = \sqrt{|\mathrm{det} g|}\epsilon_{lnm} g^{lk} F_k,$$

where F is a 1-form. This relation will prove to be a key in transformation optics. The appendix gives more details about forms and the Hodge star operator. We can now write Maxwell equations using forms:

$$d\mathscr{E} = -\partial_t \mathscr{B},\ d\mathscr{D} = \rho,\ d\mathscr{B} = 0,\ d\mathscr{H} = \mathscr{J} + \partial_t D. \tag{1.17}$$

Let us remark that the equations are written in a way independent of any particular choice of coordinates.

1.6 Dispersion Relations

1.6.1 Introduction

In physics, dispersion most often refers to materials that bring about frequency-dependent effects in wave propagation. To a large extent, all materials are dispersive, except in vacuum. This frequency dependence gives rise to numerous physical phenomena such as light refraction through a prism, rainbows or the spreading of wave packets in optical fibers. As regards to dispersion relations, they specify the link between the angular frequency and some characteristics of materials trough the so-called constitutive relations. In classical electromagnetism, these constitutive relations are encoded in the so-called permittivity ε and permeability μ, and link the mesoscopic fields between them (see Chapter 3). For not too large energies, these relations are linear and may be expressed as a convolution operator. The use of a Fourier transform in time converts this rather intricate operation into a simple operation of multiplication by a function of ω. It turns out that this function is complex valued and that its imaginary part is related to the phenomenon of leakage: Dispersion and leakage are thus the two sides of the same coin. Once the relative permittivity and permeability are obtained, the norm k of the vector $\mathbf{k}$, which is the associated variable to the position vector $\mathbf{x}$, may be expressed as a function of ω: $k = k(\omega)$. Finally, from this last relation, the different notions of velocities such as phase, group, and wave-packet velocities may be derived.

It remains to give explicitly the functions $\varepsilon(\omega)$ and $\mu(\omega)$ for a given material, which is far from being a simple task. For doing so, a quantum treatment is necessary (see Chapter 2), which is beyond the scope of this paragraph. As a consequence, the electromagnetic quantities have to be derived in a phenomenological way. By this, we mean that the mesoscopic electromagnetic quantities (see Chapter 3) are not derived from—possibly statistical—quantum physics but are derived from *ad hoc* quantities obtained from the experiments and other general concepts such as causality and some reasonable assumptions. To be more precise, the relationships given in this paragraph rely on a theorem, Titchmarsh's theorem (Titchmarsh, 1948, 1958), the principle of causality, and the fact that any material is transparent at very high frequency.

We now have to give some relations between, on the one hand, the mesoscopic quantities derived from $\mathscr{H}$ and $\mathscr{D}$ defined in the Chapter 3, and the mean values of the microscopic fields $< \mathscr{E} >$ and $< \mathscr{B} >$, on the other hand. We first begin with the simplest model: the model of perfect media. In that case, we assume that the polarization field is given by:

$$\mathscr{P} = \varepsilon_0 \chi_e < \mathscr{E} > , \qquad (1.18)$$

where χ_e is a real number (usually positive in the optical wavelength range) called electric susceptibility. We thus have:

$$\mathscr{D} = \varepsilon_0(1 + \chi_e) < \mathscr{E} > := \varepsilon_0 \varepsilon_r < \mathscr{E} > . \qquad (1.19)$$

Analogously, we assume that:

$$\mathscr{M} = \frac{\chi_m}{1+\chi_m} < \mathscr{B} > , \tag{1.20}$$

where χ_m is a real number (positive or negative) called magnetic susceptibility. We thus have:

$$\mathscr{B} = \mu_0(1+\chi_m) < \mathscr{B} > := \mu_0\mu_r < \mathscr{B} > , \tag{1.21}$$

where μ_r and ε_r are functions that can depend on the space variable $\mathbf{r}$, respectively called relative permeability and relative permittivity (note that they have no physical dimensions). From now on, in order to lighten the notation, we will denote the mesoscopic electromagnetic field $(< \mathscr{E} >, < \mathscr{H} >)$ by $(\mathscr{E}, \mathscr{H})$.

1.6.2 Causality and Kramers–Kronig Relations

It is clear that the constitutive relations have been obtained roughly: They cannot convey physical properties as important as dispersion and absorption as mentioned before. We make less restrictive assumptions concerning $\mathscr{P}$. We state the existence of an operator L such that:

$$\mathscr{P}(\mathbf{r}, t) = L(\mathscr{E}(\mathbf{r}, t)) . \tag{1.22}$$

where L is a linear and local operator. In that case, it can be proved that L is an operator of convolution. In other words, we can find a distribution S such that:

$$\mathscr{P} = S \star \mathscr{E} . \tag{1.23}$$

Moreover, we assume that S is a regular distribution associated with a sufficiently smooth function χ such that:

$$\mathscr{P} = \varepsilon_0 \chi \star \mathscr{E} . \tag{1.24}$$

For a homogeneous medium,[c] we have:

$$\hat{\mathscr{P}}(\mathbf{r}, \omega) = \varepsilon_0 \hat{\chi}(\omega) \hat{\mathscr{E}}(\mathbf{r}, \omega) . \tag{1.25}$$

Consequently, we can find the electric constitutive relation:

$$\hat{\mathscr{D}}(\mathbf{r}, \omega) = \varepsilon_0 \varepsilon_r(\omega) \hat{\mathscr{E}}(\mathbf{r}, \omega), \tag{1.26}$$

[c] For inhomogeneous media, it is enough to replace $\hat{\chi}(\omega)$ by $\hat{\chi}(\mathbf{r}, \omega)$.

with $\varepsilon_r(\omega) = 1 + \hat{\chi}_e(\omega)$. This last relation has important consequences. Causality implies that the function χ has a positive support. Indeed, the polarization vector can be written *as per*:

$$\mathscr{P}(\mathbf{x}, t) = \varepsilon_0 \left(\int_{-\infty}^{0} \chi(t - t')\mathscr{E}(\mathbf{x}, t')\, dt' + \int_{0}^{+\infty} \chi(t - t')\mathscr{E}(\mathbf{x}, t')\, dt' \right). \tag{1.27}$$

The former integral involves the electric field at times posterior to t ($t - t'$, with $t' < 0$), which violates the principle of causality unless χ vanishes for negative t. Besides, we assume that χ is of finite energy (i.e. χ is in $L^2(\mathbb{R})$). It turns out that Titchmarsh's theorem connects causality and finiteness of energy to the analytical properties of its Fourier transform and to a powerful property involving the Hilbert transform. Here, Titchmarsh's theorem is recalled (Titchmarsh, 1948), th. 95:

Theorem 1.1 (Titchmarsh). *Let f be a function in $L^2(\mathbb{R})$. The following three statements are equivalent:*

(1) The function f is causal (i.e., $f(t) = 0$ if $t \leq 0$).
(2) The Fourier transform of f, $\hat{f}(\omega) = \mathcal{F}(f) = \int_{\mathbb{R}} f(t)e^{i\omega t}\, dt$, is the limit as $y \to 0^+$ of a holomorphic function in the upper half complex plane[d] $\mathbb{C}^+$.

$$\hat{f}(\omega') = \lim_{\omega'' \to 0^+} \hat{f}(\omega' + i\omega'')\, a.e.\, ,$$

such that $\int_{\mathbb{R}} \left|\hat{f}(\omega' + i\omega'')\right|^2 d\omega' < K$, for $\omega'' > 0$.
(3) $f = i\mathcal{H}(f)$, where $\mathcal{H}$ denotes the Hilbert transform:

$$\mathcal{H}(f)(\omega) = \frac{1}{\pi} \fint_{\mathbb{R}} \frac{f(\omega')}{\omega' - \omega}\, d\omega'. \tag{1.28}$$

This last relation gives an unexpected link between the real and the imaginary parts of the Fourier transform of a finite energy causal function denoted by $\hat{\chi}'$ and $\hat{\chi}''$ in the sequel. By making use of the linearity and reality of the Hilbert transform, we have:

$$\hat{\chi}' + i\hat{\chi}'' = i\left(\mathcal{H}(\hat{\chi}') + i\mathcal{H}(\hat{\chi}'')\right), \tag{1.29}$$

[d] This condition means that the possible poles are in the lower half complex plane, $\mathbb{C}^-$. It is worth noting that this dissymmetry between the upper and the lower complex plane is due to the arbitrariness of the definition of the Fourier transform. In certain domains of physics, the choice of the direct Fourier transform is $\hat{f}(\omega) = \mathcal{F}(f) = \int_{\mathbb{R}} f(t)e^{-i\omega t}\, dt$. In that case, the condition on holomorphy would be inverted.

which leads to:

$$\hat{\chi}' = -\mathcal{H}\left(\hat{\chi}''\right) \quad \text{and} \quad \hat{\chi}'' = \mathcal{H}\left(\hat{\chi}'\right) . \tag{1.30}$$

Moreover, if the function is real valued, which is, of course, the case when dealing with the function χ, its Fourier transform for the negative and positive frequencies are related by Hermitian symmetry:

$$\hat{\chi}(-\omega) = \chi^*(\omega), \tag{1.31}$$

i.e.,

$$\hat{\chi}'(-\omega) + i\hat{\chi}''(-\omega) = \hat{\chi}'(\omega) - i\hat{\chi}''(\omega) . \tag{1.32}$$

By equating the real and the imaginary parts throughout this equality, it appears that the real part, $\hat{\chi}'(\omega)$, is an even function, whereas the imaginary part $\hat{\chi}''(\omega)$ is odd. The aforementioned integrals may be thus expressed over positive frequencies in the following manner:

$$\hat{\chi}''(\omega) = -\frac{1}{\pi} \fint_{\mathbb{R}} \frac{\hat{\chi}'(\omega')(\omega + \omega')}{\omega'^2 - \omega^2} d\omega' \tag{1.33}$$

$$= -\frac{\omega}{\pi} \fint_{\mathbb{R}} \frac{\hat{\chi}'(\omega')}{\omega'^2 - \omega^2} d\omega' + \frac{1}{\pi} \fint_{\mathbb{R}} \frac{\omega'\hat{\chi}'(\omega')}{\omega'^2 - \omega^2} d\omega' . \tag{1.34}$$

Using the parity of the two integrands, we obtain:

$$\hat{\chi}''(\omega) = -\frac{2\omega}{\pi} \fint_{\mathbb{R}^+} \frac{\hat{\chi}'(\omega')}{\omega'^2 - \omega^2} d\omega', \tag{1.35}$$

and similarly

$$\hat{\chi}'(\omega) = \frac{2}{\pi} \fint_{\mathbb{R}^+} \frac{\omega'\hat{\chi}''(\omega')}{\omega'^2 - \omega^2} d\omega' . \tag{1.36}$$

The formulae 1.35–1.36 are the so-called Kramers–Kronig relations (de L. Kronig, 1926; Hu, 1989; Kramers, 1927; van Kampen, 1961). As a consequence, giving the real part of the permittivity ε_r' in the whole frequency range allows to derive the imaginary part ε_r'' and *vice versa* :

$$\varepsilon_r''(\omega) = -\frac{2\omega}{\pi} \fint_{\mathbb{R}^+} \frac{(\varepsilon_r'(\omega') - 1)}{\omega'^2 - \omega^2} d\omega', \tag{1.37}$$

and similarly

$$\varepsilon_r'(\omega) - 1 = \frac{2}{\pi} \fint_{\mathbb{R}^+} \frac{\omega'\varepsilon_r''(\omega')}{\omega'^2 - \omega^2} d\omega'. \tag{1.38}$$

Now we would like to apply these relations in a simple situation, namely, the Lorentz model. In this framework, it is said that an electron submitted to an exterior electric field $\mathscr{E}$ bound to a nucleus has a motion described by the position vector $\boldsymbol{x}$ given by the equation:

$$m\ddot{\boldsymbol{x}} + s\dot{\boldsymbol{x}} + k\boldsymbol{x} = e\mathscr{E}. \tag{1.39}$$

Besides, the polarization vector $\mathscr{P}$ is given by:

$$\mathscr{P} = Ne\boldsymbol{x}. \tag{1.40}$$

Seeking the polarization vector $\mathscr{P}$ amounts to deriving x_i from E_i thanks to the equation:

$$\ddot{x}_i + \frac{2}{\tau}\dot{x}_i + \omega_0^2 x_i = \frac{e}{m}E_i, \tag{1.41}$$

where τ and ω_0 are, respectively, characteristic time and frequency. The uniqueness of the solution needs an extra condition, namely, causality. In other words: "If $E_i = 0$ for $t < t_0$, so shall x_i". To guarantee this hypothesis, it is enough to consider the Green's function g associated with the equation 1.41, solution of:

$$\ddot{g} + \frac{2}{\tau}\dot{g} + \omega_0^2 g = \delta, \tag{1.42}$$

where g is a *positive support function*. This equation is equivalent to the following conditions:

$$\begin{cases} g(t) = 0 & \text{if } t < 0 \\ \ddot{g} + \frac{2}{\tau}\dot{g} + \omega_0^2 g = 0 & \text{if } t > 0 \\ g(0) = 0 & \\ \dot{g}(0) = 1 & \end{cases} \tag{1.43}$$

For subcritical motions ($\omega_0 > \frac{1}{\tau}$), the general solution is the sum of two exponentials:

$$g(t) = a_- e^{-t/\tau} e^{i\Omega_0 t} + a_+ e^{-t/\tau} e^{-i\Omega_0 t}, \tag{1.44}$$

with $\Omega_0 := \sqrt{\omega_0^2 - \frac{1}{\tau^2}}$, or equivalently:

$$g(t) = a_+ e^{-i\omega_c^+ t} + a_- e^{-i\omega_c^- t}, \tag{1.45}$$

where ω_c^+ and ω_c^- are complex characteristic frequencies in the complex ω−plane: $\omega_c^+ := \Omega_0 - i\frac{1}{\tau}$ and $\omega_c^- := -\Omega_0 - i\frac{1}{\tau}$, both located in $\mathbb{C}^-$. The conditions $g(0) = 0$ and $\dot{g}(0) = 1$ lead to:

$$a_+ = -a_- = \frac{1}{-2i\Omega_0}. \tag{1.46}$$

The Green's function is, therefore,

$$g(t) = \frac{H(t)}{-2i\Omega_0}\left(e^{-i\omega_c^+ t} - e^{-i\omega_c^- t}\right) \tag{1.47}$$

$$= \frac{H(t)}{\Omega_0}\sin(\Omega_0 t)e^{-t/\tau}\,, \tag{1.48}$$

which is compatible with condition (1) of Titchmarsh's theorem. We are now in the position to give the expression of the polarization vector in time domain:

$$\mathscr{P} = \frac{Ne^2}{m} g \star \mathscr{E}\,. \tag{1.49}$$

In frequency domain, the expression is simpler as already noted:

$$\hat{\mathscr{P}} = \frac{Ne^2}{m}\hat{g}(\omega)\hat{\mathscr{E}}\,. \tag{1.50}$$

For obtaining the permittivity in that case, we use the following property of the Fourier transform: $\widehat{\dot{x}} = -i\omega\hat{x}$. Applying this property to the equation 1.41 yields:

$$\hat{\boldsymbol{x}} = \frac{e}{m}\frac{1}{-\omega^2 - 2i\omega/\tau + \omega_0^2}\hat{\mathscr{E}}, \tag{1.51}$$

whereupon,

$$\hat{\chi}_e = -\frac{Ne^2}{m\varepsilon_0}\frac{1}{\omega^2 - 2i\omega/\tau + \omega_0^2}\hat{\mathscr{E}}\,. \tag{1.52}$$

This last expression may be remastered by emphasizing the two poles at stake, ω_c^+ and ω_c^-:

$$\hat{\chi}_e(\omega) = \frac{-\omega_p^2}{(\omega - \omega_c^+)(\omega - \omega_c^-)}\,, \tag{1.53}$$

where ω_p is a characteristic frequency: $\omega_p := \sqrt{\frac{N}{m\varepsilon_0}}e$. Note that the function $\hat{\chi}_e$ verifies condition (2) of Titchmarsh's theorem. We then derive the expression of the relative permittivity:

$$\varepsilon_r(\omega) = 1 + \frac{\omega_p^2}{(\omega_0^2 - \omega^2) - 2i\omega/\tau}\,. \tag{1.54}$$

The real and the imaginary parts are, therefore:

$$\varepsilon_r'(\omega) = 1 + \frac{\omega_p^2(\omega_0^2 - \omega^2)}{(\omega_0^2 - \omega^2)^2 + 4\omega^2/\tau^2}, \tag{1.55}$$

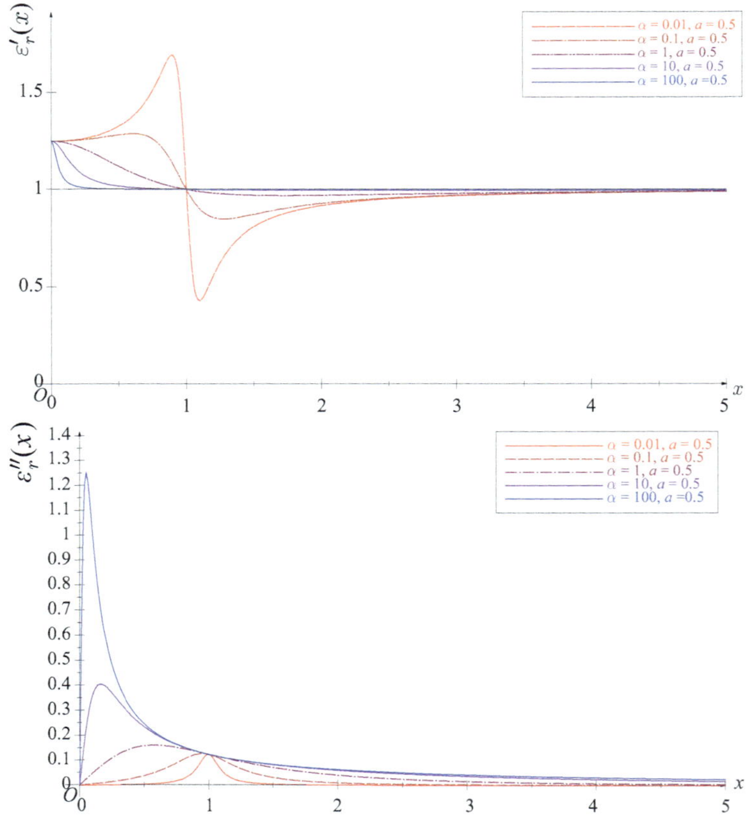

Figure 1.1 Graphs of the functions $\varepsilon_r'(x) = 1 + \dfrac{a^2(1-x^2)}{(1-x^2)^2 + 4\alpha x^2}$ and $\varepsilon_r''(x) = \dfrac{2a^2\alpha x}{(1-x^2)^2 + 4\alpha x^2}$ versus a normalized frequency x. The functions are derived from the Lorentz model. The graphs for different values of α are represented and $a = 0.5$.

and

$$\varepsilon_r''(\omega) = \frac{2\omega_p^2\,\omega/\tau}{(\omega_0^2 - \omega^2)^2 + 4\omega^2/\tau^2}\,. \tag{1.56}$$

Now, in contrast to the real part, it can be checked that the imaginary part is positive whatever the parameters ω_p, ω_0, and τ and whatever the frequency ω: This property corresponds to a passive material[e].

[e] Once again, the sign of the imaginary part depends on the very definition of the direct Fourier transform. If the other Fourier transform were chosen, ε_r'' would be negative.

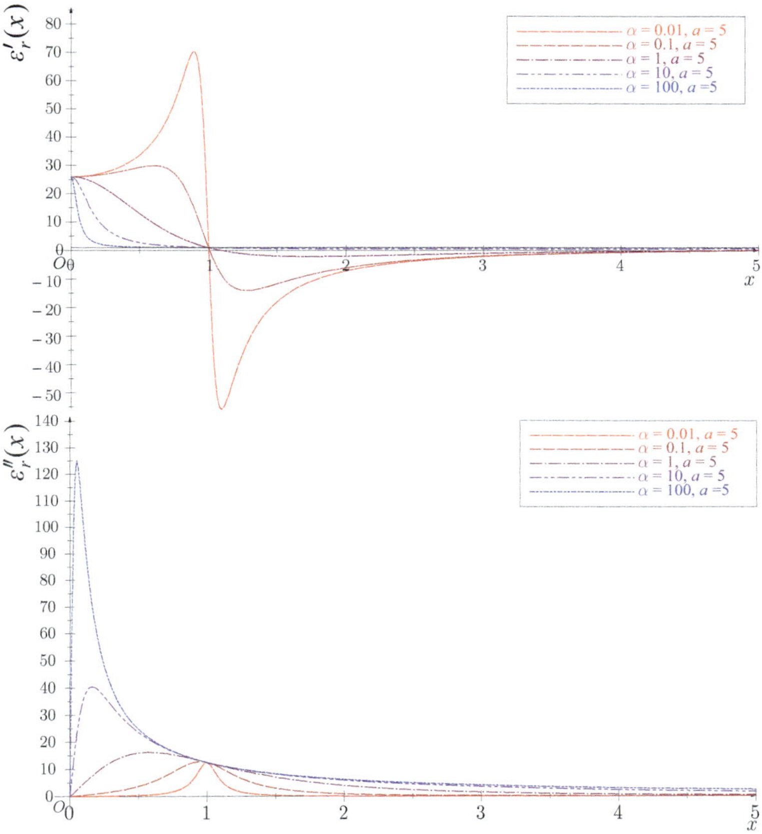

Figure 1.2 Graphs of the functions $\varepsilon_r'(x) = 1 + \dfrac{a^2(1-x^2)}{(1-x^2)^2 + 4\alpha x^2}$ and $\varepsilon_r''(x) = \dfrac{2a^2\alpha x}{(1-x^2)^2 + 4\alpha x^2}$ versus a normalized frequency x. The functions are derived from the Lorentz model. The graphs for different values of α are represented and $a = 5$.

1.6.3 Super-Convergence and Sum Rules

We have just seen that, as a consequence of causality applied to the electric polarization vector, the complex permittivity must satisfy the Kamers–Kronig relations. What happens for the derived quantities such as the refractive index? The expression (1.54) shows that $\varepsilon_r(\omega' + i\omega'') - 1$ is in $L^1(\mathbb{R})$ for $\omega'' > 0$. Therefore, $n(\omega) - 1$ fulfills the second criterion of Titchmarsh's theorem. As a consequence, the function $n(\omega) - 1$ is the Fourier transform of a function of finite energy in time domain and with a positive support. Second, the function $n - 1$ is such that $n - 1 = i\mathcal{H}(n - 1)$, which leads to the relations:

$$n''(\omega) = -\frac{2\omega}{\pi} \fint_{\mathbb{R}^+} \frac{n'(\omega') - 1}{\omega'^2 - \omega^2} \, d\omega' \tag{1.57}$$

and

$$n'(\omega) - 1 = \frac{2}{\pi} \fint_{\mathbb{R}^+} \frac{\omega' n''(\omega')}{\omega'^2 - \omega^2} \, d\omega' \, . \tag{1.58}$$

We have just seen that from causality and some rather weak hypotheses on the behavior at large frequencies on the permittivity, we have deduced the Kramers–Kronig relations for both the permittivity and the optical index. But there are further exciting discoveries to come. It is indeed possible to derive the zero-degree moment of the real part of the permittivity and the first moment of the imaginary part. These results can be obtained by taking benefit of the super-convergence theorem (Altarelli et al., 1972).

Theorem 1.2 (Super-convergence). *Let f be a continuously differentiable function with the following asymptotic behavior at infinity: $f = O\left[(\omega \ln \omega)^{-1}\right]$ and a function g defined as follows:*

$$g(\omega) = \fint_{\mathbb{R}^+} \frac{f(\omega')}{\omega^2 - \omega'^2} \, d\omega' \, , \tag{1.59}$$

then

$$\int_{\mathbb{R}^+} f(\omega) \, d\omega = \lim_{\omega' \to +\infty} \left(\omega'^2 \, g(\omega')\right) \, . \tag{1.60}$$

From 1.38, we have, indeed:

$$\frac{\pi}{2\omega} \varepsilon_r''(\omega) = \fint_{\mathbb{R}^+} \frac{(\varepsilon_r'(\omega') - 1)}{\omega^2 - \omega'^2} \, d\omega'. \tag{1.61}$$

Besides, if we assume that $\varepsilon_r = O\left[(\omega \ln \omega)^{-1}\right]$, the super-convergence theorem states:

$$\int_{\mathbb{R}^+} \varepsilon_r'(\omega') - 1 \, d\omega' = \lim_{\omega' \to +\infty} \left(\frac{\pi}{2\omega'} \omega'^2 \varepsilon_r''(\omega')\right) = 0 \, . \tag{1.62}$$

And analogously, from 1.37:

$$\frac{\pi}{2} \left(1 - \varepsilon_r'(\omega)\right) = \fint_{\mathbb{R}^+} \frac{\omega' \, \varepsilon_r''(\omega')}{\omega^2 - \omega'^2} \, d\omega', \tag{1.63}$$

leads to

$$\int_{\mathbb{R}^+} \omega' \varepsilon_r''(\omega')\, d\omega' = \lim_{\omega' \to +\infty} \left(\frac{\pi}{2} \omega'^2 (1 - \varepsilon_r'(\omega')) \right) . \tag{1.64}$$

Now if we consider that the permittivity behaves as the function $1 - \frac{\omega_p^2}{\omega^2}$ for large frequencies, which is the case for the Lorentz model, the limit of the latter limit is simply ω_p^2, which relates the first moment of the real part of the relative permittivity to a characteristic of the permittivity for large frequencies, ω_p:

$$\int_{\mathbb{R}^+} \omega' \varepsilon_r''(\omega')\, d\omega' = \omega_p^2 . \tag{1.65}$$

The two last formulae involving the moments of the real and the imaginary parts of the relative permittivity are the so-called f−sum rules. What we have done for the permittivity can be redone for the optical index *mutatis mutandis* since the starting point of the f−sum rules is nothing but the Kramers–Kronig relations. We have thus:

$$\int_{\mathbb{R}^+} n'(\omega') - 1\, d\omega' = 0 , \tag{1.66}$$

and

$$\int_{\mathbb{R}^+} \omega' n''(\omega')\, d\omega' = \omega'^2_p . \tag{1.67}$$

But

$$n(\omega) - 1 = \sqrt{1 + \chi(\omega)} - 1 \simeq \frac{\chi(\omega)}{2} \tag{1.68}$$

gives $\omega'^2_p = \frac{\omega_p^2}{2}$, so that

$$\int_{\mathbb{R}^+} \omega' n''(\omega')\, d\omega' = \frac{\omega_p^2}{2} . \tag{1.69}$$

1.6.4 Dispersion Relations Versus Mixing Laws

The homogenization techniques and mixing laws being the purpose of this book, we have to verify that these laws are consistent with causality and consequently exhibit relations of the Kramers–Kronig type. This task is even more useful in that we will see later on that seeking mixing laws is not always easy. Indeed, except in very peculiar cases, mixing laws cannot be given as closed formulae but must be numerically derived: The resulting numerical methods (often called annex problems) will be the subject of the Chapter 7. The dispersion relations may then be used to verify the mixing laws in the frequency domain. However, the reader should be aware of some subtleties arising from the two-step approach widely used in the following chapters. First, the media obtained after two-step homogenization may be dispersive, while the media making up of them are not. Second, when dealing with high-contrast media (see, for instance, the bed-of-nails structure studied in Chapter 8), the media at stake after homogenization are no longer local in space. In that case, the relations of the Kramers–Kronig type have to be modified.

1.6.5 Group Velocity

Consider a dispersive medium characterized by a permittivity tensor $\varepsilon(r, k)$. The constitutive relation between the displacement field and the electric field is given by:

$$\mathscr{D}(\mathbf{x}) = \int \varepsilon(\omega, \mathbf{k})\mathscr{E}(\mathbf{k})e^{i\mathbf{k}\cdot\mathbf{x}}d^3k.$$

Maxwell equations admit solutions in the form of monochromatic plane waves in the form:

$$\mathscr{E}_0\, e^{i(\mathbf{k}\cdot\mathbf{x}-\omega t)},\ \mathscr{H}_0\, e^{i(\mathbf{k}\cdot\mathbf{x}-\omega t)}.$$

The averaged Poynting vector is denoted by $\mathcal{P} = \frac{1}{2}\Re(\mathscr{E}_0 \times \overline{\mathscr{H}}_0)$. The following result holds:

$$w\nabla_k\omega = \mathcal{P} - \frac{1}{4}\omega|\mathscr{E}_0|^2\nabla_k\varepsilon, \tag{1.70}$$

where: $w = \frac{1}{4}\left(\mu_0|\mathscr{H}_0|^2 + \frac{\partial(\omega\varepsilon)}{\partial\omega}|\mathscr{E}_0|^2\right)$.

The rest of this section is devoted to a proof of this relation.

Putting the expression for the plane waves into Maxwell equations, we get:

$$\mathbf{k} \times \mathscr{E}_0 = \omega\mu_0\mathscr{H}_0,\ \mathbf{k} \times \mathscr{H}_0 = -\omega\varepsilon(\omega, \mathbf{k})\mathscr{E}_0.$$

We now make an infinitesimal variation on the wave vector $d\mathbf{k}$, inducing a small perturbation on the field $\mathscr{E} + d\mathscr{E},\ \mathscr{H} + d\mathscr{H}$ and on the frequency $d\omega$ and permittivity $d\varepsilon$:

$$(\mathbf{k} + d\mathbf{k}) \times (\mathscr{E}_0 + d\mathscr{E}_0) = (\omega + d\omega)\mu_0(\mathscr{H}_0 + d\mathscr{H}_0),$$
$$(\mathbf{k} + d\mathbf{k}) \times (\mathscr{H}_0 + d\mathscr{H}_0) = -(\omega + d\omega)(\varepsilon + d\varepsilon)(\mathscr{E}_0 + d\mathscr{E}_0).$$

Expanding to first order, we obtain:

$$d\mathbf{k} \times \mathscr{E}_0 + \mathbf{k} \times d\mathscr{E}_0 = \omega\mu_0 d\mathscr{H}_0 + d\omega\mu_0\mathscr{H}_0,$$
$$d\mathbf{k} \times \mathscr{H}_0 + \mathbf{k} \times d\mathscr{H}_0 = -\omega d\varepsilon\mathscr{E}_0 - d\omega\varepsilon\mathscr{E}_0 - \omega\varepsilon d\mathscr{E}_0.$$

The next step is to make the Poynting vector $\mathcal{P}$ appear. We denote $\mathcal{P}_c = \mathscr{E}_0 \times \overline{\mathscr{H}}_0$.

$$\overline{\mathscr{H}}_0 \cdot (d\mathbf{k} \times \mathscr{E}_0) + \overline{\mathscr{H}}_0 \cdot (\mathbf{k} \times d\mathscr{E}_0) = \omega\mu_0\overline{\mathscr{H}}_0 \cdot d\mathscr{H}_0 + d\omega\mu_0|\mathscr{H}_0|^2,$$
$$\overline{\mathscr{E}}_0 \cdot (d\mathbf{k} \times \mathscr{H}) + \overline{\mathscr{E}}_0 \cdot (\mathbf{k} \times d\mathscr{H}_0) = -\omega d\varepsilon|\mathscr{E}_0|^2 - d\omega\varepsilon|\mathscr{E}_0|^2 -\omega\varepsilon\overline{\mathscr{E}}_0 \cdot d\mathscr{E}_0.$$

$$d\mathbf{k} \cdot (\mathscr{E}_0 \times \overline{\mathscr{H}}_0) + d\mathscr{E}_0 \cdot (\overline{\mathscr{H}}_0 \times \mathbf{k}) = \omega\mu_0\overline{\mathscr{H}}_0 \cdot d\mathscr{H}_0 + d\omega\mu_0|\mathscr{H}_0|^2,$$
$$d\mathbf{k} \cdot (\mathscr{H}_0 \times \overline{\mathscr{E}}_0) + d\mathscr{H}_0 \cdot (\overline{\mathscr{E}}_0 \times \mathbf{k}) = -\omega d\varepsilon|\mathscr{E}_0|^2 - d\omega\varepsilon|\mathscr{E}_0|^2 -\omega\varepsilon\overline{\mathscr{E}}_0 \cdot d\mathscr{E}_0.$$

On inserting the Maxwell system, we obtain:

$$2d\mathbf{k} \cdot \mathcal{P}_c + \omega\varepsilon d\mathscr{E}_0 \cdot \overline{\mathscr{E}}_0 = \omega\mu_0\overline{\mathscr{H}}_0 \cdot d\mathscr{H}_0 + d\omega\mu_0|\mathscr{H}_0|^2,$$
$$-2d\mathbf{k} \cdot \overline{\mathcal{P}}_c - \omega\mu_0 d\mathscr{H}_0 \cdot \overline{\mathscr{H}}_0 = -\omega d\varepsilon|\mathscr{E}_0|^2 - d\omega\varepsilon|\mathscr{E}_0|^2 - \omega\varepsilon\overline{\mathscr{E}}_0 \cdot d\mathscr{E}_0.$$

Finally, we subtract the relations to obtain:

$$4d\mathbf{k} \cdot \mathcal{P} = d\omega\mu_0|\mathscr{H}_0|^2 + (\omega d\varepsilon + d\omega\varepsilon)|\mathscr{E}_0|^2,$$

$$4d\mathbf{k} \cdot \mathcal{P} = d\omega\mu_0|\mathscr{H}_0|^2 + \left(\omega\nabla_k\varepsilon \cdot dk + \omega\frac{\partial\varepsilon}{\partial\omega} + d\omega\varepsilon\right)|\mathscr{E}_0|^2,$$

$$d\mathbf{k} \cdot (4\mathcal{P} - \omega|\mathscr{E}_0|^2\nabla_k\varepsilon) = d\omega\left(\mu_0|\mathscr{H}_0|^2 + \left(\omega\frac{\partial\varepsilon}{\partial\omega} + \varepsilon\right)|\mathscr{E}_0|^2\right),$$

and *in fine* the claimed result.

References

Altarelli, M., Dexter, D. L. and Nussenzveig, H. M. (1972). Superconvergence and sum rules for the optical constants, *Phys. Rev. B* **6**, pp. 1290–1298.

Bialynicki-Birula, I. (1996). Photon wave function, *Prog. Opt.* **36**, pp. 245–294.

Bossavit, A. (1993). *Électromagnétisme, en vue de la modélisation, Mathématiques et Applications 14* (Springer-Verlag, Paris).

de L. Kronig, R. (1926). On the theory of dispersion of x-rays, *J. Opt. Soc. Am.* **12**, pp. 547–557.

do Carmo, M. P. (1994). *Differential Forms and Applications*, Universitext (Springer-Verlag, Berlin Heidelberg).

Einstein, A. (1905). Zur elektrodynamik der bewegten körper, *Annalen der Physik* **17**, pp. 891–921.

Hu, B. Y.-K. (1989). Kramers–kronig in two lines, *Am. J. Phys.* **57**, 9, p. 821, doi:10.1119/1.15901, http://link.aip.org/link/?AJP/57/821/1.

Kramers, H. A. (1927). La diffusion de la lumiere par les atomes, *Atti. Congr. Int. Fis. Como* **2**, p. 545.

Maxwell, J. C. (1873). *A Treatise on Electricity and Magnetism*, Vol. 1 (Clarendon Press).

Poincaré, H. (1905). Sur la dynamique de l'électron, *Comptes Rendus Acad. Sci. Paris* **140**, pp. 1504–1508.

Titchmarsh, E. C. (1948). *Introduction to the Theory of Fourier Integrals*, 2nd edn. (Oxford University Press, London and New York).

Titchmarsh, E. C. (1958). *Theory of Functions*, 2nd edn. (Oxford University Press, London and New York).

van Kampen, N. G. (1961). Causalité et relations de Kramers-Kronig, *J. Phys. Radium* **22**, pp. 179–191.

Wigner, E. (1939). On unitary representations of the inhomogeneous lorentz group, *Ann. Math.* **40**, 1, pp. 149–204.

Chapter 2

A Review of Natural Materials and Properties in Micro-Waves and Optics

Bernard Gil
Laboratory Charles Coulomb UMR CNRS-UM 5221, University of Montpellier, Place Bataillon, 34095 Montpellier Cedex 05, France
bernard.gil@univ-montp2.fr

2.1 Introduction

Most of the chapters following the present one, if not all, are dedicated to the resolution of Maxwell equations in specific circumstances. Beyond the scope of the description of the challenges that are faced, in each of the cases treated hereafter in this book, and beyond the pedagogical and technical descriptions of the mathematical approach required to solve them intelligibly in each case, we felt concerned with the extension of this science to situations when quantum mechanics rules the spectral behaviors of both the dielectric and magnetic susceptibilities.

The point of this chapter is, thus, to provide the elements required to bridge semiconductor and metal sciences with electro-magnetism.

Metamaterials Modeling and Design
Edited by Didier Felbacq and Guy Bouchitté

ISBN 978-981-4316-12-5 (Hardcover), 978-1-315-36500-8 (eBook)
www.panstanford.com

For semiconductors as well as metals, susceptibilities are complex quantities, exhibit a spectral dependence, and exhibit resonances centered at energies specific to the chemical element(s) assembled to form the crystal. Such energies are associated, as we will see later, to specific transitions at a given k-vector in the Brillouin zone of the crystal. The great density of electrons free to propagate in metals ($10^{29}m^{-3}$) gives rise to a bulk collective elementary excitation named plasmon, which oscillates at an energy close to 10 eV (Kittel, 1996), which is in principle larger than the resonance energies that will interest us here. Researchers in electromagnetic science are used to express dispersion relations in terms of frequency (Hz) or wavelength (nm), while solid state physicists prefer electron-volts even though this quantity does not belong to the International System of Units. As we will use this measurement unit, we provide its relationship with nanometers in Fig. 2.1. In semiconductors, while increasing the density of electrons free to propagate by the appropriate incorporation of controlled amounts of foreign atoms into pure crystals (including a group V element, for instance As, into silicon), a contamination procedure called doping, it is possible to tune the bulk plasma oscillation energy from nought to several electron-volts. We will discuss bulk plasmons in metals and semiconductors.

To give an overview of the origins of these phenomena, we will describe some band structures of various materials, detail how the spectral dependence of the dielectric constant of a material can be interpreted in terms of inter-band transitions between critical points of the band structure, introduce the concept of excitons, and conclude by giving some words regarding heavy doping of

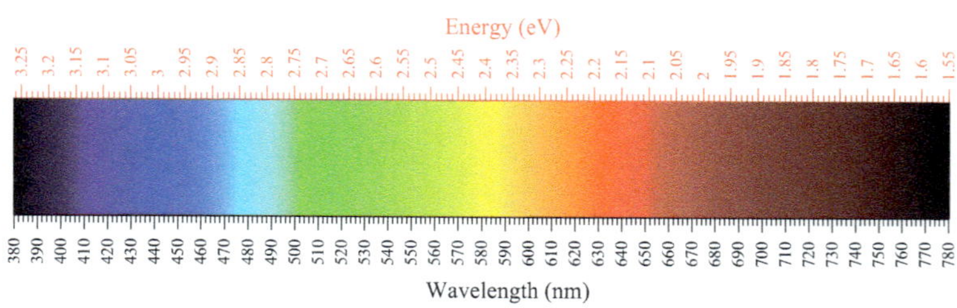

Figure 2.1 Correspondence between electron-volts and nanometers.

semiconductors and plasmons. Our philosophy has been to locate in between the strict theoretical description of the phenomena and an oversimplified one.

2.2 Metals and Non-Metals

In terms of conductivity, gold and diamond are highly different. Gold is a metal, while diamond is an insulator (or a semiconductor). Diamond is transparent, while gold is not. Diamond is a good conductor of heat, like gold. Gold is soft and can be easily molded; diamond is hard and fragile. The understanding of the puzzling physical properties of such materials, which had been known since the dawn of humanity, was achieved in the 1930s, with the simultaneous booming of quantum mechanics, atomic physics, and solid state theory (Fowler, 1993). Answers to most of the important differences between the conductivities of metals and insulators (a metal conducts at low temperature, whereas an insulator does not) were found with the development of the quantum theory of solids. To do so, a new concept of band structure, discovered by Bloch (1928) in the case of periodic and perfect crystals, was introduced with a fairly great success. This concept led to the discovery of an energy-dependent, even, and multi-dependent function of a wave number k. The birth of the concept of Brillouin zones (Brillouin, 1930a,b,c) was of paramount importance to project quantities derived from band structure effects in a sub-volume of the reciprocal space defined from group theory arguments and capable of gathering all relevant information about a periodic crystal. The concepts of Fermi wavevector k_F, Fermi energy E_F, and Fermi surfaces have revolutionized the civilization for at least a century (Ziman, 1960). The Fermi energy (resp. wave vector) at zero K is the maximum total energy (resp. wave vector) an electron can have. The spectral distribution of electron states is described by the Fermi distribution, which at zero K is the Heaviside function: The probability of occupancy of a given state is identity for energies below E_F and nought otherwise. The Fermi surface is the surface of constant energy E_F in the Brillouin zones of the crystal. It separates the occupied orbitals from non-occupied ones at zero K and is invariant under the symmetry operations of the point group of the crystal (Seitz, 1960). It is, in addition, invariant under inversion respective to the center of the Brillouin zone. It can be fully located inside the first Brillouin zone, or may spread outside it. The symmetry of the Fermi surface is broken under the application of an external electric field. Finite conductivity is then made possible at zero K when the Fermi surface falls into a partly occupied band so that the conduction electrons of the Fermi distribution, having a given wave vector, are energetically adjacent to states with different wave vectors and are available to move into. The conductivity is then ruled by the difference of populations of the energy bands in the neighborhood of the Fermi surface and by its distortion upon the application of a finite bias. When temperature increases, the Heaviside distribution is replaced by the temperature-dependent Fermi distribution. This leads to complementary distortions of the Fermi surface and modification of the conductivity with temperature. In contrast, insulating behaviors are observed when the Fermi surface coincides with the topmost of an electron band, and when no adjacent energy states are available for the conduction electrons.

The unusual conduction properties of semiconductors can also be explained by assuming that their outer band is filled, as with insulators. Therefore, they do not exhibit conductivity at zero K in the absence of any perturbation complementary to the external bias. However, their band structure is such that one or more unfilled electronic levels are close enough to the next allowed band and that electron excitation by heat from the filled band to the unfilled band can be promoted relatively easily so that conductivity appears when temperature increases. This is the qualitative interpretation of the conductivity of ideal materials during the first part of 1930s (Hoddeson et al., 1987). At that time, more quantitative interpretation faced severe difficulties in the accurate calculation of the band structure of metals, semiconductors, and insulators, and in the departure of real crystals from perfection, leading to mysterious and parasitic phenomena.

It is worthwhile to note here that, since that time, the number of solids having specific behaviors increased with the discovery of semi-metals, superconductors, topological insulators, and with the realization of complex ensembles of these differently conducting materials. We have to remind that metals have been, for a long time, very naively and wrongly distinguished from insulators in terms or their opacity to visible light. This is not true as we shall see later: It depends on the frequency of the light used to probe the material's transparency. This is the very complex issue of establishing the theory of dielectric constant through the whole energy range.

2.3 Examples of Band Structures for Monovalent Elemental Metals

Metals can furnish a large variety of band structures in line with the Bravais lattice they form and depending of the electronic structure of the element. To illustrate this, we select the simple case of monovalent atoms in the case of body-centered and face-centered cubic lattices.

Figure 2.2 represents the plot, in the first Brillouin zone, of the band structure of a monovalent crystal sodium that crystallizes in the body-centered cubic lattice (Ching and Callaway, 1975). In this textbook, the single-valence electron is in a 3s state. The

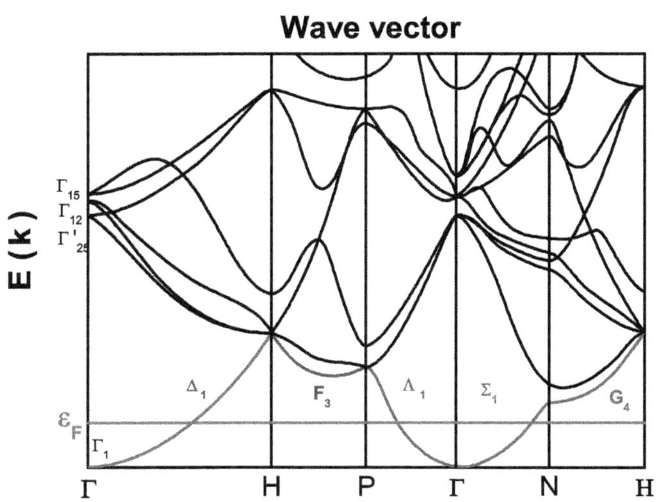

Figure 2.2 The band structure of sodium, calculated by the LCAO method.

fundamental band starts at an energy of about −0.37 Rydberg at the zone center. Then it parabolically increases when k increases in the K-N or K-P or K-H directions ($\Gamma = (0, 0, 0)$, $H = 2\pi/a(0, 0, 1)$, $N = \pi/a(1, 1, 0)$, $P = \pi/a(1/2, 1/2, 1/2)$), as it is expected for an energy-isolated s state in the absence of interactions with second atomic neighbors. The exact dispersion relation is expressed as follows versus the lattice parameter a and components k_{is} of the wave number $\cos(k_x a/2)\cos(k_y a/2)\cos(k_z a/2)$.

This low-energy band (Σ_1, Δ_1, and Λ_1 lines) is highly dispersive in the reciprocal space in the neighborhood of its minimum at the zone center (Γ_1 level). In other words, the effective mass:

$$m_{\text{eff}} = \hbar^2 \left[\frac{\partial^2 E(k)}{\partial k^2}\right]^{-1}$$

is very light. Higher in energy, this energy band interacts with contributions of the higher energy states, for instance at the zone center and the even (as well as empty) 3d states of sodium (Γ_{12} and Γ'_{25}), which gives the complicated dispersion relations at positive energies. These states being empty, the Fermi energy is computed at 3.23 eV in the first valence band and is plotted in the figure to indicate the metallic behavior of this material. For a monovalent metal, in the case of the body-centered cubic lattice, the Fermi wave number $k_F = 1.24\pi/a$, where a is the cube of the lattice constant, which indicates that the Fermi surface is totally included in the first Brillouin zone.[a] We have reproduced in Fig. 2.3 the Brillouin zone of the body-centered cubic lattice (Solyom, 2010), and the Fermi surface is plotted in blue. The semi-classical theory of the dielectric constant can explain how the lack of energy gap in the band structure of metals indicates that these materials are, at first order, opaque to visible light (thanks to the high density of mobile electrons, the real part of the dielectric constant is negative at these energies) but transparent in the ultraviolet.

Figure 2.4 plots the band structure of copper, a noble monovalent metal. The Brillouin zone states at zone center and zone boundaries

---

[a]Values of the Fermi energies are 4.72 eV for lithium (a = 0.349 nm); 3.23 eV for sodium (a = 0.423 nm); 2.12 eV for potassium (a = 0.523 nm), 1.85 eV for rubidium (a = 0.559 nm); 1.58 eV for Cesium (a = 0.605 nm) (Kittel, 1996).

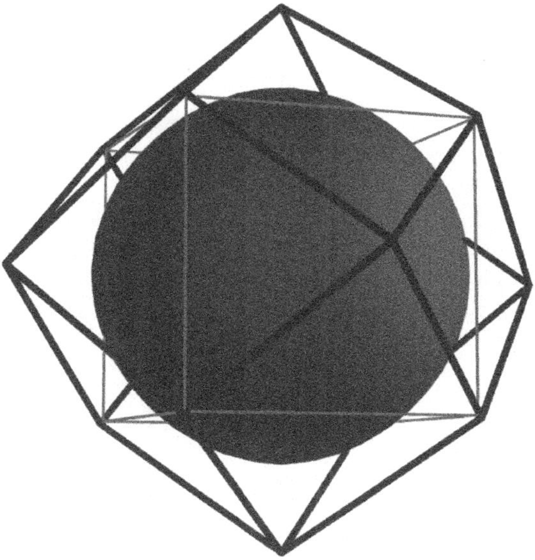

Figure 2.3 Fermi surface for the conduction electrons of sodium.

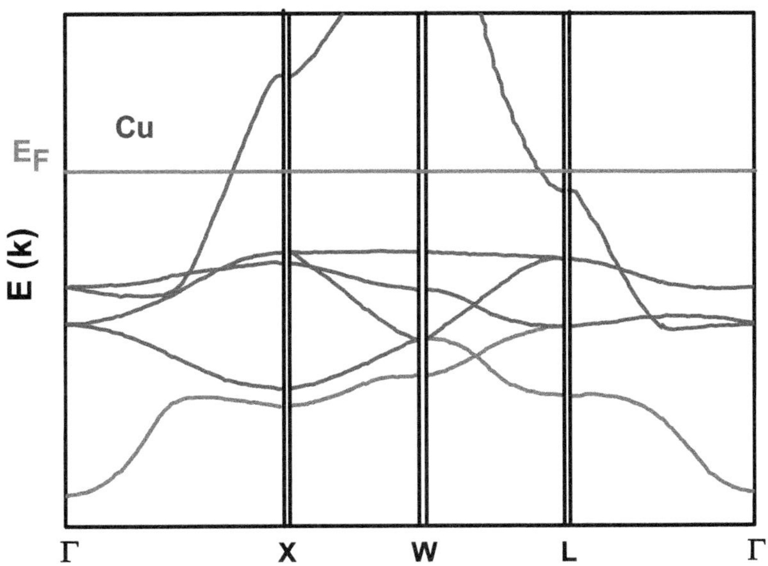

Figure 2.4 The band structure of copper along some special lines of the Brillouin zone, calculated using the APW.

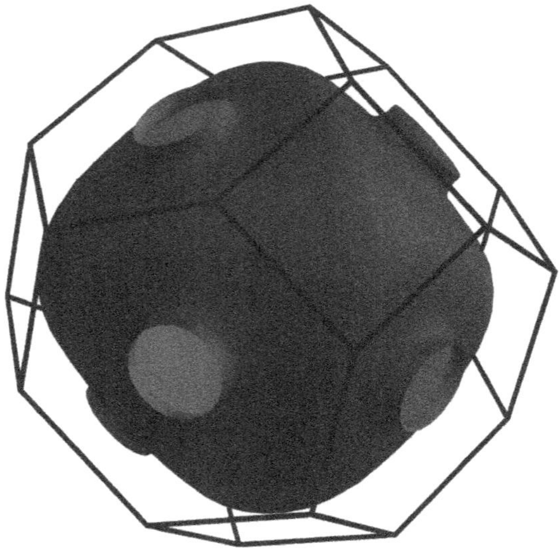

Figure 2.5 Fermi surface for the conduction electrons of copper.

are $\Gamma = (0, 0, 0)$, $X = 2\pi/a(0, 0, 1)$, $W = \pi/a(1, 1, 0)$, and $L = \pi/a(1/2, 1/2, 1/2)$. In that case, the Bravais lattice is the face-centered cubic lattice, but as another difference with respect to sodium, the 3d states of copper are occupied and hybridize with the 4s states to contribute to the chemical bonding. Then the Fermi energy ($E_F = 7$ eV) is clearly located in the fifth energy band (Burdick, 1963). Figure 2.5 plots the Fermi surface, which again is fully contained within the Brillouin zone ($k_F = 1.56\pi/a$) where a is the cube of the lattice constant. The Fermi surface (Solyom, 2010) is very different from the Fermi surface for monovalent metals in body-centered cubic lattice, which indicates how impacting 3d sates are to interpret the differences between the conductivity of copper and sodium.

2.4 Band Structures of Cubic Semiconductors

We now jump to semiconductors. There are two great families of Bravais lattices for technologically impacting semiconductors: the face-centered cubic lattice and the hexagonal one (Morkoç, 1999).

Other Bravais lattices are found for layered compounds such as GaSe, InSe, graphite, h-BN, and other exotic semiconductors, for instance lead-chalcogenides found in rock-salt (Madelung, 1992) or chalcopyrites growing as tetragonal D_{2d} crystals (Burdick and Ellis, 1917). We will not discuss them here. The semiconductors we will discuss here are tetrahedrically coordinated compounds. Silicon and germanium are the most popular of the semiconductors that crystallize in the face-centered cubic lattice. This is the case also for carbon when found as diamond. Therefore, their crystallographic structure is called diamond-like structure.

Figure 2.6 plots the band structure of undoped silicon, which is a terahedricaly bonded semiconductor that crystallizes in the diamond structure (Cardona and Pollak, 1966). The band structure of this semiconductor is particularly different from the band structure of metals: first of all, there is an energy gap in the visible portion of the electromagnetic spectrum. Second, the maximum of the valence band occurs at the Γ point (Γ_{25}), while the minimum of the conduction band occurs at the zone boundary (X_1). This band structure situation is typical of intrinsic indirect band gap

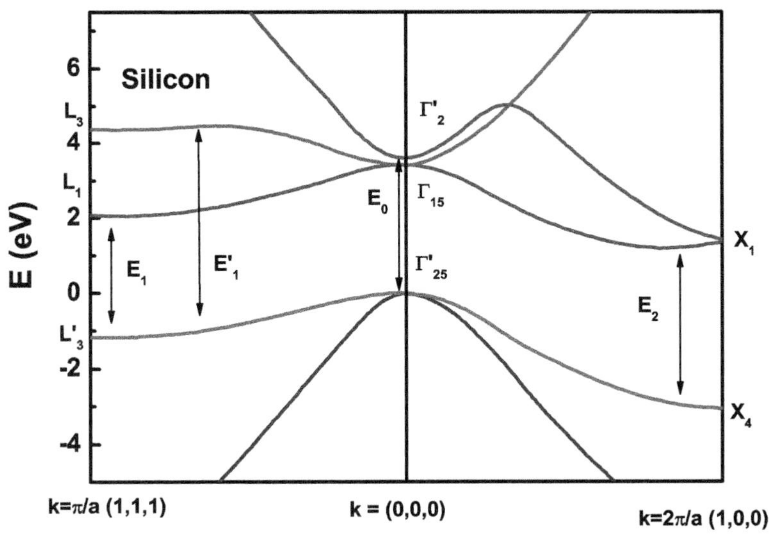

Figure 2.6 Energy bands of silicon calculated by the *k* method in (111) and (100) directions.

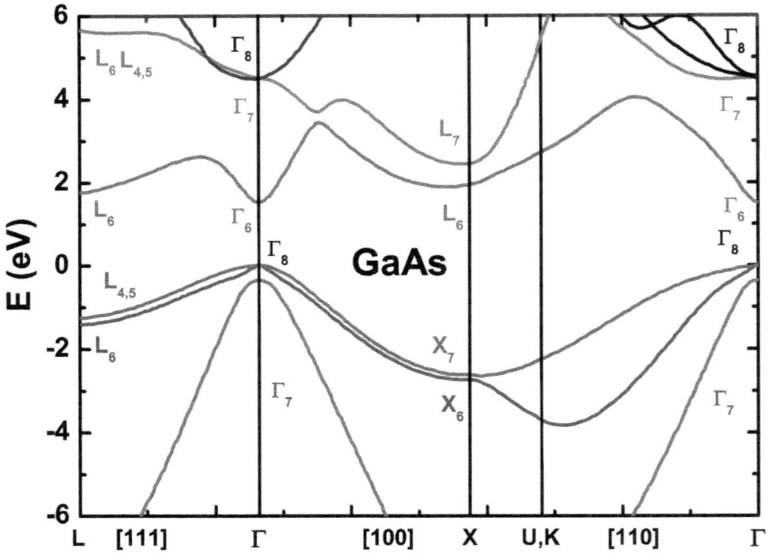

Figure 2.7 The energy bands of GaAs.

semiconductors. The theory of the electronic structure of solids indicates that the Fermi energy is located in the band gap of the undoped, perfect semiconductor. The valence band is fully occupied by electrons. The spin–orbit interaction (a moderate correction for silicon) is not included in the calculation plotted in Fig. 2.6.

Figure 2.7 plots the band structure of GaAs, a well-known semiconductor that grows in the cubic zinc blende fourfold coordinated crystalline phase (Chelikowsky and Cohen, 1976). The difference between the zinc blende and diamond structures is the loss of inversion symmetry when a III-V pair replaces two IV identical atoms. The concept of macroscopic parity is lost. The position of the deep valence band is not given in the figure. In the GaAs case, a clear band gap occurs between the p-type (Γ_8) valence band and the s-type (Γ_6) conduction band. Both extrema occur at the zone center. This is an intrinsic direct band gap semiconductor. In addition, the value of the conduction to valence splitting is minimum in that case, which we call the fundamental band gap. A higher energy band gap can also be found at the L or X boundaries of the Brillouin zone, which will be extremely important in terms of their contribution to the dielectric constant. Spin–orbit interaction (a moderate correction for GaAs) is included in this calculation.

Gallium arsenide is most commonly grown as a zinc blende crystal, such as InP or GaSb, two other interesting binary compounds obtained by assembling group III and group V elements. These so-called III-V compounds are extensively used in optoelectronics for designing optoelectronic devices operating at different wavelengths thanks to the different values of their band gaps. Close cousins of the III-Vs, the II_B-VIs have similar band structure properties when the group II element is zinc or cadmium, while the group VI element is sulfur or selenium or tellurium. Group II_A elements combined with group VI are not interesting; they are fragile, hygroscopic, and of low crystalline quality. They do not interest researchers in semiconductors. Mercury salts are very interesting but are for military application or for their behavior as topological insulators. This is something else. Oxides of group II_B elements crystallize in the hexagonal symmetry. The most popular of them is ZnO.

Nitrides of group III elements, such as GaN, AlN, and InN, generally grow as wurtzitic materials. This crystalline phase is energetically more favorable than the zinc blende one. However, it is also possible to force phosphides, arsenides, and antimonides to grow under the wurtzitic form. The band structure of wurtzitic semiconductors is strongly different from the band structure of zinc blende materials. In particular, the reduction in symmetry produces a complementary splitting of the topmost valence band at the zone center (Dingle et al., 1971; Gil, 2001; Thomas and Hopfield, 1962). The optical response, if performing an experiment in linear polarization along the six-fold symmetry axis, gives an information different from the one obtained in crossed polarization conditions: The material is birefringent (Klingshirn, 2005).

Reducing their lattice symmetry can artificially generate the anisotropy of the optical response of semiconductors. This is clearly illustrated in Figs. 2.8 and 2.9 in the specific case of gallium nitride, a wide band gap semiconductor word widely used for growing blue and white light-emitting diodes and blue solid state lasers. Here the GaN epilayer has been deposited on the anisotropic A (10-10) plane of sapphire, as shown in Fig. 2.8. The epitaxial relations are indirectly given by the orientations of the two crystals. Thanks to

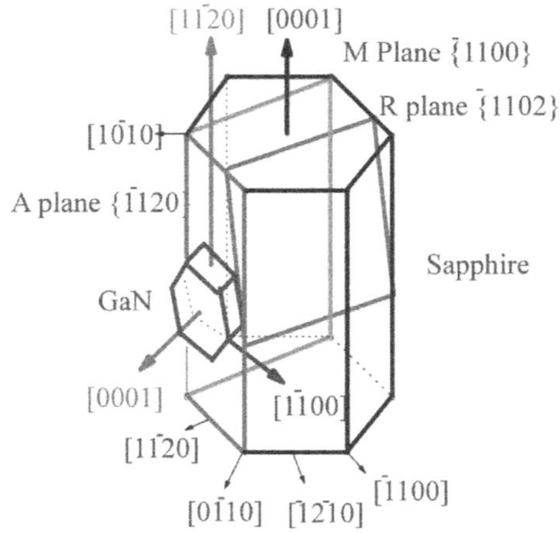

Figure 2.8 Relative orientations of the crystallographic axes of GaN in the case of heteroepitaxy on A-plane sapphire. Note that the (0001) plane of sapphire is parallel to the (11-20) plane of the GaN epilayer.

the anisotropic dilatation coefficients of the thick sapphire substrate in its A plane, the GaN epilayer experiences a strong anisotropic in-plane deformation that reduces the hexagonal symmetry to an orthorhombic one and the in-plane six-fold symmetry of GaN is lost (Alemu et al., 1998). Figure 2.9 represents the result of a reflectivity experiment, which here probes the dielectric constant of the semiconductor, and which, in this specific situation, reveals an anisotropy. As can be seen in the figure, the optical response has been made anisotropic just by choosing the growth conditions, which may sometimes be of technological value.

It is now time to correlate the dielectric constants of semiconductors and metals to the specificities of their band structure. As briefly indicated in the examples earlier, there are a lot of possible situations. Before doing so, we have to emphasize the fact that there is a large range of possibilities for assembling several group III elements with one or more group V elements and vice versa, giving ternary, quaternary, and even quinary alloys. The advantage of this possibility is often taken by device designers for tuning the band gap or the lattice parameter or for tuning the relative positions of fundamental valence and conduction bands (Bastard, 1988; Gil,

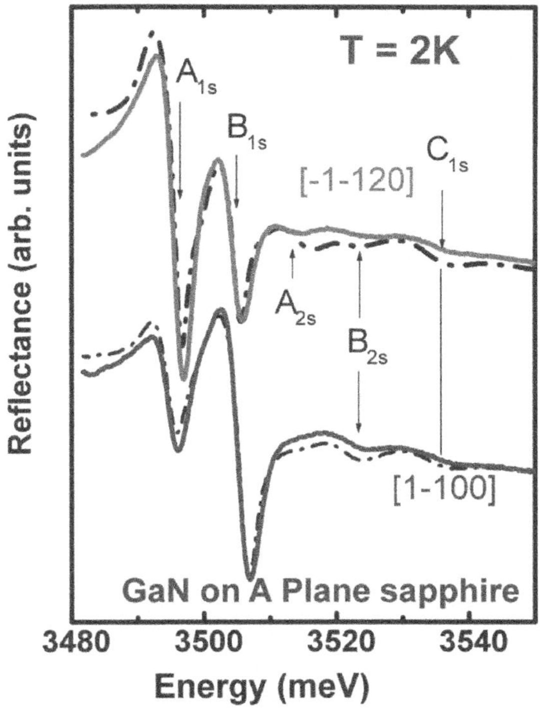

Figure 2.9 Evolution of the intensity of the reflectance structures with in-plane orientation of the electric field: (a)[10-10] orientation, (b) [-12-10] orientation. Note the increase in A_{1s} accompanied by a decrease in B_{1s}, making it easier to detect A_{2s}. Arrows indicate the average positions of transverse excitonic polaritons. The experimental data and the line-shape fitting are plotted using full and dotted lines, respectively.

2002; Weisbuch and Vinter, 1991). We also want to emphasize that the band gap of technologically relevant semiconductors varies from almost zero (for InSb, it equals about 230 meV or about 5.4 micrometers) to 6.2 eV or 200 nm for wurtzite AlN grown by epitaxy on sapphire. Finally, we would like to inform the readers that band structures shown here are not necessarily the most recent and sophisticated calculations that include spin–orbit interactions, exchange correlations, etc. Many technical improvements, although being of paramount importance for people of the skill, have not been discussed here.

2.5 Semi-Classical Theory of the Dielectric Function in Crystals

2.5.1 Intuitive Description

In the preceding section, we gave examples of band structures for two different semiconductors: silicon and gallium arsenide. In the case of silicon, the topmost valence band maximum does not coincide with the conduction band minimum in the first Brillouin zone, while they do in the case of GaAs. If the transition between the conduction and valence quantum states Γ_{6c} and Γ_{8v} does not violate a spectroscopic selection rule (it can be established from group theory arguments that the matrix element of the transition is not vanishing when $\hbar\omega \geq E_{\Gamma_{6c}} - E_{\Gamma_{8v}} = E_{g0}$). A photon with the specific energy $\hbar\omega \geq E_{g0}$ can excite an electron from the top of the filled valence band to one of the states at the bottom of the lowest conduction band, in a transition that is vertical in the reciprocal space (the lattice parameter equals 0.56 nm, while the photon wavelength is 0.8 micrometers, thus the wave number of the photon can be neglected) and called an allowed transition. Further assuming the matrix element to slightly vary with the transition energy, the absorption coefficient $\alpha(E_{g0})$ in the three-dimensional semiconductor can be naturally and intuitively written as proportional to the three-dimensional joint density of states $D\sqrt{\hbar\omega - E_{g0}}$, that is to say $\alpha(E_{g0}) = A\sqrt{\hbar\omega - E_{g0}}$ (Blakemore, 1985; Pankove, 1971).

The free electron state missing in the valence (analogous to a positive charge, named a valence band hole by Wilson in 1930 (Fowler, 1993) who established the symmetry properties between the valence missing electron and the concept of valence hole), and the excess free electrons in the conduction band share the excess energy $\hbar\omega - (E_{\Gamma_{6c}} - E_{\Gamma_{8v}})$ in such a manner that electron and hole have the same value of k. Let us assume for the sake of the lightness of equations that the degeneracy of the valence band does not exist. We take m_e and m_h as the effective masses of electron and hole. They are in the neighborhood of Γ, the curvatures of the bottom conduction band of the top valence respectively. Then the joint density of state effective mass becomes:

$$m_r = \frac{m_e m_h}{m_e + m_h}.$$

It is important to outline that the joint density of state $D\sqrt{\hbar\omega - E_{g0}}$ is proportional to $m_r^{3/2}$ according to the model of the free particle in a three-dimensional box. Thus, the absorption coefficient, for any direct transition in the reciprocal space, reads:

$$\alpha_i(\hbar\omega) = A_i m_{ri}^{3/2}\sqrt{\hbar\omega - E_{gi}}.$$

The index i indicates that for each transition E_{gi}, we attribute a given matrix element A_i, and a given joint density of states:

$$m_{ri}^{3/2}\sqrt{\hbar\omega - E_{gi}}.$$

The next important point to outline here is the expression of the joint density of state effective mass:

$$m_{ri} = \frac{m_{ei} m_{hi}}{m_{ei} + m_{hi}},$$

which rules the amplitude of the absorption. In particular, these m_{ri}s vary with band index i and wave number. Thus, one can again intuitively imagine that these conditions may lead to specific situations featuring the absorption coefficient, which we write:

$$\sum_i \alpha_i(\hbar\omega)$$

or

$$\sum_k \sum_{i=n_c-n_v} \alpha_i(\hbar\omega),$$

where the summation is extended for all vectors k of the Brillouin zone and for all couple of valence (n_v) and conduction bands (n_c). In silicon, which has an indirect energy gap, electron hole pairs of the minimum creation energy cannot be directly created: The momentum conservation rule is not fulfilled in the case of transition between a Γ hole and an X electron, although it can be produced if the photon has an energy high enough to boost the electron in the non-fundamental conduction state. An indirect (non-vertical) transition can occur via a virtual state where photon absorption is accompanied by either absorption or creation of a phonon. This model requires either a photon of energy $E_{\text{gindirect}} + \hbar\omega_{\text{phonon}}$ for creating a phonon with energy $\hbar\omega$ and momentum $k_c - k_v$ or a photon of energy $E_{\text{gindirect}} - \hbar\omega_{\text{phonon}}$ for absorbing a phonon with energy $\hbar\omega$ and momentum $k_c - k_v$. Thus, absorption can occur at energies sitting at both sides of energy $E_{\text{gindirect}}$. Phonons of several branches of the lattice vibration spectrum may contribute. The absorption coefficient is, roughly speaking, one hundred times smaller than that for direct transitions and was difficult to analyze. It is worthwhile to notice that Bose–Einstein statistics prescripts the density of existing phonons to very small at low temperature, thus making the probability to annihilate one is extremely small compared to the probability to create one (Pankove, 1971). The absorption spectra are not symmetrical with respect to the creation–annihilation process and their shapes may be tuned when changing temperature. In silicon, these indirect absorption processes are found at 1.1 eV, while the strong direct absorption at the zone center occurs at 3.4 eV as shown later.

2.5.2 Microscopic Theory of the Dielectric Constant

When a solid is shined by an electromagnetic radiation, part of this radiation is reflected, part of it is transmitted, part of it is absorbed, and part of it is scattered inelastically or elastically. The solid may also emit some fluorescence. In this section, we will consider the three first processes that are directly connected with the semi-classical theory of the dielectric constant.

It is well established that a vector potential $\mathscr{A}(\mathbf{x}, t)$ and a scalar potential ϕ have to be added to the classical one-electron hamiltonian to account for the influence of the radiation field on the electronic structure of the solid (Haug and Koch, 1990; Messiah, 1961). After some quantum mechanics textbook manipulation, the electron-radiation interaction Hamiltonian appears:

$$H_{er} = \frac{e}{mc}\mathscr{A} \cdot \mathbf{p},$$

which is written versus impulsion operator $\mathbf{p}$, when second-order nonlinear effects proportional to $\mathscr{A}^2$ are neglected. Applying time-dependent perturbation theory, the transition probability per unit volume R for an electron in an initial state $|i\rangle$ with energy E_i and wave number $\mathbf{k}_i$ toward a final state $|f\rangle$ with energy E_f and wave number k_f requires calculating the matrix element:

$$|\langle i|\, H_{er}\,|f\rangle|^2 = \frac{e^2}{m^2c^2}\,|\langle i|\,\mathscr{A}\cdot\mathbf{p}\,|f\rangle|^2 = \frac{A^2e^2}{m^2c^2}\,|\langle i|\,\mathbf{u}\cdot\mathbf{p}\,|f\rangle|^2 .$$

We have defined **u** as the unit vector in the direction of $\mathscr{A}$. The Bloch states of the crystal are written as

$$|\beta\rangle = u_{\beta,\mathbf{k}_\beta(\mathbf{x})} exp(i\mathbf{k}_\beta\cdot\mathbf{x})$$

for both initial and final states. Further introducing the amplitude of the electric field $\mathscr{E}(q,\omega)$, $\mathscr{A}$ can be written as the sum of two complex conjugate quantities:

$$\mathscr{A} = \frac{E}{2q}\left[e^{i(\mathbf{q}\cdot\mathbf{x}-\omega t)} + e^{-i(\mathbf{q}\cdot\mathbf{x}-\omega t)}\right].$$

Time integration of $\langle i|\,\mathscr{A}\cdot\mathbf{p}\,|f\rangle$ gives two terms $\delta(E_i(\mathbf{k}_i) - E_f(\mathbf{k}_f) - \hbar\omega)$ and $\delta(E_i(\mathbf{k}_i) - E_f(\mathbf{k}_f) + \hbar\omega)$ for the two previous quantities, which respectively correspond to the absorption of the photon and the excitation of the electron from state to state $|i\rangle$ with energy E_i toward a final state $|f\rangle$ with energy E_f, and to stimulated emission of a photon from energy relaxation from state $|f\rangle$ to $|i\rangle$. The second term is ignored hereafter since we disregard emission processes. Assuming wave vector conservation $\mathbf{q} = \mathbf{k}_i - \mathbf{k}_f$ and finally noting the matrix element $|\langle i|\,\mathbf{u}\cdot\mathbf{p}\,|f\rangle|^2$ as $\left|P_{if}\right|^2$, some mathematical manipulations lead us to obtain the transition probability for photon absorption per unit of time:

$$R = \frac{2\pi}{\hbar}\Sigma_{\mathbf{k}_i,\mathbf{k}_f}\,|\langle i|\, H_{er}\,|f\rangle|^2\,\delta(E_i(\mathbf{k}_i) - E_f(\mathbf{k}_f) - \hbar\omega)$$

as:

$$R = \frac{2\pi}{\hbar}\frac{e^2}{m^2c^2}\left|\frac{E(\omega)}{2}\right|^2\Sigma_{\mathbf{k}}\left|P_{if}\right|^2\delta(E_i(\mathbf{k}_i) - E_f(\mathbf{k}_f) - \hbar\omega).$$

where the k summation is restricted to the **ks** allowed per unit volume of the crystal.

The power lost per unit volume is by the incident beam I of the field due to absorption per unit volume is $-\frac{di}{dt}$ or $R\hbar\omega$. Further introducing the refractive index n, the imaginary part ε'' of the dielectric constant and the absorption coefficient α, one writes:

$$R = -\frac{dI}{dt} = \frac{\varepsilon''}{n^2}c\omega\frac{n^2}{8\pi}|E(\omega)|^2,$$

which straightforwardly provides the expression of the imaginary part of the dielectric constant:

$$\varepsilon'' = |\frac{2\pi e}{m\omega}|^2 \sum_k |P_{if}|^2 \delta(E_i(\mathbf{k}_i) - E_f(\mathbf{k}_f) - \hbar\omega).$$

A straightforward application of the Kramers–Krönig analysis furnishes the expression of the real part:

$$\varepsilon' = 1 + \frac{4\pi^2 e^2}{m} \sum_k \left(\frac{2}{m[E_i(\mathbf{k}) - E_f(\mathbf{k})]} \right) \frac{|P_{if}|^2}{[E_i(\mathbf{k}) - E_f(\mathbf{k})]^2 - \hbar^2\omega^2}.$$

In the above equation $\left(\frac{2}{m[E_i(\mathbf{k}) - E_f(\mathbf{k})]} \right)$ is the number of oscillators with energy $E_i(\mathbf{k}) - E_f(\mathbf{k})$ and is called oscillator strength and noted f_{if}.

Most of the dispersion in ε'' comes from the summation over the delta function. This summation can be replaced by an integral over the energy by defining a joint density of state that reads as follows for the doubly degenerated (spin is included) conduction and valence band:

$$D_j(E_{if}) = \frac{1}{4\pi^3} \int \frac{dS_{\mathbf{k}}}{|\nabla \mathbf{k} E_{if}(\mathbf{k})|},$$

where $E_{if}(\mathbf{k}) = E_i(\mathbf{k}) - E_f(\mathbf{k})$, and $S_{\mathbf{k}}$ is the energy surface $E_{if}(\mathbf{k}) = cte$. Then $\sum_{\mathbf{k}}$ is replaced by $\int D_i(E_{if})dE_{if}$.

It has been shown that the density of state displays singularities when $\nabla_{\mathbf{k}} E_{if}(\mathbf{k})$ vanishes in general at high symmetry points of the Brillouin zone. These singularities are the so-called van Hove singularities (van Hove, 1953). The imaginary part of the dielectric constant exhibits a series of resonances peaking at these specific energies. From the above equations, which give the real and imaginary parts of the dielectric function, one can now calculate the absorption and reflectivity using well-known formulae. The equations intuitively established in the preceding section are recovered.

2.5.3 Experimental Values of the Spectral Dependence of the Dielectric Constants of Semiconductors and Metals

Figures 2.10 and 2.11 report the real and imaginary parts of the dielectric constant of silicon measured by Lautenschlager et al. (1987)

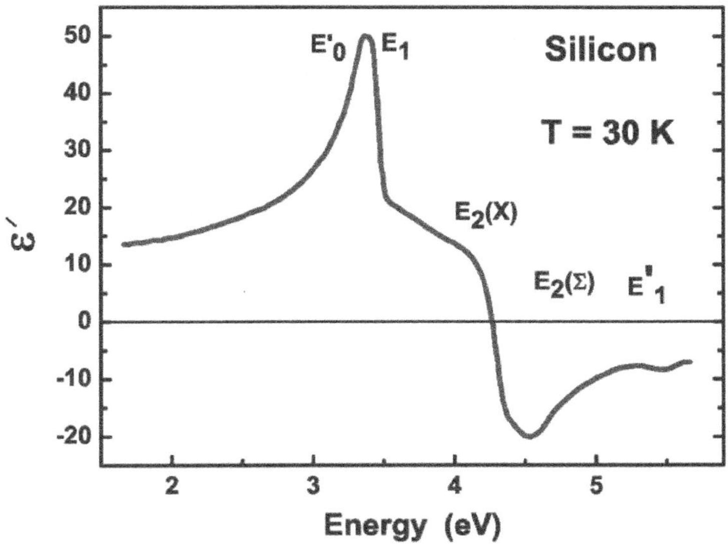

Figure 2.10 Real part of the dielectric function of Si measured at several temperatures.

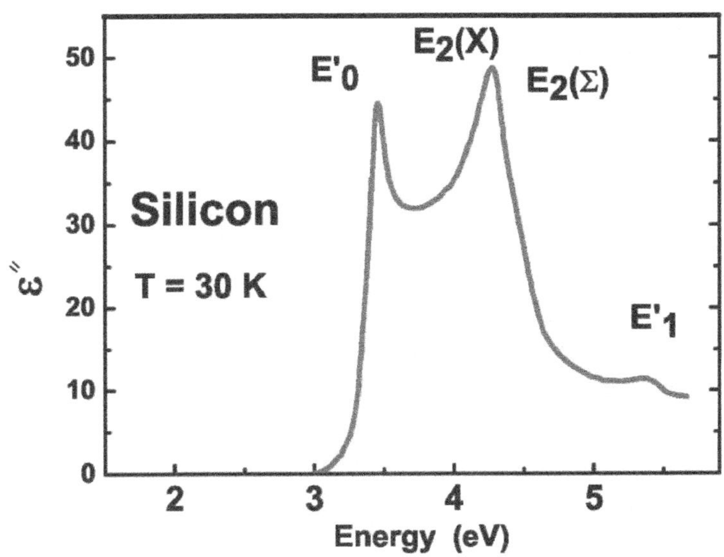

Figure 2.11 Imaginary part of the dielectric function of Si measured at several temperatures.

for different lattice temperatures. The imaginary part is positive, and it exhibits a series of peaks centered at the energies of the direct transitions between the valence and conduction bands of the band structure of silicon (see Fig. 2.6). These peaks overlap. When the temperature increases, they all red-shift. This is a general trend for semiconductors. The real part of the dielectric constant is measured to be positive and negative with an asymptotic trend to zero at high energies. Semiconductors, more or less, exhibit similar shapes, with differences correlated to different values of their direct band gap transitions at the zone center or at other critical points of the Brillouin Zone (Aspnes and Studna, 1983).

Figures 2.12 and 2.13 give the real and imaginary parts of the dielectric function and the loss function $Im(\epsilon^{-1})$ for copper and silver, respectively (Ehrenreich and Philipp, 1962). We note that the real part of the dielectric function is negative at low energy before it gets positive values. This behavior is opposite to what is found for silicon, GaAs, and is typical of metals. We also remark that the imaginary part of the dielectric constant is very large at low energy. This (and the negative value of the real part of the dielectric constant at these energies too) is due to the absorption by the free electron gas. It decreases monotonously before to peak again (near 3.9 eV for silver, 2.1 eV for copper) due to the onset of contributions of inter-band transitions from the Fermi surface to the next higher empty band or from a lower lying band to the Fermi surface. The inter-band contributions are correlated to the shape of the band structure of these metals and to the joint density of states.

The spectral dependence of the absorption coefficients of silver and copper is given in Fig. 2.14 and correlates the observed features with the band structure. The broad structures at energies above 10 eV are attributed to either the excitation of the core shell d electrons to the Fermi surface or from the excitation of electrons at the Fermi surface to the next empty conduction band (Kubo et al., 1976). Peaks of losses are found at 3.8 eV, 7.5 eV, and 18 eV for silver, and 4.1 eV, 7.5 eV, and 20 eV for copper. The 7.5 eV features are attributed to bulk plasma oscillations.

Bulk plasma oscillations are observed at energy $\hbar\omega_{\text{plasma}}$ when:

$$\varepsilon(\hbar\omega_{\text{plasma}}) = 0.$$

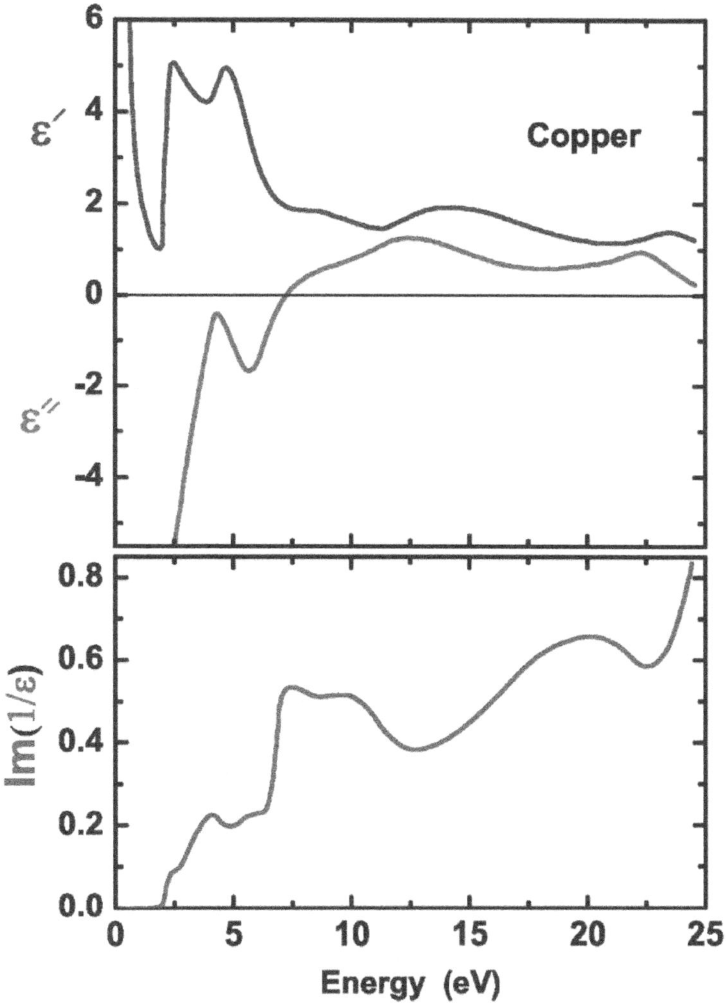

Figure 2.12 Dielectric function (real, imaginary parts) and loss function of Cu from top to bottom, respectively.

The solution to this equation leads to a complex frequency, the real part of which corresponds to the plasma frequency $\frac{2\pi}{\hbar\omega_{\text{plasma}}}$ and the imaginary part to the damping of the plasma oscillation. In the limit of small dampings (to get a simple snapshot of the situation), this is simplified to $\varepsilon'(\hbar\omega_{\text{plasma}}) = 0$. A simple calculation gives plasma

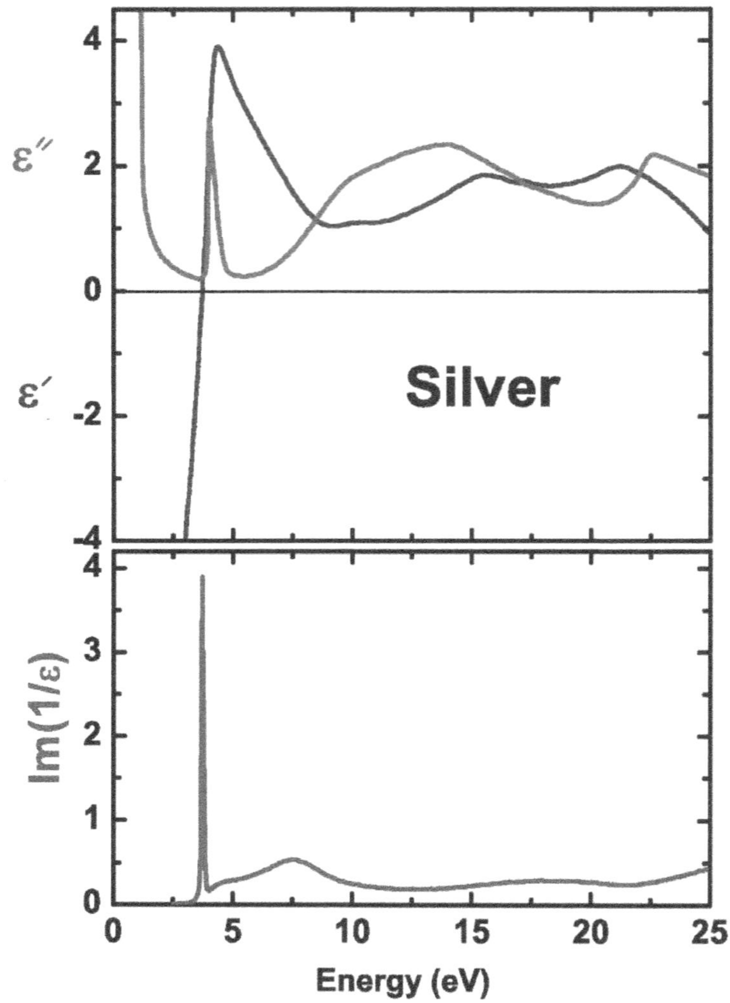

Figure 2.13 Dielectric function (real, imaginary parts) and loss function of Ag from top to bottom, respectively.

oscillation energies at 9.2 eV and 9.3 eV in Ag and Cu, respectively, which correspond to the 7.5 eV features of the loss function. This energy discrepancy is attributed to the influence of d electrons and of the rest of the band structure.

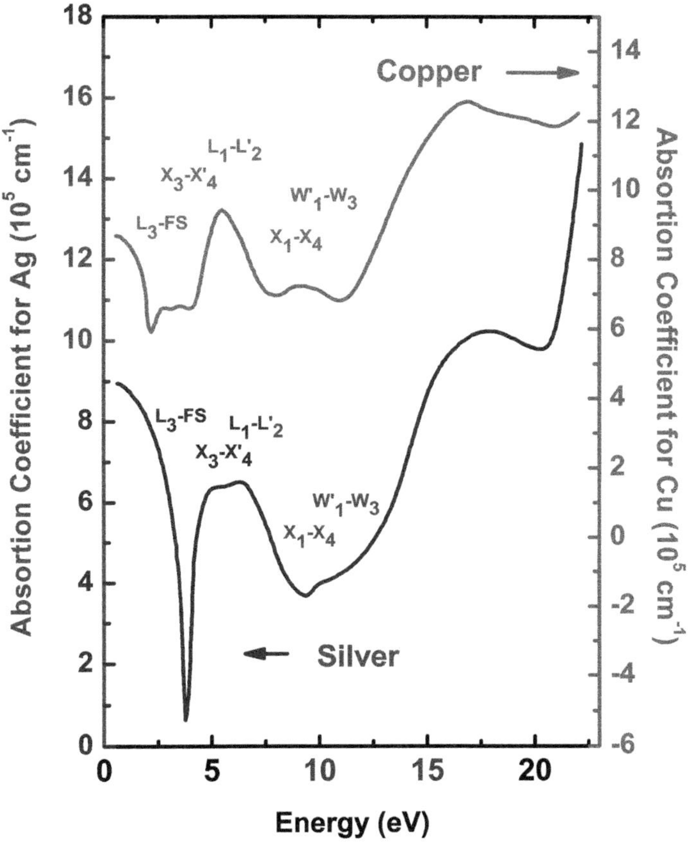

Figure 2.14 Spectral dependence of the absorption coefficients of Ag and Cu.

2.6 Excitonic Effects

The equations above account neither for the experimental behavior in terms of the exact shape for the measured dielectric constant, nor for the absorption coefficient or the reflectance properties. In the case of direct band gap III-V and II-VI semiconductors or group IV element (silicon, germanium, diamond) semiconductors, the more intense lowest energy peaks reported for the imaginary part of the dielectric constant arise from the fundamental (E_0) band gap at the zone center or from the excited one (E_0'). In the language of the van Hove singularities, these transitions arise at M_0 singularities of the band structure. Then the absorption coefficient is zero at energies below the band gap, and it increases with the square root of the energy measured relatively to the band gap. The experimental situation is very different from what measured in the 1940s in the case of ZnO by Mollwo (1943). He observed a rapid onset of the absorption, which saturated, then decreased, and increased again forming some kind of plateau at about 200,000 cm^{-1}. Excitonic effects invented by Frenkel (1931) and Peierls (1932), also described more quantitatively by Wannier (1937) and Mott (1938), are responsible for this behavior (Bassani and Parravicini, 1975). In the simplest picture, they can be viewed as the quantum of electrostatic interaction between the photo-created conduction electron (a negative charge) and the valence-missing electron (a positive charge named hole by Wilson). The long-range Coulomb interaction between these two particles makes them behave like a hydrogen-like quantum system with bound and un-bound states, electron–hole short-range spin exchange interaction (analogous to the contact interaction in the hydrogen atom). The absorption coefficient consists in a discrete hydrogen series of absorption line energy split according to the hydrogen series and to contribution of the continuous unbound states, which superimpose to the band-to-band process. The absorption onset starts at an energy lower than the energy gap E_g by an amount E_R, which is the exciton binding energy. E_R is expressed as a function of the hydrogen atom Rydberg R_y as:

$$E_R = \frac{\mu}{\varepsilon^2} R_y.$$

In these equation, the excitonic reduced mass μ is defined as:

$$\frac{1}{\mu} = \frac{1}{m_e} + \frac{1}{m_v},$$

and the dielectric constant is averaged through the low-frequency $\varepsilon(0)$ and high-frequency $\varepsilon(\infty)$ dielectric constants as follows (Mahan and Berland, 2011):

$$\frac{1}{\varepsilon} \approx \frac{11}{16\varepsilon(0)} + \frac{5}{16\varepsilon(\infty)}.$$

In the strictest sense, the amplitudes of the absorption of the excited excitonic states vary like ν^{-3}, where ν is the principal quantum

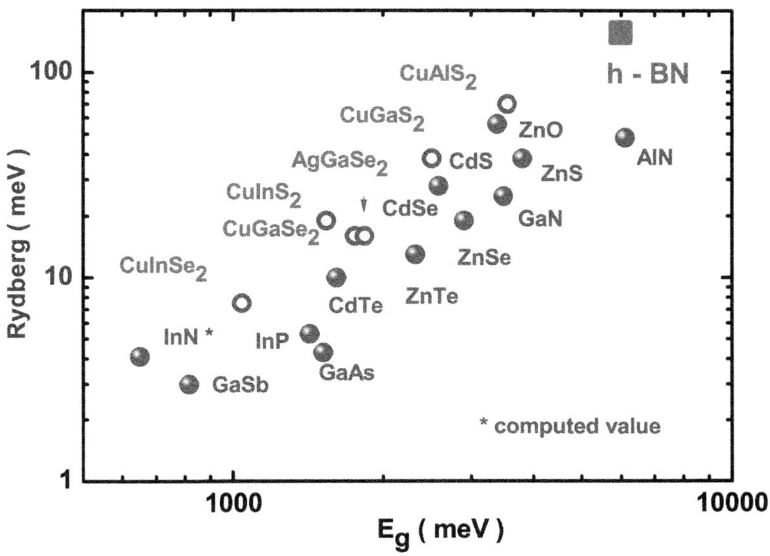

Figure 2.15 Plot of the excitonic Rydberg versus energy gaps for fourfold coordinated semiconductors.

number relative to the absorption at the energy of ground state exciton. Excitonic levels have a finite lifetime. The excitonic lines have a finite line-width, and one generally does not observe the whole excitonic series except in high-quality zinc blende semiconductors (Sturge, 1962), or in wide band gap semiconductors (see Fig. 2.9) or in chalcopyrite semiconductors. The reason for this is the large value of the exciton binding energy in the latter case (see Fig. 2.15) (Gil et al., 2012). This is illustrated in Fig. 2.9, which displays excitonic energy splitting larger than the natural broadening of the exciton line. When the lattice temperature is increased, thermal broadening leads to the overlap between excitonic levels and the sharp absorption peaks are sometimes no longer resolved. The spectral dependence of the absorption coefficient according to the band-to-band model is never observed, as shown in Fig. 2.16, including at high temperatures. From the mathematical point of view, the absorption coefficient including excitonic effect has been published by Elliot (1957); it contains excitonic contributions: the contributions of bound excitonic hydrogen-like states and the

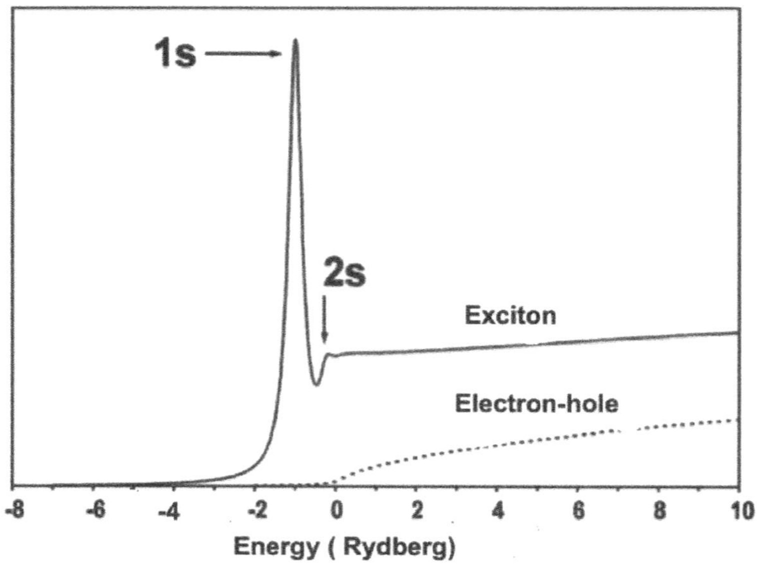

Figure 2.16 Optical density within the context of excitonic model or electron-hole one.

contribution of unbound states. The contribution of bound states consists in a series Lorentzian shaped peaks centered at the energies of the hydrogen series, of amplitudes that vary accordingly. For large values of the principal quantum number, the density of excitonic state increases and they merge and amalgamate to form a pseudo continuum of density $D_\nu(E) = \frac{\nu^3}{2E_R}$ at energies below the energy of the band-to-band gap. The excitonic contribution to the absorption for high values of the principal quantum number is the product of the oscillator strength ($\sim\nu^{-3}$) timed by the density of state of the pseudo-continuum ($\sim\nu^3$). They compensate each other. The absorption coefficient has a finite value. Now regarding unbound states, the joint density of states is:

$$D_{3D} = \frac{1}{\sqrt{2}h^3}(\hbar\omega - E_g)\mu^{3/2}\sqrt{\hbar\omega - E_g},$$

which is rewritten as a function of the excitonic Rydberg, of its Bohr radius a_B, and as a function of a dimensionless quantity $\gamma = \sqrt{\frac{E_R}{\hbar\omega - E_g}}$

as follows:

$$D_{3D} = \frac{1}{4\pi^2 a_B^3 R_y \gamma},$$

and:

$$\alpha_{3D}(\hbar\omega) = \frac{1}{4\pi^2 a_B^3 R_y} \frac{e^{\pi\gamma}}{\sinh \pi\gamma}.$$

Beyond the scope of shifting the energy onset of absorption, Coulomb interaction and excitons introduce an enhancement of the absorption regarding band-to-band process. This enhancement factor:

$$S(\hbar\omega - E_g) = \frac{K_{\text{continuum}}(\hbar\omega - E_g)}{K_{\text{interbands}}(\hbar\omega - E_g)} = \pi\gamma \frac{e^{\pi\gamma}}{\sinh \pi\gamma}$$

is called the Sommerfeld factor. This ratio between the absorption, including excitons, and the band-to-band result is always larger than 1, as shown in Fig. 2.16.

We have chosen here to discuss absorption coefficient, but similar contributions of excitonic effect are expected at the scale of real and imaginary parts of the dielectric constant, which can be found in Yu and Cardona (1996).

The impact of the excitonic corrections is important far away from the fundamental band gap energy but is impacting the whole shape of the dielectric function. Figure 2.17 illustrates this quantitatively: the imaginary part of the dielectric function of wurtzitic InN for Coulomb-correlated electron–hole pairs (solid lines) and independent quasiparticles (dashed lines). The ordinary and extraordinary functions are given in the upper (a) and lower (b) panel, respectively (Furthmuller et al., 2005). In the low-energy region ($E < 4$ eV), they report a moderate influence of the excitonic effects. A substantial red shift of about 0.5 eV is found in the 4–9 eV energy. In addition, a redistribution of spectral strength from higher to lower photon energies happens. This effect is obvious for the extraordinary dielectric function. Similar effects have been observed for the E_1 and E_2 peaks of diamond and zinc blende semiconductors.

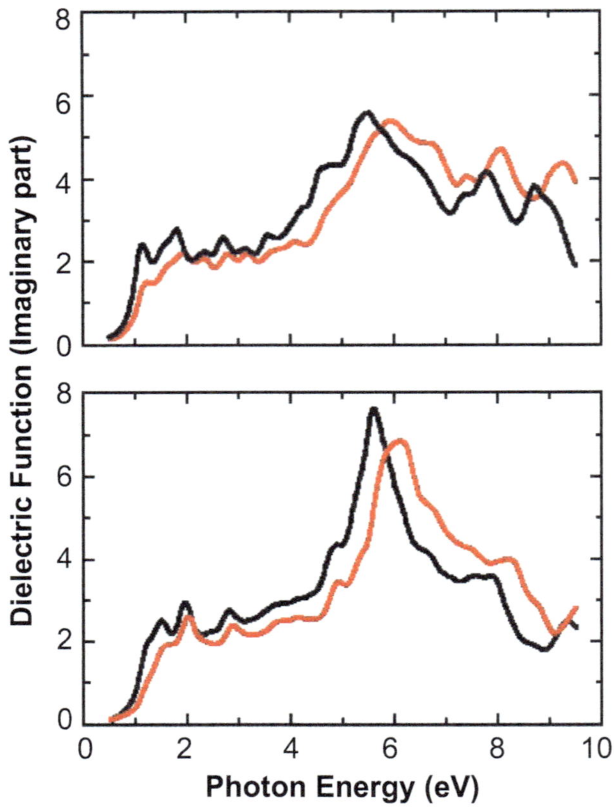

Figure 2.17 Imaginary part of the dielectric function of w-InN for Coulomb-correlated electron–hole pairs (black lines) and independent quasiparticles (red lines). The ordinary and extraordinary functions are given in the upper (a) and lower (b) panel, respectively.

2.7 Influence of Doping and Alloying

Semiconductor alloying is realized by sometimes substituting one atom of the lattice by another one belonging to the same column of Mendeleev's table. An example of this is replacing some gallium atoms of GaAs by indium or aluminum to form a ternary compound. This substitution leads to a random distribution of both elements; if the proportion of substitution equals some specific proportion like 0.25 or 0.5, then one can obtain either the ordered or the disordered phase, depending on the growth conditions. Both anions and cations can be simultaneously alloyed to form a quaternary alloy like $Ga_{1-x}In_xAs_{1-y}P_y$. By doing that, device designers try to achieve band gap engineering (to tune the color of the emitted light); they engineer built-in strain field in strained layer heterostructures, or they tune band offset, or they try to realize waveguides (Burstein and Weisbuch, 1995). The random distribution of atoms washes out the clear value of the bang gap, which loses its meaning at the strictest sense simultaneously with the vanishing of translation symmetry. Tails of states occur below the average energy of the conduction and above the average energy of the valence band. One rather speaks of spectral dependence of densities of states rather than of bands. These local fluctuations of compositions may trap carriers or photons, depending on their extensions in real space and they are either deleterious or profitable for device designers.

Doping is achieved by the appropriate incorporation of controlled amounts of foreign atoms into a pure crystal (for instance, including a group V element, As, in silicon increases the density of free electrons). As is called a donor, while B is called an acceptor. The first effect of this is to modify the Fermi energy as soon as the semiconductor is doped. Residual impurities will be impacting the transport properties and sometimes the optical ones. There is a great zoology of dopants, and we do not intend to review them, but we consider one important consequence (we believe it) of doping. Consider an n-type dopant in a semiconductor giving a residual electron density n. According to the classical Drude model for the dielectric constant, this electron gas in excess behaves collectively and oscillates with a plasma energy:

$$\hbar\omega_{\text{plasma}} = \hbar\left(\frac{ne^2}{m\varepsilon_0}\right)^{1/2}.$$

We note that a better approximation gives the longitudinal oscillations of the plasmon to occur at pulsation

$$\omega_{\text{plasma}}(\mathbf{k}) \approx \omega_{\text{plasma}}\left(1 + \frac{3v_F^2}{10\omega_{\text{plasma}}^2}\mathbf{k}^2 + o(\mathbf{k}^4)\right),$$

where v_F is the electron Fermi velocity (Pines, 1963). Volume plasmons are collective eigen-modes of the free-electron gas inside a metal. Because of their longitudinal character and the transversal nature of light, the photoexcitation of volume plasmons is forbidden in classical electrodynamics. However, they can couple with longitudinal phonons and thus contribute to the dielectric constant in the infrared portion of the electromagnetic spectrum (Perlin et al., 1995). We also would like to emphasize the fact that the whole dielectric constant is changed by doping (Viña and Cardona, 1984). It is worthwhile to notice that doped silicon has been proposed (Soref et al., 2008) to tune surface plasmon oscillation frequencies (Pitarke et al., 2007) and that surface plasmons are also used to enhance the spontaneous emission rate in modern devices, among which are nitride-based blue light emitters (Gontijo et al., 1999; Lin et al., 2010; Mohammadi et al., 2008; Okamoto et al., 2005, 2007; Singh et al., 2010a,b).

2.8 Conclusion

In this review, I tried to give the readers an overview of the optical properties of metals and semiconductors, by highlighting their similarities and differences in an identical manner. This is a great field. Hundreds of thousands of pages have been written on these compounds. In this chapter, I tried to give a rapid overview. I am sure that I failed to address important issues, but if you got the flavor to tune the browser of your computer to the topics you would like to know more about, my failure will be, I hope, almost forgiven.

References

Alemu, A., Gil, B., Julier, M., and Nakamura, S. (1998). Optical properties of wurtzite gan epilayers grown on a-plane sapphire, *Phys. Rev. B* **57**, 3761.

Aspnes, D. E. and Studna, A. A. (1983). Dielectric functions and optical parameters of si, ge, gap, gaas, gasb, inp, inas, and insb from 1.5 to 6.0 ev, *Phys. Rev. B* **27**, pp. 985–1009.

Bassani, F. and Parravicini, G. P. (1975). *Electronic States and Optical Transitions in Solids* (Pergamon Press, New York).

Bastard, G. (1988). *Wave Mechanics Applied to Semiconductor Heterostructures* (EDP Sciences, Paris).

Blakemore, J. S. (1985). *Solid State Physics* (Cambridge Press, Cambridge).

Bloch, F. (1928). Uber die quantenmechanik der elektronen in kristallgittern, *Zs. Phys* **52**, pp. 555–600.

Brillouin, L. (1930a). Les électrons dans les métaux et le classement des ondes de de broglie correspondantes, *C.R.* **191**, pp. 292–294.

Brillouin, L. (1930b). Les électrons dans les métaux et le rôle des conditions de réflexion sélective de bragg, *C.R.* **191**, pp. 198–201.

Brillouin, L. (1930c). Les électrons libres dans les métaux et le rôle des réflexions de bragg, *J. Phys. Radium* **1**, pp. 377–400.

Burdick, C. L. and Ellis, J. H. (1917). The crystal structure of chalcopyrite determined by x-rays, *Proc. Natl. Acad. Sci. USA* **3**, p. 644.

Burdick, G. A. (1963). Energy band structure of copper, *Phys. Rev.* **129**, p. 138.

Burstein, E. and Weisbuch, C. (1995). *Confined Electron and Photon: New Physics and Applications,* NATO ASI Series (Plenum Press, New York).

Cardona, M. and Pollak, F. H. (1966). Energy-band structure of germanium and silicon: The k p method, *Phys. Rev.* **142**, p. 530.

Chelikowsky, J. R. and Cohen, M. L. (1976). Nonlocal pseudopotential calculations for the electronic structure of eleven diamond and zincblende semiconductors, *Phys. Rev. B* **14**, p. 556.

Ching, Y. and Callaway, J. (1975). Energy bands, optical conductivity and compton profile of sodium, *Phys. Rev. B* **11**, p. 1324.

Dingle, R., Sell, D. D., Stokowski, S. E., and Ilegems, M. (1971). Absorption, reflectance, and luminescence of gan epitaxial layers, *Phys. Rev. B* **4**, p. 1211.

Ehrenreich, H. and Philipp, H. R. (1962). Optical properties of ag and cu, *Phys. Rev.* **128**, p. 1622.

Elliot, R. J. (1957). Intensity of optical absorption by excitons, *Phys. Rev.* **108**, p. 1384.

Fowler, A. B. (1993). A semicentury of semiconductors, *Phys. Today* **46**, 59–62, doi:dx.doi.org/10.1063/1.881385.

Frenkel, J. (1931). On the transformation of light into heat in solids. II, *Phys. Rev.* **37**, 17, p. 1276.

Furthmuller, J., Hahn, P. H., Fuchs, F., and Bechsted, F. (2005). Band structures and optical spectra of inn polymorphs: Influence of quasiparticle and excitonic effects, *Phys. Rev. B* **72**, p. 205106.

Gil, B. (2001). Oscillator strengths of A, B, and C excitons in ZnO films, *Phys. Rev. B* **64**, 201310.

Gil, B. (2002). *Low-Dimensional Nitride Semiconductors* (Oxford University Press Oxford, Oxford).

Gil, B., Felbacq, D., and Chichibu, S. F. (2012). Exciton binding energies in chalcopyrite semiconductors, *Phys. Rev. B* **85**, p. 075205.

Gontijo, I., Borodtiski, M., Yablonovitch, E., Keller, S., Mishra, U. K., and Baars, S. P. D. (1999). Coupling of InGan quantum-well photoluminescence to silver surface plasmons, *Phys. Rev. B* **60**, pp. 11564–11567.

Haug, H. and Koch, S. W. (1990). *Quantum Theory of Optical and Electronic Properties of Semiconductors* (World Scientific, Singapore).

Hoddeson, L., Baym, G., and Eckert, M. (1987). The development of the quantum mechanical theory of metals, *Rev. Mod. Phys.* **59**, p. 287.

Kittel, C. (1996). *Introduction to Solid State Physics* (John Wiley and Sons, New York).

Klingshirn, C. (2005). *Semiconductor Optics* (Springer Verlag, Berlin).

Kubo, Y., Wakoh, S., and Yamashita, J. (1976). Energy loss spectrum of copper, *J. Phys. Soc. Jpn* **41**, pp. 1556–1561.

Lautenschlager, P., Garriga, M., Vina, L., and Cardona, M. (1987). Temperature dependence of the dielectric function and interband critical points in silicon, *Phys. Rev. B* **36**, pp. 4821–4830.

Lin, J., Mohammadizia, A., Neogi, A., Morkoc, H., and Ohtsu, M. (2010). Surface plasmon enhanced UV emission in AlGaN/GaN quantum well, *Appl. Phys. Lett.* **97**, p. 221104.

Madelung, O. (1992). *Data in Science and Technology: Semiconductors other than Group IV Elements and III-V Compound* (Springer Verlag, Berlin, Heidelberg).

Mahan, G. D. and Berland, K. (2011). Theory of polar corrections to donor binding, *Phys. Rev. B* **84**, p. 235203.

Messiah, A. (1961). *Quantum Mechanics* (North Holland, Amsterdam).

Mohammadi, A., Sandoghdar, V., and Agio, M. (2008). Gold nanorods and nanospheroids for enhancing spontaneous emission, *New J. Phys.* **10**, p. 105015.

Mollwo, E. (1943). Gassorption un photolumineszenz von zinkoxid, *Reichsber. der Physik* **1**, p. 1.

Morkoç, H. (1999). *Nitride Semiconductors* (Springer Verlag, Berlin, Heidelberg).

Mott, N. F. (1938). Conduction in polar crystals. II. The conduction band and ultra-violet absorption of alkali-halide crystals, *Trans. Faraday Soc.* **34**, p. 500.

Okamoto, K., Niki, I., Scherer, A., Narukawa, Y., Mukai, T., and Kawakam, Y. (2005). Surface plasmon enhanced spontaneous emission rate of In-GaN/GaN quantum wells probed by time-resolved photoluminescence spectroscopy, *Appl. Phys. Lett.* **87**, p. 071102.

Okamoto, K., Niki, I., Shvartser, A., Maltezos, G., Narukawa, Y., Mukai, T., Kawakami, Y., and Scherer, A. (2007). Surface plasmon enhanced bright light emission from InGaN/GaN, *Phys. Status Solidi (a)* **204**, p. 2103.

Pankove, J. I. (1971). *Optical Processes in Semiconductors* (Dover Publication Inc.).

Peierls, R. E. (1932). Zur theorie der absorptionsspekten fester kormper, *Annalen der Physik* **13**, p. 905.

Perlin, P., Camassel, J., Knap, W., Taliercio, T., Chervin, J. C., Suski, T., Grzegory, I., and Porowsk, S. (1995). Investigation of longitudinal optical phonon plasmon coupled modes in highly conducting bulk GaN, *Appl. Phys. Lett.* **67**, p. 2524.

Pines, D. (1963). *Elementary Excitations in Solids*, Vol. A (Benjamin).

Pitarke, M., Silkin, V. M., Chulkov, E. V., and Echenique, P. M. (2007). Theory of surface plasmons and surface-plasmon polaritons, *Rep. Prog. Phys.* **70**, pp. 1–87.

Seitz, F. (1960). *Théorie Moderne des Solides* (Masson, Paris).

Singh, A., Grczynski, K. G., McDaniel, F. D., Park, S. Y., Kim, M., and Neogi, A. (2010a). Localized surface plasmon polariton enhanced radiative recombination in ion-implanted silicon emitters, *Appl. Phys. Express* **3**, p. 102201.

Singh, A., Grczynski, K. G., McDaniel, F. D., Park, S. Y., Kim, M., and Neogi, A. (2010b). Localized surface plasmon polariton enhanced radiative recommbination in ion-implanted silicon emitter, *Appl. Phys. Express* **3**, p. 102201.

Solyom, J. (2010). *Fundamentals of the Physics of Solids: Volume II: Electronic Properties* (Springer, Berlin).

Soref, R., Peale, R. E., and Buchwald, W. (2008). Longwave plasmonics on doped silicon and silicides, *Opt. Express* **16**, 9, pp. 6507–6514.

Sturge, M. D. (1962). Optical absorption of gallium arsenide between 0.6 and 2.75 ev, *Phys. Rev.*, pp. 768–773.

Thomas, D. G. and Hopfield, J. J. (1962). Optical properties of bound exciton complexes in cadmium sulfide, *Phys. Rev.* **128**, p. 2135.

van Hove, L. (1953). The occurrence of singularities in the elastic frequency distribution of a crystal, *Phys. Rev.* **89**, pp. 1189–1193.

Viña, L. and Cardona, M. (1984). Effect of heavy doping on the optical properties and the band structure of silicon, *Phys. Rev. B* **29**, p. 6739.

Wannier, G. H. (1937). The structure of electronic excitation levels in insulating crystals, *Phys. Rev.* **52**, 191.

Weisbuch, C. and Vinter, B. (1991). *Quantum Semiconductor Structures: Fundamentals and Applications* (Academic Press, London).

Yu, P. and Cardona, M. (1996). *Fundamentals of Semiconductors* (Springer Verlag, Berlin, Heidelberg).

Ziman, J. M. (1960). *Electrons and Phonons* (Oxford University Press, Oxford).

Chapter 3

From Microphysics to Mesophysics: Obtaining Effective Properties from Microscopic Behaviors

Alexandrù Cabuz
PEERGATE SRL, Str. Stirbei Vodă nr. 97, Bl. 25C, Apt. 15, Bucharest Sector 1, 010108 Romania
alexcabuz2@gmail.com

The awareness that matter is not smooth and homogeneous but grainy and discrete has come to us gradually, over a period of time spanning centuries. Perhaps the first indication of the discrete nature of matter came from the study of chemical reactions. Certain discrete ratios between the masses of reactants involved in certain reactions were observed. This remained at a fairly vague level until the end of the 19th century, when experimental methods advanced to the point where the notion of atom emerged as the indivisible unit making up commonly known substances.

The reason why science took so long to realize that matter is grainy, and why only sufficiently advanced experimental techniques were able to determine that it is so, is that previously humans only observed matter *on a large scale* compared to the graininess. In the vast majority of situations, scale imposes an impenetrable barrier to observation, which is why it took us so long to realize

Metamaterials Modeling and Design
Edited by Didier Felbacq and Guy Bouchitté

ISBN 978-981-4316-12-5 (Hardcover), 978-1-315-36500-8 (eBook)
www.panstanford.com

that there is any structure "at the bottom" at all. Even today, phenomena spanning different scales are among the most difficult to model, understand, and predict: chaos, complexity, theoretical biology, neuroscience, and econophysics. In Nobel laureate P. W. Anderson's words, "More is different" (Anderson, 1972).

The notion of scale is the essential concept in the study of the physics of materials in general, and of electromagnetic metamaterials in particular. In the context of electromagnetic metamaterials, we are relatively fortunate due to the fact that the same kind of physics (classical electrodynamics) governs the behavior on the smaller scale and also on the larger scale. It is only the material description that changes from a detailed, complex description to a general, simple description, respectively.

We have two reasons for our interest in homogenization. First, it allows us to *simplify* a model, keeping only the essential features. Second, it makes it possible to *design* materials that are not naturally available. The actual computation techniques that may be used to *model and design* various structures will be discussed in subsequent chapters. Here, we only discuss and explain the general concepts that emerge when dealing with phenomena spanning different scales.

In Section 3.1, we highlight the central importance of the relationship between scales in determining how we model a given structure. The three scales (wavelength, period, and charge mobility scales) determine the qualitative behavior of the medium. A fourth scale (which has become increasingly important recently, particularly in nanophotonics), the size of the objects a, is also briefly mentioned in Section 3.6. In Section 3.2, we discuss the averaging method, which makes it possible to define the macroscopic effective fields (electric **E** and magnetic **B**) and the macroscopic polarization fields (only the electric polarization **P** is treated in detail). Sections 3.3 and 3.4 introduce the susceptibility (including the Clausius–Mossotti relation), the permittivity, permeability, and refractive index, including a discussion on the negative index of refraction. Sections 3.5 and 3.6 will discuss the two physical origins of the phenomenon of spatial dispersion: the periodicity in Section 3.5, and free charges in Section 3.6. In the last section of the chapter, we will bring together the various concepts in order to highlight the flexibility that effective medium theory offers us: the tradeoff between the homogeneity and the nonlocality of the material description.

3.1 Metamaterials and Scales

In an electromagnetic material, three scales determine its behavior: the wavelength λ of the harmonic field, the scale d on which the medium is structured (e.g., the period in the case of periodic structures), and the distance l over which the charges are free to move within the structure. Thus, we impose no restriction on the medium as to whether it contains "free charges" or not, because charges are always free to some extent: It is just the scale l that changes. The relationship between these scales will govern the kind of model we will obtain.

The most familiar case is the homogeneous dielectric: $l \approx d \ll \lambda$. We may be talking about a naturally occurring dielectric, where l and d are on the atomic scale, i.e., several angstroms, while λ is in the optical domain, hundreds of nanometers. Or we may be talking about cermets or composites (Aspnes, 1986; Bergman, 1980; Milton, 1980), where l and d are on the scale of tens of nanometers, while the wavelength is also optical, in the hundreds of nanometers. Or, also, we may be talking about microwave delay lenses (Kock, 1948), made of metallic scatterers, where the size of the scatterers (l) is similar to the distances separating them (d), while the wavelength (λ) is at least an order of magnitude larger. The two latter situations are examples of metamaterials: They are artificial structures engineered so as to exhibit certain properties not available in naturally occurring materials.

Due to the very large wavelength, in this situation we lose sight of the graininess on the d scale, obtaining a homogeneous dielectric with an effective permittivity and permeability. The microscopic inhomogeneity on the a scale is averaged to give a *homogeneous* description of the medium as represented by the permittivity ε and permeability μ. Thus, at this scale, *all* traces of the medium's true underlying discreteness are lost, and it looks smooth, continuous, and homogeneous.

Different behaviors emerge as the wavelength is reduced, and it starts to approach either l or d or both. When this happens, the medium starts to exhibit spatial dispersion, also known as nonlocal behavior. Light propagation is more complicated, and the waves are no longer plane waves. This regime has been studied, for instance, in the context of photonic crystals, which exhibit photonic band gaps, guiding, localization, autocollimation, and superprism effects. Light propagation in photonic crystals is strongly dependent on their inhomogeneous nature, and it is through careful design that a wealth of phenomena and applications has been uncovered. Spatial dispersion can also occur when the medium contains free charges, and we have $d \ll l$.

It is important to distinguish between the nonlocal behavior due to the proximity between λ and d, which may be called "structural" dispersion (discussed in Section 3.5), and the proximity between λ and l, which may be called "free-charge" dispersion (discussed in Section 3.6). The first appears when the wavelength approaches the period of a crystal (for instance) and it will occupy the main part of our attention, while the second appears when charges are mobile on distances comparable to the wavelength, i.e., in conducting structures. The two involve quite different physical mechanisms.

The objective of any homogenization procedure is twofold. First, given a certain structure, we would like to know what the effective optical parameters are: the index and impedance, or the permittivity and permeability. Second, given the optical parameters, we would like to know what structure may be constructed to realize them. In order to realize this objective, we must first define a general homogenization procedure.

In the following sections, we will treat effective media in such a way as to include naturally occurring materials made of atoms. Thus, the arguments will also discuss issues related to the random thermal motion and the quantum mechanics that makes it possible for atoms to *exist* in the first place. When considering atoms, l will correspond to the size of the atom. When considering metamaterials made by stacking arrays of conducting scatterers, l will correspond to the size of the scatterers. Note that nothing prevents scatterers from being larger than the distances separating them. For instance, star-shaped or thin and long objects may be arranged in this way. In these cases, we have $l > d$ and the properties of each object may be strongly modified by the presence of nearby objects. The case of large or strongly asymmetric charge mobility regions is a relative newcomer to effective medium theory, and we only briefly mention it in Section 3.6. In this case, a scatterer may not be considered independently, but only as a part of its "electromagnetic neighborhood".

3.2 Averaging—Time and Space

The most important notion discussed in this chapter is the notion of the observation horizon, or barrier, which is implicit when observing a system on a certain scale. With each scale is associated a given barrier, below which one can no longer distinguish features and events. Things that are *too small* and things that happen *too fast* fall below the barrier and thereby become uncertain. The way we deal with uncertainty is to take averages, because an average can be described also as a "best fit" to noisy data.

The key word in the previous paragraph is "uncertainty." This is the first thing that must be dealt with, and a probability distribution function is required for this. By this means, it is possible to perform a statistical average to eliminate the low-level thermal and quantum agitation, leaving a stable, *time-invariant* system.

The time-invariant feature is absolutely fundamental for our purposes, since it is the only way in which one may meaningfully speak of a set of *material parameters* that characterize a certain structure. When, in addition, the medium is linear, we are able to marshall the full power of the theory of *linear time-invariant systems*, in order to extract and understand the *transfer functions* we are interested in: permittivity and permeability. Importantly, therefore, the statistical average may be seen as a low-pass filter. Out of the complex and frantic dynamics on a microscopic scale, it extracts the time-invariant component, or "DC" component. In a single word, what the statistical average brings us is *reproducibility*.

The details of the statistical average are technically involved and rarely discussed in the literature mainly because the debate about the correct way to go about it is mostly academic. We know what we are supposed to obtain: a time-invariant system. Thus, we do not really have much freedom, which makes it a mostly philosophical issue. A particular approach is superior to another not according to whether it works (it *must* work), but according to how well the author is able to explain *why* it works. This is somewhat analogous to remark, which has been attributed (Laughlin, 2005) to Lev Landau: we could *calculate* the properties of water, but in practice, it makes so much more sense to just measure them. The main reason this is true is because we cannot *design* water. It works just fine already.

Exactly the contrary is true of the spatial average. There are many ways to do it; we do not know ahead of time what we will obtain, and there are non-trivial advantages and disadvantages in the different ways to go about it, some of which lead to novel and even exotic new *applications and designs*. Some of these will be discussed in the following chapters.

For the time being, we will simply state the features we require of the statistical average and we direct the interested reader toward the relevant literature (de Groot, 1969; Mazur, 1958; Robinson, 1973; van Vleck, 1932) in order to focus on the spatial average.

All we require for the present purpose is that statistical averaging has the following key features:

(1) Makes it possible to treat the structure as a linear time-invariant system, in particular allowing us to define inputs, outputs, and transfer functions. Transfer functions are particularly useful since they contain the macroscopic time-invariant information we seek about the *behavior* (as opposed to simply the *state*) of the system.
(2) Incorporates the symmetry properties of the medium into the statistical distribution function by eliminating random spatial fluctuations so as to bring the periodic structures within the reach of Bloch theorem. In other words, the statistical averaged charge distribution of a periodic medium must be periodic.
(3) Includes the effect of the interactions between the particles and the microscopic fields as well as the interactions (classical and quantum) among the particles themselves on the sub-atomic scale. Both interactions within the same atom and with particles in neighboring atoms must be accounted for. These effects would be reflected in the statistical distribution function.

The first point was discussed earlier. Periodic media will be discussed in Section 3.5, and treated in detail in Chapter 5. The third point is essential when considering naturally occurring media where the basic unit is the atom. This is because in classical electromagnetics, atoms do not exist. As a matter of fact, this failure of Maxwell's theory was one of the major motivations for the later development of quantum theory. We, therefore, have two options. Either we stick to purely classical considerations, or we attempt to include, even if only phenomenologically, quantum effects. If we want to remain strictly within the classical domain, then, since atoms are outside its scope, no study of lossless dielectric media is possible, and we are limited to the study of collision-less plasmas, since they can be treated completely using only Maxwell's equations (Dendy, 1990). Otherwise, we must include quantum effects, even if only phenomenologically, through an ad hoc relation introduced at the appropriate moment. That moment will come toward the end of the following section, on the spatial average.

3.2.1 The Spatial Average as Truncation

This section presents the first step leading from the microscopic description of matter, involving point charges moving in empty space, to macroscopic materials described by indexes of refraction. It is a way of averaging the microscopic fields **e** and **b** and the distribution of charge and current $\eta(\mathbf{x}, t)$ and $\mathbf{j}(\mathbf{x}, t)$ to obtain the macroscopic fields **E**, **B**, **P**, and **M**, and the macroscopic charge and current densities $\rho(\mathbf{x}, t)$ and $\mathbf{J}(\mathbf{x}, t)$. If one assumes that the statistical average has been applied and the medium behaves as a linear time-invariant system, one can define permittivity and permeability tensors through the relations

$$\bar{\varepsilon}\mathbf{E} = \varepsilon_0\mathbf{E} + \mathbf{P}$$
$$\bar{\mu}\mathbf{B} = \mu_0(\mathbf{B} + \bar{\mu}\mathbf{M}).$$

The formal asymmetry between these definitions has the benefit of leading to a highly symmetrical formulation of the *macroscopic* Maxwell's equations. It also results in a simple relationship between permittivity and permeability on one hand, and the phenomenological parameters of refractive index n and the impedance Z on the other hand, which in an isotropic medium are defined as

$$n = \sqrt{\mu\varepsilon}$$
$$Z = \sqrt{\frac{\mu}{\varepsilon}}.$$

P and **M** are called the macroscopic polarization and magnetization, respectively, and they represent the overall macroscopic effect of the microscopically complicated distribution of charges and currents.

We start, therefore, with the microscopic Maxwell's equations:

$$\begin{array}{ll} \operatorname{div}\mathbf{b} = 0 & \mathbf{curl}\,\mathbf{e} + \frac{\partial \mathbf{b}}{\partial t} = 0 \\ \operatorname{div}\mathbf{e} = \eta/\varepsilon_0 & \frac{1}{\mu_0}\mathbf{curl}\,\mathbf{b} - \varepsilon_0 \frac{\partial \mathbf{e}}{\partial t} = \mathbf{j} \end{array} \tag{3.1}$$

and with an averaging procedure. Various approaches to the averaging have been put forward: spatial, temporal, or statistical averaging. Russakoff (1970) argued that only spatial averaging was truly necessary in order to consistently define the macroscopic *fields*. However, as argued in the previous section, in order to define macroscopic *parameters* such as the relative permittivity, an additional statistical average is required. The usual macroscopic quantities we are familiar with are, therefore, *both* spatial and statistical averages.

Spatial averaging can be seen from two points of view: as a spatial "sliding average" or as a low-pass filter in reciprocal space or **k**-space. We explain by considering a generic space- and time-dependent quantity $\xi(\mathbf{x}, t)$ though in what follows the time is fixed and we will omit it to avoid cluttering the equations.

In the moving average view, the macroscopic quantity $[\xi(\mathbf{x})]$ is defined at each point by taking the average of the original $\xi(\mathbf{x})$ over a small region centered at **x**. We write

$$[\xi(\mathbf{x})] = \int d^3x' f(\mathbf{x}')\xi(\mathbf{x} - \mathbf{x}') \tag{3.2}$$

where the function $f(\mathbf{x})$ is real, its support is larger than l, it contains the origin where it is nonzero, is normalized to 1: $\oint f(\mathbf{x})dV = 1$, and is radially symmetric in order to preserve the symmetry properties of ξ: $f = f(r)$.[a] This corresponds to a generalized version of our

[a]We treat here only the case where $l < d$. For a discussion on large l, asymmetric charge mobility, and finite size objects, where some of these conditions have to be removed, see Section 3.6.

intuitive notion of a sliding average. The form of the integral above is also known as a *convolution*, and we can rewrite the equation as

$$[\xi(\mathbf{x})] = f(\mathbf{x}) \star \xi(\mathbf{x}) \tag{3.3}$$

where the small circle denotes convolution.

In the low-pass filter view, the average is seen as a truncation of the spatial Fourier transform of the quantity $\xi(\mathbf{x})$ whereby all components with $|\mathbf{k}| > k_0$ are excluded. We apply the convolution theorem to Eq. (3.2):

$$\begin{aligned}[\xi(\mathbf{x})] &= f(\mathbf{x}) \star \xi(\mathbf{x}) \\ &= \mathcal{F}^{-1}(\mathcal{F}(f(\mathbf{x}))\mathcal{F}(\xi(\mathbf{x}))) \\ &= \mathcal{F}^{-1}(\tilde{f}(\mathbf{k})\tilde{\xi}(\mathbf{k}))\end{aligned} \tag{3.4}$$

where the Fourier transform of ξ is denoted as $\mathcal{F}(\xi(\mathbf{x})) = \tilde{\xi}(\mathbf{k})$ and has the specific form

$$\mathcal{F}(\xi(\mathbf{x})) = \int \xi(\mathbf{x})e^{-i\mathbf{k}\cdot\mathbf{x}}d^3x \qquad \mathcal{F}^{-1}(\tilde{\xi}(\mathbf{k})) = \frac{1}{(2\pi)^3}\int \tilde{\xi}(\mathbf{k})e^{i\mathbf{k}\cdot\mathbf{x}}d^3k. \tag{3.5}$$

It is clear that $\tilde{f}(\mathbf{k})$ plays the role of a filter on the frequency components of $\xi(\mathbf{x})$. In our case, we want to remove the microscopic features of ξ, which is equivalent to removing its high-frequency components. $\tilde{f}(\mathbf{k})$ must then be a low-pass filter, a point of view emphasized by Robinson (1973). Moreover, from the well-known general properties of the Fourier transform, we know that if f is well behaved, normalized to 1, and symmetrical, then $\tilde{f} \approx 1$ and $\mathbf{grad}_\mathbf{k}\,\tilde{f} \approx \mathbf{0}$ in some neighborhood of $\mathbf{k} = \mathbf{0}$ and the approximations can be made arbitrarily good in the right neighborhood. The importance of these facts will become clear in the following paragraphs.

Since the convolution commutes with space and time differentiation, when we apply the brackets to Maxwell's equations, we obtain directly

$$\begin{array}{ll}\mathrm{div}[\mathbf{b}] = 0 & \mathbf{curl}[\mathbf{e}] + \frac{\partial[\mathbf{b}]}{\partial t} = 0 \\ \mathrm{div}[\mathbf{e}] = [\eta]/\varepsilon_0 & \frac{1}{\mu_0}\mathbf{curl}[\mathbf{b}] - \varepsilon_0\frac{\partial[\mathbf{e}]}{\partial t} = [\mathbf{j}]\end{array}. \tag{3.6}$$

The macroscopic fields $\mathbf{E}$ and $\mathbf{B}$ are then defined as $\mathbf{E} = [\mathbf{e}]$ and $\mathbf{B} = [\mathbf{b}]$, and in order to obtain the macroscopic equations, we need to write the average charge and current densities, $[\eta]$ and $[\mathbf{j}]$. We will write only the charge density in detail.

We now make two simplifying assumptions.

The first and relatively innocuous one is that the medium as a whole is neutral. This assumption is due to the fact that electromagnetic interactions are so strong compared to the masses of the objects involved that electrical charges will quickly pair up, such that even over microscopic distances (say, several unit cells) most media of interest are all but almost perfectly neutral.

The second assumption, which we call the *atomic assumption* (the medium is composed of *stable atoms*), was discussed above on page 67, and is a direct consequence of the third feature of statistical averaging.

Thus, the whole charge distribution of the medium can be considered a sum over the charge distributions of individual atoms $\eta = \sum_n \eta_n(\mathbf{x} - \mathbf{x}_n)$. In other words, there are no surplus charges. Note that the individual atoms need not be neutral, only collectively.

We now apply Eq. (3.4) to $[\eta(\mathbf{x})]$:

$$\begin{aligned}[\eta(\mathbf{x})] &= \mathcal{F}^{-1}(\tilde{f}(\mathbf{k})\tilde{\eta}(\mathbf{k})) \\ &= \mathcal{F}^{-1}(\tilde{f}(\mathbf{k})\sum_n \tilde{\eta}_n(\mathbf{k})) \\ &= \sum_n \mathcal{F}^{-1}(\tilde{f}(\mathbf{k})\tilde{\eta}_n(\mathbf{k})). \end{aligned} \tag{3.7}$$

Since we have seen that the multiplication by $\tilde{f}(\mathbf{k})$ has the role of a filter that passes only frequency components with $\mathbf{k}$ close to $\mathbf{0}$, it is reasonable to attempt to represent $\tilde{\eta}_n(\mathbf{k})$ as a Taylor series around $\mathbf{k} = \mathbf{0}$ and hope that we may only need to keep a few terms. We have

$$\begin{aligned}\tilde{\eta}_n(\mathbf{k}) &= \tilde{\eta}_n(\mathbf{k})|_{\mathbf{k}=\mathbf{0}} + \mathbf{k}\cdot \mathbf{grad}_{\mathbf{k}}\, \tilde{\eta}_n(\mathbf{k})\big|_{\mathbf{k}=\mathbf{0}} + \mathbf{k}\cdot \widetilde{\mathbf{R}}_n(\mathbf{k}) \\ &= \tilde{\eta}_n(\mathbf{k})|_{\mathbf{k}=\mathbf{0}} + \mathbf{k}\cdot \left(\mathbf{grad}_{\mathbf{k}}\, \tilde{\eta}_n(\mathbf{k})\big|_{\mathbf{k}=\mathbf{0}} + \widetilde{\mathbf{R}}_n(\mathbf{k})\right) \end{aligned} \tag{3.8}$$

where the first term is easily seen as the total net charge q_n, the second term is the dipolar term, while the $\mathbf{k}$-dependent term $\mathbf{R}_n$ collects all the rest of the higher order multipolar terms, which we hope are small; the above equation is, therefore, not an approximation but a true equality. Before going any further, let us try to get a feel for the physical meaning of the quantity in parentheses. Let us assume that the $\widetilde{\mathbf{R}}_n$ term is negligible, and write the gradient term. The interpretation is facilitated if we take the Fourier transform around $\mathbf{x}_n$. From Eq. (3.5), we have

$$\begin{aligned}\mathbf{grad}_{\mathbf{k}}\,\tilde{\eta}_n(\mathbf{k})\big|_{\mathbf{k}=\mathbf{0}} &= \int \eta_n(\mathbf{x})e^{-i\mathbf{k}\cdot(\mathbf{x}-\mathbf{x}_n)}(-i(\mathbf{x}-\mathbf{x}_n))d^3x\bigg|_{\mathbf{k}=\mathbf{0}} \\ &= -i\int(\mathbf{x}-\mathbf{x}_n)\eta_n(\mathbf{x})d^3x \\ &= -i\mathbf{p}_n\end{aligned}$$

where we have introduced $\mathbf{p}_n$, the *equivalent point dipole moment* of the atom, in the limit of $\mathbf{k} \to \mathbf{0}$. We now define the *generalized electric moment*

$$\widetilde{\mathbf{p}}_n(\mathbf{k}) = i\ \mathbf{grad}_{\mathbf{k}}\,\tilde{\eta}_n(\mathbf{k})\big|_{\mathbf{k}=\mathbf{0}} + i\widetilde{\mathbf{R}}_n(\mathbf{k}) = \mathbf{p}_n + i\widetilde{\mathbf{R}}_n(\mathbf{k}) \tag{3.9}$$

Note that the charge distribution of a given atom at any given time t need not be symmetrical, even when there is no external field applied. The dipolar term $\mathbf{grad}_{\mathbf{k}}\,\tilde{\eta}_n(\mathbf{k})\big|_{\mathbf{k}=\mathbf{0}}$, therefore, need not be zero. Summarizing:

$$\tilde{\eta}_n(\mathbf{k}) = q_n - i\mathbf{k}\cdot\widetilde{\mathbf{p}}_n(\mathbf{k}). \tag{3.10}$$

We now write the Taylor expansion of $\tilde{f}(\mathbf{k})$:

$$\begin{aligned}\tilde{f}(\mathbf{k}) &= 1 + \mathbf{k}\cdot\mathbf{grad}_{\mathbf{k}}\,\tilde{f}_n(\mathbf{k})\big|_{\mathbf{k}=\mathbf{0}} + \mathbf{k}\cdot\widetilde{\mathbf{R}}_{\mathrm{f}}(\mathbf{k}) \\ &= 1 + \mathbf{k}\cdot\widetilde{\mathbf{R}}_{\mathrm{f}}(\mathbf{k})\end{aligned} \tag{3.11}$$

where we have used the symmetry of f as mentioned above. The remainder terms $\widetilde{\mathbf{R}}_n$ and $\widetilde{\mathbf{R}}_{\mathrm{f}}$ are by definition null at the origin: $\widetilde{\mathbf{R}}_n(\mathbf{k})\big|_{\mathbf{k}=\mathbf{0}} = \widetilde{\mathbf{R}}_{\mathrm{f}}(\mathbf{k})\big|_{\mathbf{k}=\mathbf{0}} = 0$ and continuous there. Moreover, it is important to note that the $\mathbf{k}$-dependent rest terms $\widetilde{\mathbf{R}}_n(\mathbf{k})$ and $\widetilde{\mathbf{R}}_{\mathrm{f}}(\mathbf{k})$ are not on the same footing, from a physical point of view. While the $\widetilde{\mathbf{R}}_n$ term is related to the microscopic configuration of the medium at the given time, the $\widetilde{\mathbf{R}}_{\mathrm{f}}$ term is related to the properties of the function f, which is a mathematical construct that we can choose as suits our needs. We can, therefore, constrain f to be such that $\widetilde{\mathbf{R}}_{\mathrm{f}}(\mathbf{k})$ be arbitrarily small compared to the other terms in Eq. (3.7). We shall see what this constraint entails in the following sections.

The generic term of Eq. (3.7) takes the form:

$$\begin{aligned}\mathcal{F}^{-1}(\tilde{f}(\mathbf{k})\tilde{\eta}_n(\mathbf{k})) &= \mathcal{F}^{-1}\left(q_n\tilde{f}(\mathbf{k}) - i\mathbf{k}\cdot\widetilde{\mathbf{p}}_n(\mathbf{k})\tilde{f}(\mathbf{k})\right) \\ &= q_n\delta(\mathbf{x}-\mathbf{x}_n)\star f(\mathbf{x}) - \delta'(\mathbf{x}-\mathbf{x}_n)\star\mathcal{F}^{-1}\left(\tilde{f}(\mathbf{k})\widetilde{\mathbf{p}}_n(\mathbf{k})\right) \\ &= q_n f(\mathbf{x}-\mathbf{x}_n) - \mathrm{div}\left(f(\mathbf{x})\star\mathbf{p}_n(\mathbf{x}-\mathbf{x}_n)\right)\end{aligned} \tag{3.12}$$

What is the physical meaning of this result? For the interpretation of the first term, it is sufficient to look at the definition of the smoothing process, Eq. (3.3). We have

$$q_n f(\mathbf{x}-\mathbf{x}_n) = q_n \delta(\mathbf{x}-\mathbf{x}_n) \star f(\mathbf{x}) = [q_n \delta(\mathbf{x}-\mathbf{x}_n)]$$

So from a macroscopic point of view, the net charge of the atom is seen as if the atom consisted of a single point charge q_n localized at the center of the atom, $\mathbf{x}_n$. Even though the actual charge distribution within the atom may be complicated, with many individual point charges spread over a finite volume, the smoothing process wipes out all the detailed information leaving only two aspects: the net charge q_n and the mean position $\mathbf{x}_n$.

The interpretation of the second term is not quite as straightforward. We write the position-dependent polarization vector:

$$\begin{aligned}\mathbf{p}_n(\mathbf{x}-\mathbf{x}_n) &= \mathcal{F}^{-1}(\widetilde{\mathbf{p}}_n(\mathbf{k})) \\ &= \mathcal{F}^{-1}(\mathbf{p}_n + i\widetilde{\mathbf{R}}_n(\mathbf{k})) \\ &= \mathbf{p}_n\delta(\mathbf{x}-\mathbf{x}_n) + i\mathbf{R}_n(\mathbf{x}-\mathbf{x}_n). \end{aligned} \tag{3.13}$$

Note that since $\widetilde{\mathbf{R}}_n(\mathbf{k})$ is null at the origin by definition, this means that $\mathbf{R}_n(\mathbf{x} - \mathbf{x}_n)$ integrates to zero over all space. The electric polarization of the atom, therefore, has two components. One of them is singular, the ideal dipole localized at the center of the atom, while the other is regular and decreases to zero quickly with distance. When the $\mathbf{k}$ dependence of $\widetilde{\mathbf{p}}_n(\mathbf{k})$ is negligible, the homogenization process reduces the atom to a smoothed version of a point dipole $\mathbf{p}_n$ localized at $\mathbf{x}_n$. In the more general case, however, we must write div$[\mathbf{p}_n(\mathbf{x}-\mathbf{x}_n)]$ where the electric moment of the atom cannot be idealized as a point dipole but is smeared out, in a sense, over a finite region of space. We now sum over all the atoms to obtain the total smoothed charge density

$$[\eta(\mathbf{x})] = -\operatorname{div}\mathbf{P}(\mathbf{x}) \tag{3.14}$$

where the macroscopic polarization $\mathbf{P}(\mathbf{x})$ is defined

$$\mathbf{P}(\mathbf{x}) = \left[\sum_n \mathbf{p}_n(\mathbf{x}-\mathbf{x}_n)\right] \tag{3.15}$$

which in the limit of $\mathbf{k} \to \Gamma$ (the origin in reciprocal space) becomes

$$\mathbf{P}^{\Gamma}(\mathbf{x}) = \left[\sum_n \mathbf{p}_n\delta(\mathbf{x}-\mathbf{x}_n)\right].$$

Note that among the consequences of the spatial averaging is the fact that a *discrete* microscopic quantity (the atomic polarization **p**) has been transformed into a *continuous* macroscopic one (the polarization field **P**). It has been argued that this is a general feature of the universe: Macroscopic continuity hides microscopic discreteness. Many examples of this phenomenon can be found in solid state physics in particular (Laughlin, 2005).

It is important to point out that until now in this section, we have made no mention of time. The statistical average discussed in the previous section has not been used at any point. In fact, everything in this section remains valid even if we are dealing with a "snapshot" of a structure with point charges suspended in space at some random positions. However, in this case no connection can be established between any external fields and the state of the system because, in a sense, the system has not had time to "react." We have come to the point where we must introduce such a connection, and the only thing that makes this possible is the fact that we assume the statistical average has already been performed. The relation in question is the one connecting the microscopic electric field and the microscopic polarization, i.e., the *definition of polarizability*:

$$\mathbf{p}_n(\mathbf{x} - \mathbf{x}_n) = \varepsilon_0 \gamma_n^{\mathrm{e}}(\mathbf{x} - \mathbf{x}_n)\mathbf{e}(\mathbf{x}) \tag{3.16}$$

This equation would be meaningless if the statistical average had *not* been taken, because we want γ_n^{e} to be a *time-invariant property* of the internal structure of atom *n*, not just the (randomly fluctuating in time) factor of proportionality between the polarization and the electric field at some particular instant in time *t*. Thus, each atom will be treated as a black box, and the interactions between the particles inside will be inaccessible to us. Some of these interactions may be electrical in nature, but some will clearly not be.[b] Thus, the internal dynamics of the atom, as well as the modification in its internal dynamics due to the presence of neighboring atoms' electron clouds, are encapsulated within the parameter γ_n^{e} such that the total polarization can be written as a sum over the atoms, and

[b]As Robinson (1973) observes, "[...] for all we are concerned atoms could equally well be held together with glue and rubber bands".

Eq. (3.15) may be written:

$$\mathbf{P}(\mathbf{x}) = \left[\sum_n \mathbf{p}_n(\mathbf{x} - \mathbf{x}_n)\right] = [\mathbf{p}(\mathbf{x})] = [\varepsilon_0 \gamma^{\mathrm{e}}(\mathbf{x})\mathbf{e}(\mathbf{x})] \qquad (3.17)$$

where

$$\mathbf{p}(\mathbf{x}) = \sum_n \mathbf{p}_n(\mathbf{x} - \mathbf{x}_n) \text{ and } \gamma^{\mathrm{e}}(\mathbf{x}) = \sum_n \gamma_n^{\mathrm{e}}(\mathbf{x} - \mathbf{x}_n).$$

We see that the effect of taking a statistical average in an atomic medium is to replace the two fundamental interacting quantities **e** and η with the quantities **e** and **p**. The charges are displaced by an electric field according to Eq. (3.16), while the polarization charge density $\eta(\mathbf{x}) = -\operatorname{div}\mathbf{p}(\mathbf{x})$ produces an electric field according to Coulomb's law,

$$\mathbf{e}_{\mathrm{pol}}(\mathbf{x}) = \frac{1}{4\pi\varepsilon_0}\,\mathbf{grad}_{\mathbf{x}} \int \frac{\operatorname{div}_{\mathbf{x}'} \mathbf{p}(\mathbf{x}')}{|\mathbf{x} - \mathbf{x}'|} d^3x'. \qquad (3.18)$$

The net result of statistical averaging is, therefore, that it is now possible to define a time-independent polarizability, such that the time dependence of the polarization **p** (and, therefore, the charge distribution η) is tied directly to the time dependence of the electric field. The two equations $\operatorname{div}\mathbf{e}(\mathbf{x}, t) = \eta(\mathbf{x}, t)/\varepsilon_0$ and $\mathbf{F}_j(t) = q_j\mathbf{e}(\mathbf{x}_j, t)$ have been replaced by Eqs. (3.18) and (3.16), respectively. There are no more forces and no more point charges, only two position- and time-dependent continuous fields **e** and **p**, whose time dependence is synchronous if the atomic polarizability is real. The total electric field then satisfies the equation

$$\begin{aligned}\mathbf{e}(\mathbf{x}) &= \mathbf{E}_{\mathrm{ext}}(\mathbf{x}) + \mathbf{e}_{\mathrm{pol}}(\mathbf{x}) \\ &= \mathbf{E}_{\mathrm{ext}}(\mathbf{x}) + \frac{1}{4\pi\varepsilon_0}\,\mathbf{grad}_{\mathbf{x}} \int \frac{\operatorname{div}_{\mathbf{x}'} \gamma^{\mathrm{e}}(\mathbf{x}')\mathbf{e}(\mathbf{x}')}{|\mathbf{x} - \mathbf{x}'|} d^3x'.\end{aligned}$$

Before moving on, we must point out a very important and deep aspect. Equation (3.16) introduces a new quantity, the polarizability, which is designed to encapsulate (i.e., hide from view) all the complex quantum interactions taking place within a unit cell. Since this is only possible once the statistical average has been performed, this immediately implies that Maxwell's equations are *not* invariant with respect to statistical averaging.[c] A priori, the microscopic Maxwell's equations (3.1) and the macroscopic Maxwell's equations (3.32–3.35) are completely different objects: The former do not allow atoms to exist, while the latter depend crucially on their existence! Obtaining the macroscopic equations requires including quantum phenomena, so there is no obvious reason why they *should* have the same form as the microscopic equations. The fact that they **do** is extremely surprising and mysterious.

3.3 Polarizability and Susceptibility

Dielectric media are linear systems when the field intensity is not too large. As with any linear system, there are a number of degrees of freedom, some of which are of interest and some of which are either not of interest or in any case inaccessible or unobservable (Antsaklis and Michel, 2006). In the case of dielectric media, which are composed of extremely large numbers of extremely small particles, the unobservable parameters are those related to the microscopic degrees of freedom of the particles. The external description (Antsaklis and Michel, 2006) of dielectric media, also known as the macroscopic description, therefore, must be obtained by averaging over a large number of inaccessible microscopic degrees of freedom. Distances on this microscopic scale are, therefore, meaningless from the point of view of the macroscopic description. This leads to a spatial smearing, which implies that the observable properties of the material at a given position x are in fact the result of a large number of individual microscopic interactions over a whole region surrounding the point x. What happens at x, therefore, depends to some extent on the conditions prevailing in a certain volume v *surrounding* x. This

[c]This is a fact with very deep implications. It implies that the inhomogeneous Maxwell's equations in matter are physically hybrid because they would not exist without quantum mechanics. Moreover, since *they have the exact same form in matter as in vacuum*, this has led some to speculate about space itself as a fundamentally inhomogeneous "material," and the familiar continuous Maxwell's equations emerging in the limit of large wavelengths from some yet unknown underlying physics (Laughlin, 2005; Rovelli, 2004).

is what we refer to as the "electromagnetic neighborhood" of x (also mentioned at the end of Section 3.1. If we apply, therefore, some position-dependent stimulus to the material, the macroscopic response of the medium at x will depend on the value of the stimulus over the whole volume v. When the stimulus is approximately constant over that volume, we will say the medium looks local because the very *existence* of an "electromagnetic neighborhood" (not to mention its size and properties) is hidden from view. When the stimulus varies non-negligibly over that volume, the medium will look nonlocal, because the behavior starts to be affected by the size and properties of the volume v. It is important to emphasize, therefore, that locality or nonlocality refers not, strictly speaking, to a property of the medium but mostly to how it is probed. The same medium may be local at a given frequency and nonlocal at another. In other words, nonlocality is not a feature of an object, but of an *interaction*.

3.3.1 The Master Equations: Electric and Magnetic

If the stimulus is the macroscopic electric field noted $\mathbf{E}$ and the response is the polarization $\mathbf{P}$, then we can define the transfer function of the medium through the relation

$$\mathbf{P}(\mathbf{x}) = \varepsilon_0 \chi^{\mathrm{e}}(\mathbf{x}) \star \mathbf{E}(\mathbf{x}) = \varepsilon_0 \chi^{\mathrm{e}}(\mathbf{x}) \star [\mathbf{e}(\mathbf{x})]. \tag{3.19}$$

The microscopic field inside the structure is, in turn, composed of two contributions, one external and one internal, due to the polarization charge density in the medium, given by $\mathbf{e}_{\mathrm{pol}}$ of Eq. (3.18):

$$\mathbf{e}(\mathbf{x}) = \mathbf{E}_{\mathrm{ext}}(\mathbf{x}) + \mathbf{e}_{\mathrm{pol}}(\mathbf{x}).$$

The nonlocal nature of the transfer function might seem peculiar given that the interaction between charged particles and electromagnetic fields is local, according to the electrostatic Lorentz force equation $\mathbf{F}_n = q_n \mathbf{E}(\mathbf{x}_n)$, where $\mathbf{F}_n$ is the force on particle n located at $\mathbf{x}_n$. The force on a particle at $\mathbf{x}_n$ depends only on the electromagnetic fields at $\mathbf{x}_n$.

This, however, is a purely classical view, a view that, as we have seen in the section on statistical averaging, is incompatible with any consideration of a medium composed of stable atoms. The statistical averaging is the procedure that accounts for these quantum effects, resulting in a continuous charge distribution and rendering the classical notion of *position* of any given charge henceforth meaningless. The spatial averaging required to eliminate the oscillations of the electric field due to the periodicity of the lattice contributes even further to this blurring. Since both averages involve a loss of information about the positions and velocities of particles, they implicitly render the description nonlocal.

From a historical point of view, the definition of the response function of a dielectric medium as nonlocal in direct space (and therefore local in reciprocal space) can be understood by considering the fact that from a classical perspective, where light is seen as a wave, it does not make much sense to insist on the notion of position of the wave, and rather more on the frequency (temporal or spatial). It was far easier to fix the wavelength of a light wave in an experiment than its exact position; indeed, the very notion of the position of a wave seemed meaningless. Transfer functions local in direct space are typical of particle-like behavior, while transfer functions local in reciprocal space are typical of wave-like behavior. Since the dual wave/particle nature of light became known rather late, *after* the work on the photoelectric effect in the early years of the 20th century, the nonlocal wave-like description was (and for most purposes remains) the most natural.

We now compare Eq. (3.19) with Eq. (3.17) obtaining the master equation of the dielectric medium

$$\varepsilon_0 f(\mathbf{x}) \star (\gamma^{\mathrm{e}}(\mathbf{x})\mathbf{e}(\mathbf{x})) = \varepsilon_0 \chi^{\mathrm{e}}(\mathbf{x}) \star [\mathbf{e}(\mathbf{x})] \tag{3.20}$$

where we have assumed zero intrinsic polarization at zero field (no ferro-electricity). The susceptibility χ^{e} is, therefore, a macroscopic quantity *defined as a relationship between two macroscopic quantities* rather than as an average of some microscopic quantity. As such it is a macroscopic property whose relationship to the microscopic description is indirect and intuitively slippery. While χ^{e} is obviously fundamentally dependent on γ^{e}, it is far from clear in what way (if any) this dependence may be made more explicit or straightforward in the general case. The difficulty resides in the fact that in this equation, we see an intricate interplay of the micro-

and macro-, worlds that are intuitively and physically apart and no straightforward or smooth transition is possible.

For the magnetic case, one starts from the magnetic analog of Eq. (3.14), which has a somewhat more complex form:

$$[\mathbf{j}(\mathbf{x})] = \frac{\partial[\mathbf{p}(\mathbf{x})]}{\partial t} + \mathbf{curl}[\mathbf{m}(\mathbf{x})] \text{ where } \mathbf{m}(\mathbf{x}) = \sum_n \mathbf{m}_n(\mathbf{x} - \mathbf{x}_n) \tag{3.21}$$

and $\mathbf{m}_n$ is the atomic magnetic moment. By noting the macroscopic magnetization $\mathbf{M}(\mathbf{x}) = [\mathbf{m}(\mathbf{x})]$ and defining the magnetic field $\mathbf{H}(\mathbf{x})$

$$\mathbf{H}(\mathbf{x}) = \frac{\mathbf{B}(\mathbf{x})}{\mu_0} - \mathbf{M}(\mathbf{x}) \tag{3.22}$$

it is straightforward (Section 1.5 of Ref. (Cabuz, 2007)) to obtain the magnetic analog of Eq. (3.20)

$$f(\mathbf{x}) \star (\gamma^{\mathrm{m}}(\mathbf{x})\mathbf{b}(\mathbf{x})) = \chi^{\mathrm{m}}(\mathbf{x}) \star \mathbf{H}(\mathbf{x}). \tag{3.23}$$

Equations (3.20) and (3.23) provide the starting point for calculating the susceptibilities of a given medium in the semi-classical approximation. More rigorously, however, in order to obtain the polarizability of atoms, their internal dynamics must be considered and one cannot avoid a detailed quantum mechanical analysis. A detailed understanding is, therefore, a very ambitious enterprise (Delerue and Lannoo, 2004; Haug and Koch, 2004; Mahan, 2000). Fortunately, this will not be the case for artificial materials since both the unit cell and the macroscopic description are governed by the same equations, *macroscopic* Maxwell's equations. An exact description is, therefore, more easily accessible.

The simplest illustration of Eq. (3.20) is for the case of isolated independent atoms, with no near-field coupling or electron cloud overlap. The field seen by any given atom is no longer the *total* field due to all the other atoms but is dominated by the dipolar *radiated* field due to all the other atoms. The distinction between the total field and the radiated field of a distribution of charges (an atom, or any other scatterer) is important because the radiated field excludes the fields over the region *occupied* by the said distribution. A multipole expansion, for instance, is only given with respect to some closed surface that must completely enclose the charge and is valid only *outside* of it. If we consider the field of a point dipole placed at the origin, we have

$$\mathbf{e}_{\text{dipole}}(\mathbf{x}) = \frac{1}{4\pi\varepsilon_0}\,\text{div}\left(\mathbf{p}\,\mathbf{grad}\,\frac{1}{\|\mathbf{x}\|}\right)$$
$$= \frac{1}{4\pi\varepsilon_0}\frac{(3\hat{\mathbf{x}}(\hat{\mathbf{x}}\cdot\mathbf{p})-\mathbf{p})}{\|\mathbf{x}\|^3} - \frac{\mathbf{p}}{3\varepsilon_0}\delta(\mathbf{x}) \qquad (3.24)$$

whereas the radiated field (in the static limit) of the same dipole is

$$\mathbf{e}^*_{\text{dipole}}(\mathbf{x}) = \frac{1}{4\pi\varepsilon_0}\frac{(3\hat{\mathbf{x}}(\hat{\mathbf{x}}\cdot\mathbf{p})-\mathbf{p})}{\|\mathbf{x}\|^3} \qquad (3.25)$$

since the singular term encapsulates the localized fields that are not seen by neighboring scatterers. $\hat{\mathbf{x}}$ is the unit vector in the direction of $\mathbf{x}$. In what follows, we will distinguish radiated fields from total fields with a star in the superscript. The field $\mathbf{e}^*(\mathbf{x})$ seen by any particular dipole can, therefore, be written in terms of the total field as

$$\mathbf{e}^*(\mathbf{x}) = \mathbf{e}(\mathbf{x}) + \sum_n \frac{\mathbf{p}_n\delta(\mathbf{x}-\mathbf{x}_n)}{3\varepsilon_0}. \qquad (3.26)$$

Therefore, for this case, Eq. (3.20) must be written as

$$f(\mathbf{x}) \star (\gamma^{\text{e}}(\mathbf{x})\mathbf{e}^*(\mathbf{x})) = \chi^{\text{e}}(\mathbf{x}) \star [\mathbf{e}(\mathbf{x})]. \qquad (3.27)$$

We begin by writing the total polarizability $\gamma^{\text{e}}(\mathbf{x})$ as a sum over the atomic polarizabilities $\gamma_n^{\text{e}}(\mathbf{x})$

$$f(\mathbf{x}) \star (\gamma^{\text{e}}(\mathbf{x})\mathbf{e}^*(\mathbf{x})) = f(\mathbf{x}) \star \left(\sum_n \gamma_n^{\text{e}}(\mathbf{x})\mathbf{e}^*(\mathbf{x})\right). \qquad (3.28)$$

Since we have assumed that $\gamma_n^{\text{e}}(\mathbf{x})$ is very localized, its Fourier transform is well represented by the zeroth term of its Taylor expansion, such that $\widetilde{\gamma}_n^{\text{e}}(\mathbf{k}) \approxeq \widetilde{\gamma}_n^{\text{e}}(\mathbf{0}) \equiv \gamma_{\text{s}}^{\text{e}}$ (all atoms are identical) and we have $\gamma_n^{\text{e}}(\mathbf{x}) \approxeq \gamma_{\text{s}}^{\text{e}}\delta(\mathbf{x}-\mathbf{x}_n)$ where the s subscript stands for the singular or DC component of the polarizability. The left side of Eq. (3.20) becomes

$$f(\mathbf{x}) \star (\gamma^{\text{e}}(\mathbf{x})\mathbf{e}^*(\mathbf{x})) = f(\mathbf{x}) \star \left(\mathbf{e}^*(\mathbf{x})\sum_n \gamma_{\text{s}}^{\text{e}}\delta(\mathbf{x}-\mathbf{x}_n)\right).$$

We take the Fourier transform of the above equation to obtain

$$\tilde{f}(\mathbf{k})\left(\widetilde{\mathbf{e}}^*(\mathbf{k}) \star \frac{\gamma_{\text{s}}^{\text{e}}}{V_{\text{uc}}}\sum_n \delta(\mathbf{k}-\mathbf{G}_n)\right) \qquad (3.29)$$

where the Fourier transform of a Dirac comb is also a Dirac comb in reciprocal space, $\mathbf{G}_n$ are the reciprocal lattice vectors, and V_{uc} is the volume of the unit cell in real space (see Eq. 2.12 of Ref. (Kittel, 1996)). We note $N = 1/V_{\mathrm{uc}}$, the atomic number density. The electric field is quasiperiodic as per Bloch theorem, such that its Fourier transform can be written as

$$\widetilde{\mathbf{e}}^*(\mathbf{k}) = \sum_m \mathbf{E}_m^* \delta(\mathbf{k} - \mathbf{k}_{\mathrm{B}} - \mathbf{G}_m)$$

and we can rewrite expression (3.29) as

$$N\gamma_{\mathrm{s}}^{\mathrm{e}} \tilde{f}(\mathbf{k}) \left(\sum_{n,m} \mathbf{E}_m^* \delta(\mathbf{k} - \mathbf{k}_{\mathrm{B}} - \mathbf{G}_n - \mathbf{G}_m) \right).$$

We must now recall that the function $\tilde{f}(\mathbf{k})$ has been *designed* in order to filter out all spatial frequencies that are not in the first Brillouin zone. Consequently, in the above sum, only those terms will survive where $\mathbf{G}_n + \mathbf{G}_m = \mathbf{0}$ such that the delta function is at $\mathbf{k}_{\mathrm{B}}$ and, therefore, within the first Brillouin zone. This relation is satisfied when $m = -n$; therefore, for any m nonzero, there will be exactly one n such that the term survives. The $\mathbf{E}_0$ term is not filtered out because it is already by default in the first Brillouin zone. We can, therefore, rewrite the above expression as

$$N\gamma_{\mathrm{s}}^{\mathrm{e}} \tilde{f}(\mathbf{k}) \left(\mathbf{E}_0^* \delta(\mathbf{k} - \mathbf{k}_{\mathrm{B}}) + \sum_{n \neq 0} \mathbf{E}_n^* \delta(\mathbf{k} - \mathbf{k}_{\mathrm{B}}) \right).$$

Now, the $\mathbf{E}_n$ coefficients are the Fourier coefficients of a Bloch wave, and as such they have some dependence on the Bloch vector $\mathbf{k}_{\mathrm{B}}$. When this vector is large (approaching the edges of the Brillouin zone), this dependence is strong; the quantity in parentheses becomes a function of $\mathbf{k}_{\mathrm{B}}$, leading to a spatially dispersive medium. However, when this vector approaches $\mathbf{0}$, the dependence diminishes and the Fourier coefficients become, to a good approximation, independent of $\mathbf{k}_{\mathrm{B}}$. Moreover, when the medium is highly symmetrical (cubic symmetry), one can show (Jackson, 1999) that the sum over $\mathbf{E}_n^*$ reduces to zero and we obtain

$$\begin{aligned} f(\mathbf{x}) \star (\gamma^{\mathrm{e}}(\mathbf{x}) \mathbf{e}^*(\mathbf{x})) &= \mathcal{F}^{-1} \left(N\gamma_{\mathrm{s}}^{\mathrm{e}} \tilde{f}(\mathbf{k}) \left(\mathbf{E}_0^* \delta(\mathbf{k} - \mathbf{k}_{\mathrm{B}}) \right) \right) \\ &= N\gamma_{\mathrm{s}}^{\mathrm{e}} \left[\mathbf{e}^*(\mathbf{x}) \right] \end{aligned}$$

This only holds for cubic crystals. The averaged radiation field can be written in terms of the averaged total field by using Eqs. (3.24) and (3.25) and the fact that the unit cell contains only one atom:

$$\begin{aligned} N\gamma_s^e\,[\mathbf{e}^*(\mathbf{x})] &= N\gamma_s^e\left[\mathbf{e}(\mathbf{x}) + \sum_n \frac{\mathbf{p}_n\delta(\mathbf{x}-\mathbf{x}_n)}{3\varepsilon_0}\right] \\ &= N\gamma_s^e\left([\mathbf{e}(\mathbf{x})] + \frac{1}{3\varepsilon_0}[\mathbf{p}(\mathbf{x})]\right) \\ &= N\gamma_s^e\left([\mathbf{e}(\mathbf{x})] + \frac{1}{3}\chi^e(\mathbf{x}) \star [\mathbf{e}(\mathbf{x})]\right). \end{aligned}$$

Putting this back into the definition of the susceptibility, we have

$$N\gamma_s^e\left([\mathbf{e}(\mathbf{x})] + \frac{1}{3}\chi^e(\mathbf{x}) \star [\mathbf{e}(\mathbf{x})]\right) = \chi^e(\mathbf{x}) \star [\mathbf{e}(\mathbf{x})] \tag{3.30}$$

and by grouping the terms containing χ, we obtain

$$\chi^e(\mathbf{x}) \star [\mathbf{e}(\mathbf{x})] = \frac{N\gamma_s^e}{1 - N\gamma_s^e/3}[\mathbf{e}(\mathbf{x})].$$

The susceptibility is, therefore, singular:

$$\chi^e(\mathbf{x}) = \frac{N\gamma_s^e}{1 - N\gamma_s^e/3}\delta(\mathbf{x}) \tag{3.31}$$

and we recognize *the Mossotti–Clausius relation*. A similar relation holds in the magnetic case, where the "e" superscripts are replaced by "m" (see Section 1.5.1 of Ref. (Cabuz, 2007)).

It is important to note that this relation is only valid for isolated atoms and in the cubic symmetry case. Its widespread use and generality, however, come from the ability to use an effective polarizability, which includes some of these other effects. When this is done, however, it becomes a reverse-engineered relation, in the sense that the quantities are defined in such a way that it holds, and it is no longer useful for purposes of *design* but only as a descriptive tool.

3.4 Permittivity and Permeability: Index and Impedance

Now that we have obtained macroscopic parameters that characterize the behavior of the material, we would like to write the full macroscopic Maxwell's equations in order to obtain the plane wave solutions. The spatial and statistical averaged divergence equation

$$\operatorname{div}[\mathbf{e}(\mathbf{x})] = [\eta(\mathbf{x})]/\varepsilon_0$$

becomes (using Eq. (3.14))

$$\varepsilon_0 \operatorname{div} \mathbf{E}(\mathbf{x}) = -\operatorname{div} \mathbf{P}(\mathbf{x}).$$

We now introduce the electric displacement vector

$$\mathbf{D}(\mathbf{x}) = \varepsilon_0 \mathbf{E}(\mathbf{x}) + \mathbf{P}(\mathbf{x}) = \varepsilon_0(\delta(\mathbf{x}) + \chi^{\mathrm{e}}(\mathbf{x})) \star \mathbf{E}(\mathbf{x}) = \varepsilon_0 \varepsilon(\mathbf{x}) \star \mathbf{E}(\mathbf{x})$$

where the relative permittivity (also often referred to abusively as the dielectric "constant") is defined in general by $\varepsilon(\mathbf{x}) = \delta(\mathbf{x}) + \chi^{\mathrm{e}}(\mathbf{x})$ which reduces to $\varepsilon(\mathbf{x}) = 1 + \chi^{\mathrm{e}}(\mathbf{x})$ for local media. The electric divergence equation in a medium with no free charges now takes the simple form

$$\operatorname{div} \mathbf{D}(\mathbf{x}) = 0.$$

If we use the newly introduced electric displacement field, the macroscopic Maxwell–Amèpre equation is then written as

$$\mathbf{curl}\, \mathbf{H}(\mathbf{x}) - \frac{\partial \mathbf{D}(\mathbf{x})}{\partial t} = 0.$$

By rewriting Eq. (3.22), we have

$$\mathbf{B}(\mathbf{x}) = \mu_0 \mathbf{H}(\mathbf{x}) + \mu_0 \chi^{\mathrm{m}}(\mathbf{x}) \star \mathbf{H}(\mathbf{x}) = \mu_0 \mu(\mathbf{x}) \star \mathbf{H}(\mathbf{x})$$

where the relative permeability is defined in general by $\mu(\mathbf{x}) = \delta(\mathbf{x}) + \chi^{\mathrm{m}}(\mathbf{x})$ which reduces to $\mu(\mathbf{x}) = (1 + \chi^{\mathrm{m}}(\mathbf{x}))\delta(\mathbf{x})$ in local media.

We can now write the complete source-free macroscopic Maxwell's equations:

$$\mathbf{curl}\, \mathbf{H}(\mathbf{x}) + i\omega \mathbf{D}(\mathbf{x}) = 0 \tag{3.32}$$

$$\operatorname{div} \mathbf{B}(\mathbf{x}) = 0 \tag{3.33}$$

$$\mathbf{curl}\, \mathbf{E}(\mathbf{x}) - i\omega \mathbf{B}(\mathbf{x}) = 0 \tag{3.34}$$

$$\operatorname{div} \mathbf{D}(\mathbf{x}) = 0 \tag{3.35}$$

and the corresponding constitutive relations

$$\mathbf{B}(\mathbf{x}) = \mu_0 \mu(\mathbf{x}) \star \mathbf{H}(\mathbf{x})$$

$$\mathbf{D}(\mathbf{x}) = \varepsilon_0 \varepsilon(\mathbf{x}) \star \mathbf{E}(\mathbf{x})$$

We see immediately one of the benefits of the homogenization procedure. Whereas on a microscopic scale the electric and magnetic phenomena are inevitably asymmetric, from a macroscopic point of view, electric and magnetic phenomena in source-free regions are, *at least formally*, perfectly symmetric, as seen by inspecting the above equations.

3.4.1 The Negative Index of Refraction

An important set of solutions to these equations is the plane waves

$$\mathbf{E}(\mathbf{x}) = \begin{pmatrix} 0 \\ E_y \\ E_z \end{pmatrix} e^{ikx - i\omega t}. \tag{3.36}$$

This represents a plane wave of frequency ω propagating in the positive x direction. k is the wavevector, and by the Helmholtz equation, it is related to ω through the relation $k^2 = \mu_0\varepsilon_0\mu\varepsilon\omega^2$ where if we denote the speed of light in a vacuum $c = 1/\sqrt{\mu_0\varepsilon_0}$ and we introduce the index $n^2 = \mu\varepsilon$, then we can rewrite it as $k^2 = n^2\omega^2/c^2$. It is interesting to consider the lossy case. If the electric field and the polarization of the medium are not exactly in phase, then the permittivity and/or the permeability must have a nonzero imaginary part. It is easily seen that this imaginary part is positive for lossy media and negative for gain media when the time dependence is of the form $e^{-i\omega t}$, because if k has a positive imaginary part, the wave is attenuated as it propagates along the x axis. We write $k = \beta + i\alpha$ with α, β real and $\alpha > 0$. We have

$$\begin{aligned} \mu &= \mu' + i\mu'' \\ \varepsilon &= \varepsilon' + i\varepsilon'' \\ n^2 &= \mu'\varepsilon' - \mu''\varepsilon'' + i(\mu'\varepsilon'' + \varepsilon'\mu'') \end{aligned}$$

with μ'', $\varepsilon'' > 0$. Once the permittivity and the permeability have been determined for a given material, then the wave propagates with a wavevector that can be determined from the equations

$$\beta^2 - \alpha^2 = \mu'\varepsilon' - \mu''\varepsilon'' \tag{3.37}$$

$$2\alpha\beta = \mu'\varepsilon'' + \varepsilon'\mu'' \tag{3.38}$$

Keep in mind that $\beta = \text{real}(n)c^2/\omega^2$ and $\alpha = \text{real}(n)c^2/\omega^2$. In particular, if the imaginary parts are much smaller than the real parts of the permeability and permittivity, but the real parts of both are negative, then β must be negative also, as seen from the second equation. Consequently, *a material with negative real parts of permittivity and permeability will exhibit a negative real part of the index*. It is important to note, however, that this condition is sufficient but not necessary, at least not in lossy media. In other words, the real part of the index can be negative even when one of the real parts of either the permittivity or permeability is positive. This can be seen from Eq. (3.38) if one keeps in mind that α, ε'', and μ'' must all be positive in a passive medium with the sign conventions chosen in this work.

To understand what it means to have a negative index of refraction, let us consider the propagation of a linearly polarized wave in the region of space $0 < z < Z$. We assume this region contains no sources, so that the field satisfies the homogeneous Helmholtz equation. We write the total field (either the electric or magnetic, indifferently, we will denote it V):

$$V(x, y, z, t) = U(x, y, z)e^{-i\omega t}$$

and the spatial part $U(x, y, z)$ satisfies

$$\left(\triangle + k^2\right) U(x, y, z) = 0$$

where we have written $k^2 = \mu_0\varepsilon_0\mu\varepsilon\omega^2 = \mu\varepsilon\frac{\omega^2}{c^2} = \mu\varepsilon k_0^2$. The field in a plane $z =$ constant can be represented as a Fourier integral

$$U(x, y, z) = \int\int_{-\infty}^{\infty} \widetilde{U}(u, v; z)e^{i(ux+vy)}dudv. \tag{3.39}$$

Replacing this in the Helmholtz equation, we obtain

$$\int\int_{-\infty}^{\infty} \left(\triangle + k^2\right)\left(\widetilde{U}(u, v; z)e^{i(ux+vy)}\right) dudv = 0$$

or

$$\int\int_{-\infty}^{\infty} \left(\left(k^2 - u^2 - v^2\right)\widetilde{U}(u, v; z) + \frac{\partial^2\widetilde{U}(u, v; z)}{\partial z^2}\right) e^{i(ux+vy)}dudv = 0$$

This is the Fourier development of the null function, so each coefficient must be null independently. We have

$$\frac{\partial^2\widetilde{U}(u, v; z)}{\partial z^2} + \left(k^2 - u^2 - v^2\right)\widetilde{U}(u, v; z) = 0.$$

If we introduce $w^2 = k^2 - u^2 - v^2$, then the general solution to this equation takes the form

$$\widetilde{U}(u, v; z) = A(u, v)e^{iwz} + B(u, v)e^{-iwz}.$$

If we assume that the sources of the fields are all in the $z < 0$ half-space, then all waves must propagate in the positive z direction and $B(u, v) = 0$. We obtain

$$\widetilde{U}(u, v; z) = A(u, v)e^{iwz} \tag{3.40}$$

It is, therefore, clear that in the plane $z = 0$, the Fourier components of the field distribution are given by $A(u, v)$. As the field propagates in the z direction, the Fourier composition in the plane $z = Z$ is $\widetilde{U}(u, v; Z) = A(u, v)e^{iwZ}$. The propagation in an isotropic homogeneous medium, therefore, has the effect of transforming the Fourier components of the field distribution according to the factor e^{iwz}. In the following, we shall refer to it as the evolution operator, by analogy to the quantum mechanical time evolution operator. The parameter w, therefore, seems to be of paramount importance. Recall that it is defined as

$$w^2 = \mu\varepsilon k_0^2 - u^2 - v^2.$$

In this relation, all quantities are real except possibly μ, ε, and w. As in Section 3.4, we introduce the index $n^2 = (\beta + i\alpha)^2 = \mu\varepsilon$ where β and α are real, α is positive, and they are given by Eqs. (3.37) and (3.38). As before, the imaginary part α is only introduced in order to determine the signs of the real parts, and it is then made to tend to zero. We also write $w = b + ia$. Since no active media are present, all imaginary parts must be positive. We write the real and imaginary parts of the above relation:

$$\begin{aligned} b^2 - a^2 &= (\beta^2 - \alpha^2)k_0^2 - u^2 - v^2 \\ ab &= \alpha\beta k_0^2 \end{aligned}$$

By letting α tend to zero in the second equation, it results that either a or b must also go to zero. It is the first equation that will determine which. For vanishing α, we have $b^2 - a^2 = \beta^2 k_0^2 - u^2 - v^2$. When $u^2 + v^2 < \beta^2 k_0^2$, the quantity is positive and a must be the one that vanishes along with α. On the other hand, if $u^2 + v^2 > \beta^2 k_0^2$, then b must vanish. The sign of the remaining quantity is determined by the sign of β from the second relation above. We have already seen in Section 3.4 that when the medium is double positive, β is positive, but that when it is double negative, β is negative. The different possibilities are summarized in Table 3.1.

From the table it is clear that features corresponding to large spatial frequencies u, v correspond to evanescent waves, while low-frequency features correspond to propagating waves. In the course of propagating between the planes $z = 0$ and $z = Z$, the low-frequency components have undergone a unitary transformation,

Table 3.1 Table of the possible behaviors of w

	$u^2+v^2<\beta^2k_0^2$	$u^2+v^2>\beta^2k_0^2$
$\beta>0$	w real positive	w imaginary positive
	$w=\sqrt{\beta^2k_0^2-u^2-v^2}$	$w=+i\sqrt{u^2+v^2-\beta^2k_0^2}$
$\beta<0$	w real negative	w imaginary negative
	$w=-\sqrt{\beta^2k_0^2-u^2-v^2}$	$w=-i\sqrt{u^2+v^2-\beta^2k_0^2}$

or a change of phase. In the case of the high spatial frequencies, however, the phase does not evolve, but the amplitude does. When β is positive, this amplitude is attenuated, while when β is negative, this amplitude is amplified. If we note w_p for a double positive medium and w_n for the corresponding double negative medium, then from the above table we have

$$w_\mathrm{p}=-w_\mathrm{n}.$$

The evolution of the field in the z direction in the material with $\beta=-1$ is exactly the reverse from the evolution of the field when $\beta=1$ if losses are ignored. In fact, it is as if the time runs backwards. This is not an accident. If we consider the Maxwell curl equations (3.32) and (3.34), then changing the signs of the constitutive parameters μ and ε is formally equivalent to taking the inverse of the time-dependent term $e^{-i\omega t}$.

It is tempting now to consider a system of two such complementary slabs by simply multiplying the exponential evolution operators. However, this is in general not correct. The reason is that we have assumed that the sources of all fields were to the left of the region of interest and that consequently all fields propagate in the same direction, the positive z direction. But this can only be the case if the medium is homogeneous *and infinite* in the z direction. If an interface or a scattering element of any kind is present, then this is no longer true. In such cases, we must consider both left- and right-going waves and the way they couple at interfaces. This is done by calculating the transmission and reflection of each wave at each interface, by employing the notion of impedance. Without going into the details, we will only point out that these reflections are absent when the materials are matched. If the two media have the same impedance, then our assumption is justified and it is possible to simply multiply the evolution operators.

It is then possible to consider a region of free space of width z as a filter with a transfer function given by the evolution operator e^{iwz}, while a similar region filled with $\mu = \varepsilon = -1$ placed next to the first provides the *inverse* filter, e^{-iwz}. After propagation through the two layers, the field is reproduced *exactly*. It is, however, well known in the theory of linear systems that inverse filtering is sensitive to noise. If the initial filter has reduced the amplitude of some frequency components to values close to the noise amplitude at those frequencies, then when the inverse filter re-amplifies them, it amplifies the noise as well, resulting in a very noisy reconstructed signal. A way to avoid this problem is to avoid small-signal amplitudes. Since extinction and amplification are given by a term exponential in the distance z, then it may be advantageous to use many thin alternating regions of double positive and double negative media, rather than two thick ones. As long as they both occupy equal volumes, the signal will be reproduced exactly. The noise limitation also places an upper bound on the distance between an object and the surface of the lens, for any given required resolution. The higher the resolution we seek, the higher the k components that must be resolved. Higher k components, in turn, attenuate faster with distance, which means that the object must be placed closer to the lens surface in order for the signal level at the large k to be larger than the noise. The alternative is cooling the lens to very low temperatures. Note also that noise can be not only of a physical origin but also numerical, due, for instance, to the discretization employed in a computer simulation (Rao and Ong, 2003).

The signal-reproducing device described earlier is known as a superlens because it is capable of reproducing an image, including the high spatial frequencies, which in normal optical systems are inevitably lost. The loss of high-frequency components of an image is known as the "diffraction limit," and it is often said in the literature that the superlens can overcome the diffraction limit. The fact that high-frequency components of a signal are carried by waves that are attenuated in space, or evanescent waves, has also led some workers to say that the superlens can "focus" the evanescent waves, in addition to focusing the propagating waves. This is, however, subject to the noise limitation mentioned earlier, even when absorption is ignored.

We must also mention that the possibility of having media with negative constitutive parameters was investigated for the first time in a speculative article by Victor Veselago in 1967, translated in English in 1968 (Veselago, 1968). Veselago showed that such media would have many exotic and unexpected properties such as a reversed Doppler shift, reversed Cerenkov radiation, and negative refraction. He did not, however, point out that evanescent waves would also be transmitted by such a medium. This was done over 30 years later by John Pendry in the now famous *Physical Review Letters* article (Pendry, 2000), which can be said to have truly launched the field of negative index metamaterials. In order to picture these ideas, it is convenient to illustrate them by plotting the transmission coefficient of a homogeneous slab as a function of the tangential component of the incident wavevector. It is especially interesting to observe the behavior for tangential components that are larger than the wavevector. The plots below show the magnitude of the transmission coefficient through various slab media as a function of the parameter $\alpha = \frac{\sqrt{u^2+v^2}}{k_0} = \frac{\sqrt{k^2-w^2}}{k_0}$ for $H_{\parallel}$ polarization. Though it may sound paradoxical, α is the tangential component of the incident wavevector in units of the magnitude of the said wavevector. When $\alpha > 1$, the "incident" wave in question is an evanescent wave (also known as inhomogeneous because equiphase planes and equiamplitude planes do not coincide). A large α corresponds to high spatial harmonics of the object in front of the lens. When $\alpha > 1$, the wave is evanescent in free space, though it may propagate in media with a high enough index. In other words, w is not necessarily imaginary when $\alpha > 1$ but only when $\alpha > |k| / |k_0|$.

We begin by comparing the transmission of a regular dielectric slab, with both μ and ε positive with the transmission of a superlens, i.e., a slab with $\mu = \varepsilon = -1$. On the left side of Fig. 3.1, we have plotted the transmission through a slab of thickness $d = 1$au with $\mu = 1, \varepsilon = 12$ for $\lambda = 5$au, while on the right side we have plotted the transmission of free space in blue and of the superlens with $\mu = \varepsilon = -1$ in green.

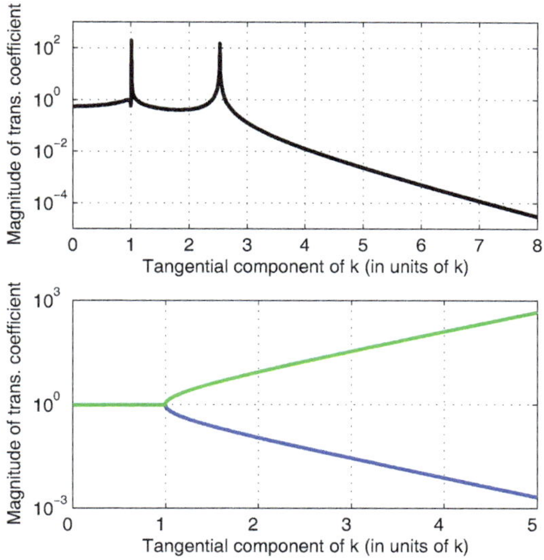

Figure 3.1 Transmission of dielectric slab with thickness $d = 1$ au, $\mu = 1$, $\varepsilon = 12$, and wavelength $\lambda = 5$ au as a function of the tangential component of the wavevector (top). In the lower figure, we compare the transmission of the free space slab (blue) and the superlens (green).

There are several important differences between the three situations. Propagating waves, i.e., for which $\alpha < 1$, are perfectly transmitted by the two media in the right plot, though not quite in the left plot. The transmission is exponentially decreasing with increasing α for the case of the dielectric slab as well as for free space, though the decay is faster in the dielectric slab. In addition, the transmission of the slab exhibits poles or values of α for which the transmission diverges. This happens close to $\alpha = 1$ and $\alpha = 2.5$ for our choice of parameters. No such divergences appear in the right plot. These poles correspond to guided modes in the dielectric slab. The absence of poles in the right plot indicates that neither free space nor the superlens support guided modes. The reason for this difference is that in the right plot, neither of the two media exhibits total internal reflection at the interface with free space. Light cannot be guided using these media.

It would be tempting to explain the absence of guided modes by the fact that these media are impedance matched to free space.

However, this is not correct. It is possible to have guided modes in structures that are perfectly impedance matched to free space. In order to see this, we plot the transmission for three slabs with $\mu = \varepsilon = \{-1.1, -1.01, -1.001\}$, traced in blue, green, and red, respectively. The superlens transmission is plotted in black. All three slabs support a single guided mode, though the mode is shifted to higher α the closer we get to the superlens condition. In fact the superlens can be seen as having a guided mode at infinity on the α axis. Total internal reflection and partial reflection at transmission through an interface are two physically distinct and unrelated phenomena. Partial reflection is related to the impedance mismatch, while total internal reflection is an effect related to the translation symmetry of the interface and the conservation of the tangential component of the wavevector, which the symmetry requires. *Partial reflection is an impedance phenomenon, while total internal reflection is a symmetry and index phenomenon.*

One interesting aspect that can also be seen in Fig. 3.2 is that in principle one does not need a perfect superlens in order to observe the amplification of evanescent waves across a slab. If one can obtain a medium with $\mu = \varepsilon = -1.01$, corresponding to the green curve, then spatial harmonics up to about $\alpha = 4$ will be transmitted accurately across the slab. Higher spatial harmonics will still be lost, but speaking very loosely one may say that the traditional diffraction limit has been beaten very roughly by a factor of 4. Of course, this is

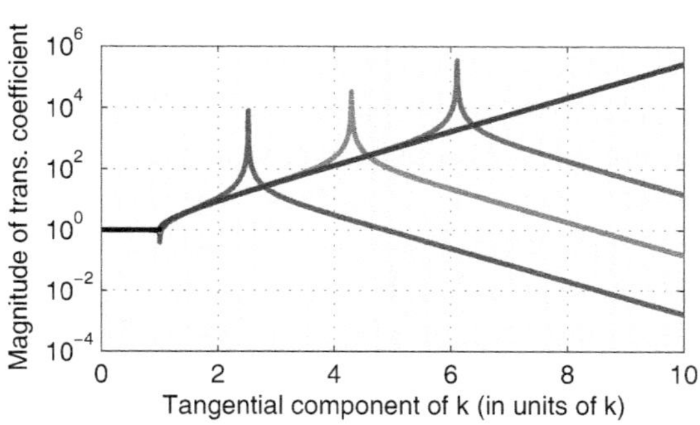

Figure 3.2 As the permittivity and permeability of the slab approach -1, the guided mode is shifted to infinity and the transmission approaches that of the superlens, the black curve, and also the green curve in Fig. 3.1.

nothing revolutionary, since resolutions far better than this can be achieved using widespread near-field optical microscopy techniques (Girard, 2005; Hecht et al., 2000; Kulzer and Orrit, 2004; Novotny and Stranick, 2006). What is novel in this case is the means used: a double negative medium.

Before discussing periodic media, let us emphasize once more that *the remarkable properties of the superlens reside essentially in its response to an incident evanescent field, i.e., for* $\alpha > 1$ *and that negative refraction is a phenomenon that pertains to fields with* $\alpha < 1$. Consequently, while the superlens may exhibit negative refraction, negative refraction does not imply a superlens.

3.5 Periodic Media: Structural Nonlocality

In this section, we consider one of the two causes of spatial dispersion: the proximity between the λ scale and the structure scale d in the special case of periodic media. In this section, we assume that $l \ll d$ (for the case $\lambda \approx l$, see the next section).

We begin by pointing out that the spatial filtering (truncation) viewpoint discussed in the preceding sections must be applied to the divergence equation $\operatorname{div} \mathbf{e} = \eta/\varepsilon_0$, in which is present the charge distribution (on the right side) and the wavelength of the electric field (on the left side). Thus, whenever we truncate one of them (to eliminate small scale, i.e., high frequency, details), we must automatically truncate the other. As we will see below, this has major consequences.

The Bloch theorem tells us that a wave at a single temporal frequency ω propagating in a periodic lattice has a space-dependent part of the form

$$\mathbf{e}(\mathbf{x}) = \mathbf{U}(\mathbf{x})e^{i\mathbf{k}_B\mathbf{x}} \tag{3.41}$$

where $\mathbf{U}(\mathbf{x})$ is a function with the periodicity of the lattice. If we consider a one-dimensional lattice of period a, then the function $\mathbf{U}(\mathbf{x})$ contains spatial frequencies of the form $2\pi n/a$ with $n \in \mathbb{Z}$. Thus, aside from the "DC" component corresponding to $n = 0$, the lowest frequency is $K = 2\pi/a$. Thus, the microscopic field $\mathbf{e}$, which results when a wave at a *single* temporal frequency ω propagates in a periodic lattice, contains *more than one* spatial frequency; specifically, it contains the frequencies: $\mathbf{k}_B + n\mathbf{K}$, $n \in \mathbb{Z}$. However, in a homogeneous material, only one spatial frequency would be present, the one corresponding to the wavevector $\mathbf{k}_B$. The smoothing procedure must, therefore, remove the harmonics that are due to the periodic structure, namely the harmonics corresponding to $n \neq 0$, leaving only the Bloch phase harmonic $\mathbf{k}_B$.

Since this filter is applied to both sides of the electric divergence equation simultaneously, this means that *the frequencies filtered out of* $\tilde{\mathbf{e}}(\mathbf{k})$ *must also be filtered out of* $\tilde{\eta}(\mathbf{k})$. In what follows, we will refer to $K/2$ or the "edge of the Brillouin zone" interchangeably, with the first being the one-dimensional version of the second, and useful for purposes of illustration. We begin by considering the various options for the choice of filter function $\tilde{f}(\mathbf{k})$.

It is simplest to consider the homogenization of a one-dimensional system. A reasonable first try for $\tilde{f}(\mathbf{k})$ is the Gaussian function, whose Fourier transform is also a Gaussian. This choice works fine if the wavelength is very large, so that $\mathbf{k}_B \approx \mathbf{0}$ (see Fig. 3.3, with $K = 2$[au]). When the wavelength is very large, the $\mathbf{k}_B$ term in Eq. (3.41) can be virtually ignored and the Bloch harmonics

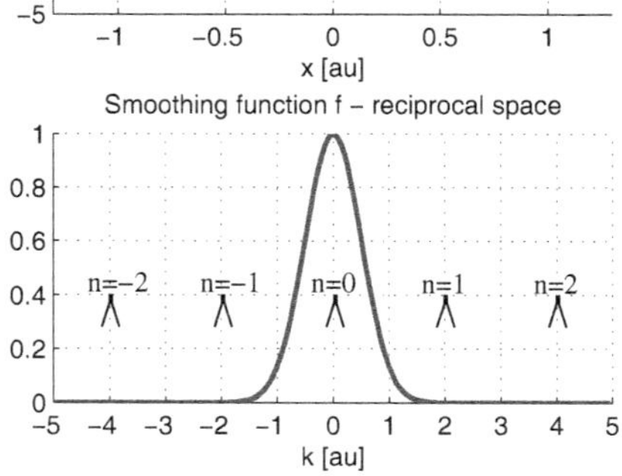

Figure 3.3 One-dimensional smooth.

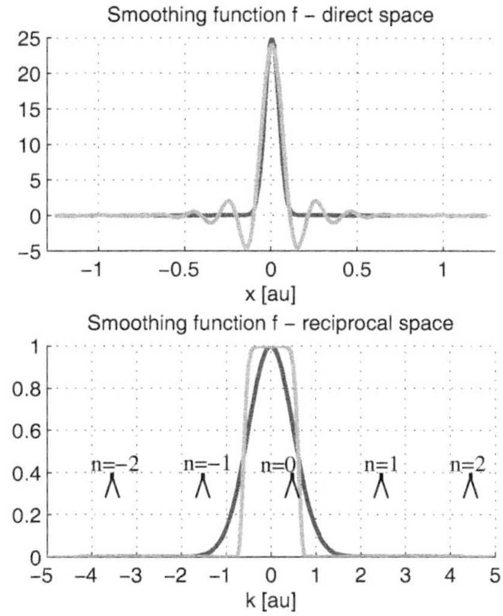

Figure 3.4 One-dimensional smooth.

coincide with the reciprocal lattice vectors of the periodic medium. The homogenization process can then be seen to consist of filtering out all but the lowest harmonic corresponding to $n = 0$.

However, as the wavelength becomes smaller, $\mathbf{k}_\mathrm{B}$ increases, and the Bloch harmonics no longer coincide with the reciprocal lattice vectors of the medium. This is illustrated on the right side of Fig. 3.4. The $n = 0$ component is partially filtered by the blue Gaussian curve, and we, therefore, require an improved, *flatter* filter function such as

$$\mathrm{real}(\tilde{f}(\mathbf{k})) = \exp\left(-\frac{1}{2\nu}\left(\frac{2\mathrm{k}}{\mathrm{K}}\right)^{2\nu}\right). \tag{3.42}$$

The green curve of Fig. 3.4 corresponds to $\nu = 5$ in Eq. (3.42). This shape change has a very important consequence in real space: The averaging volume becomes significantly larger as can be seen in the left plot.

Thus, we are witnessing a process whereby as the wavelength decreases, $\mathbf{k}_\mathrm{B}$ increases, and the size of the $f(\mathbf{x})$ in direct space (the left plot of Fig. 3.4) becomes larger. Eventually, the size of $f(\mathbf{x})$ will become comparable to the wavelength, at which point the macroscopic electric field will become dependent on the phase and direction of propagation, which corresponds to the onset of spatial dispersion or nonlocality.

This is not an intuitive phenomenon. One way of approaching it is to think of temporal rather than spatial frequencies. The job of the function $\tilde{f}$ is to tell apart two spatial frequencies k_B and K. Imagine the two as "sounds" starting in phase. If the two frequencies are quite different, it takes a small fraction of a wavelength for the two signals to go out of phase. Consequently, it takes only a short time to tell them apart, or in the case of spatial frequencies, a short distance. When the two frequencies are closer, however, they will stay in phase for longer, perhaps several wavelengths. Consequently, it takes a longer time to tell them apart, or in the case of spatial frequencies, a longer distance. This is why a sharper filter $\tilde{f}$ in k-space requires a more sprawled out smoothing function f in real space.

The size of the smoothing volume, in turn, is important because it is responsible for the macroscopically nonlocal behavior of the medium. In fact, the spatial averaging implicitly results in nonlocal macroscopic quantities in the sense that the macroscopic polarization at any given point does not depend only on the electric field at that point but over a whole region surrounding it: the smoothing volume f. In most cases of interest, however, the wavelength within the medium is much larger than this volume, the electric field being roughly constant over it. This, in effect, hides the nonlocality of the macroscopic description, making the medium response seem local.

The above argument can be extended in a straightforward way to the two-dimensional case. This is illustrated in Fig. 3.5. From left to right, we have a homogeneous model, a one-dimensional model, and the identity model ($f(\mathbf{x}) = \delta(\mathbf{x})$ the Dirac delta). The smoothing function $f(\mathbf{x})$ can be tuned to obtain different types of models. In the frequency region where nonlocal effects appear, *this idea can be used to trade off nonlocality against in-homogeneity* (Cabuz et al., 2008).

This duality is the direct consequence of the appearance in the same equation (the electric divergence equation) of the field **E** and the charge distribution. When the smoothing is applied, this equation imposes a constraint in that one must keep the *macro*scopic oscillations of the *field*, but filter out the *micro*scopic

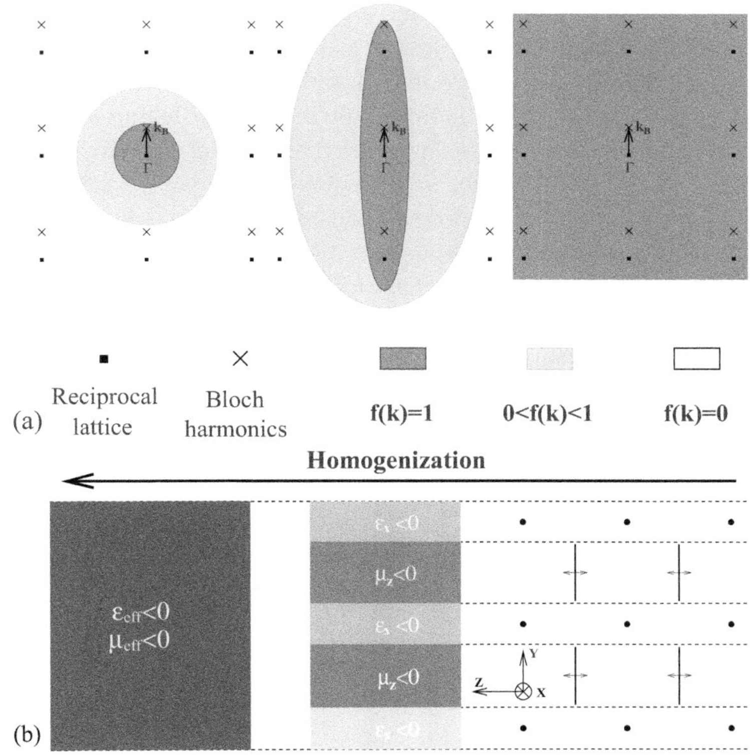

Figure 3.5 (a) Three possible models corresponding to three $f(\mathbf{k})$ functions are illustrated. The first is a homogeneous model; only the lowest harmonic is kept, and the rest are averaged over. The model on the right side is the one where all harmonics are kept, corresponding in real space to $f(\mathbf{x}) = \delta(\mathbf{x})$ the Dirac delta. The case in the middle is an intermediate model, where two of the y harmonics are kept. All three models are effective medium models, but only the one on the left is homogeneous. But is it also local? (b) Schematic illustrating the result of the intermediate homogenization step in the case of a composite material made of wires (parallel to the x axis) and magnetic resonators (dipole moments parallel to the z axis). For a more detailed discussion, see Ref. (Cabuz et al., 2008).

oscillations of the *charge density*. As these two spatial frequencies become closer, the smoothing volume f, which is required to separate them, becomes wider.

What we are observing here is a very general feature of physical systems: one which is generally associated to quantum mechanics, but which we observe here in a completely classical setting. The model of the medium depends both on its intrinsic features (charge distribution) *and* on choices made by the observer (the wavelength of the electric field, and the volume f which we have total control over as it is a purely theoretical construct). When λ is large, the model we build is mostly dependent on the intrinsic features of the medium (the choice of f is irrelevant), which gives the impression of an independent, intrinsic reality underlying its behavior. However, when λ becomes smaller, the model becomes increasingly complex, and the choice of f becomes increasingly critical. Our model of the medium, i.e., our perception of what it *is*, starts to depend not only on features intrinsic to *it* (i.e., on its charge distribution) but also on *us* (i.e., on our choice of f)!

3.6 Conductors: Free Charge Nonlocality

Most of the discussion until now has focused on media where charges are free to move only within the atoms or scatterers and, in any case, not far beyond the confines of the unit cell (in the case of periodic media). This excludes any possibility of dealing with conductors, or with metamaterials containing conducting objects, and allowing charges to travel much farther than the unit cell. Strictly speaking, in fact, we refer not only to classical currents but also to displacement currents, which may span large distances within the material. An example is, for instance, the metamaterial pictured in Fig. 3.6. While the charges are strictly confined to the unit cell, displacement currents can easily influence far away regions via the large capacitances at the interfaces between the scatterers. Similar situations involving inductance (rather than capacitance) can be constructed.

Thus, whether we are talking about an electrical current or a displacement current, the region of mobility of the charges corresponds to the "electromagnetic neighborhood" mentioned on page 3.3 and to the l scale of Section 3.1. Until now, the l scale has been assumed to be smaller than the period, or at best similar to the

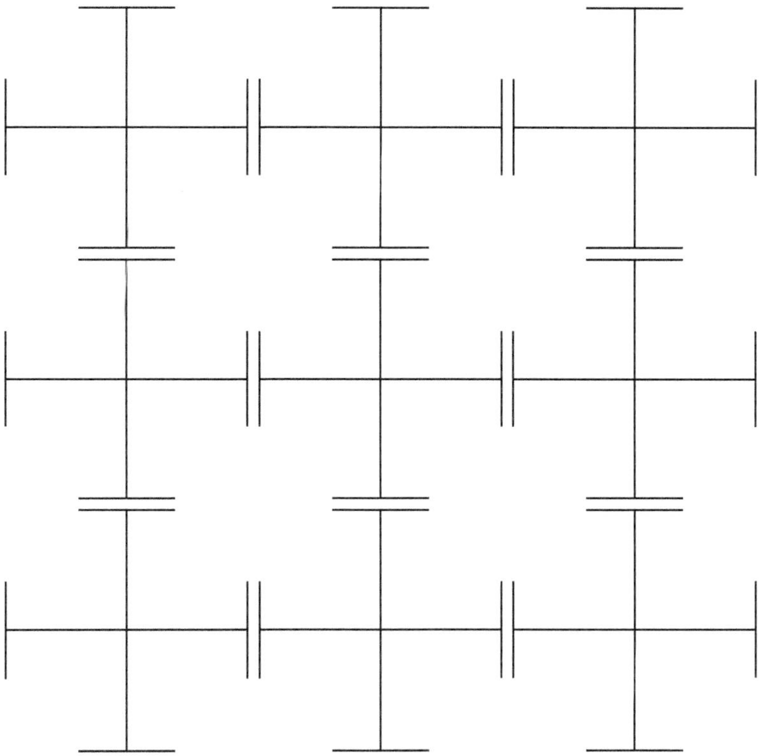

Figure 3.6 Example of metamaterial where charges are not physically free, but they strongly influence each other over large distances via displacement currents.

period. However, in a conducting material (gold) or metamaterial (Fig. 3.6), the l scale can be much larger than the period. In gold, for instance, l will generally correspond to the electron mean free path, which is on the order of tens of nanometers. This is orders of magnitude larger than the atomic scale, which is on the order of Angstroms.

The correspondence between conducting metamaterials (i.e., $l > a$) and metals is important because the latter have been studied for decades, and the results, at least the general concepts, may be transferred directly in the context of metamaterials, especially the classical or quasi-classical approaches. This situation is not very different from the way concepts associated with the physics of semiconductors were transferred to the emerging field of photonic crystals in the late 1980s (Yablonovitch, 1987). In the present case, we will be considering *the hydrodynamic model* of conducting media (Forstmann and Gerhardts, 1986).

3.6.1 The Hydrodynamic Model

According to the hydrodynamic model, one assumes that the medium under study supports two types of waves: the standard divergence-free transverse electromagnetic waves, plus the longitudinal plasma waves for which the divergence is nonzero, because they consist of waves of compression and dilation of the electron gas. Surprisingly, this seemingly minor increase in complexity can qualitatively and often even quantitatively account for a broad range of metal optics phenomena, so it is largely worth the effort to understand the basic idea.

If we consider Maxwell's equations in a medium that fills all space, we can Fourier transform the solutions, with an $e^{i\mathbf{kr}-i\omega t}$ dependence. Then the Helmholtz equation takes the form

$$-\mathbf{k}(\mathbf{k}\cdot\mathbf{E}) + k^2\mathbf{E} = \omega^2\mu\varepsilon(\mathbf{k},\omega)\mathbf{E}.$$

For transverse waves, we obtain the dispersion relation $k^2/\omega^2 = \mu\varepsilon_T(\mathbf{k},\omega)$, while for longitudinal waves where $\mathbf{k} \parallel \mathbf{E}$, the left side is identically zero, so non-trivial solutions require $\varepsilon_L(\mathbf{k},\omega) = 0$.

Initially, starting in the 1940s, these longitudinal waves, also referred to as plasma waves or plasmons, were probed and studied by sending energetic electrons through thin metallic films and measuring the absorption spectrum (Marton and Leder, 1954). These workers were concerned with the study of so-called "bulk plasmons" because the Fourier transform above assumes the medium fills all space. Much work was done starting in the 1950s on the properties of the electron gas, leading to a series of increasingly complex and general models (see Chapter 5 of Ref. (Mahan, 2000)).

However, since we are interested not in electron beams but in electromagnetic waves interacting with these materials, we are working in the small $\mathbf{k}$-region, in which all of these various models take the same simple form (in materials with inversion symmetry):

$$\varepsilon_L(k, \omega) = 1 - \frac{\omega_p^2}{\omega(\omega + i\gamma) - \beta k^2}. \tag{3.43}$$

It turns out that this form can also be obtained in a rather more intuitive way directly from an equation of motion whereby, in addition to the electric field and a damping term, one adds an electron pressure term directly proportional to the gradient of the electron density (Forstmann and Gerhardts, 1986). This is the origin of the name "hydrodynamic approximation": The electron gas is seen as a fluid that supports longitudinal waves analogous to sound.

There is an additional complication, however, when considering electromagnetic waves interacting with the material (as opposed to an electron beam): The idea of a bulk plasmon is useless because we cannot ignore the effect of the interface. The waves cannot "teleport" into the bulk. They have to cross the interface first, and unfortunately there is no automatic way to match the fields in free space with the fields in the bulk. The surface introduces *new physics*, which is reflected in the coupling efficiency between the incident field, on one hand, and the transmitted fields in the *two* available modes, the transverse (standard) and the longitudinal (plasmonic) modes, on the other hand. The continuity conditions usually associated with interfaces in the electromagnetics of local media are no longer sufficient, and we need to introduce *additional boundary conditions (ABCs)* on the fields at the surface. These ABCs basically encapsulate the additional physics of the surface, required for a complete description of the system.

The problem of the ABCs is a vexing one, and the debate surrounding the appropriate conditions is ongoing even in the older and better established study of metals. This is because there is no systematic, ab initio way to prove one ABC superior to another. Strictly speaking, what would be required would be a quantum computation of the surface region, which, besides requiring *vast* amounts of computing power, may not, however, be reducible in any straightforward way to a simple boundary condition to be satisfied by the macroscopic fields. Moreover, in the metamaterial context, there are several additional complications; for instance, we have the issue of anisotropy. In this case, we have to consider a nonlocal

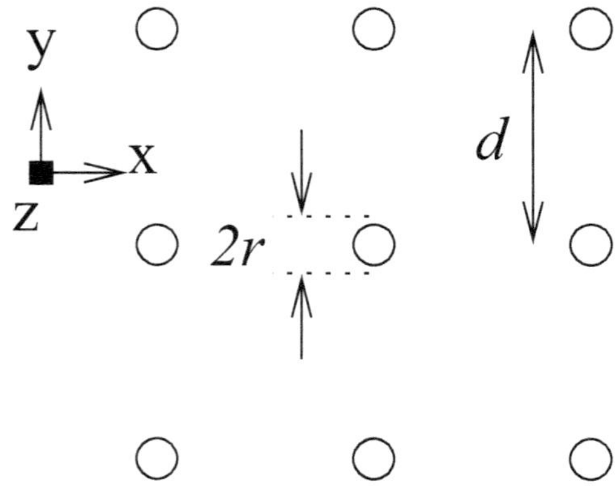

Figure 3.7 The parallel wire metamaterial. The length in the z direction is generally much larger than the period.

permittivity matrix $\tilde{\varepsilon}_L(\mathbf{k}, \omega)$ and note that this depends on the vector $\mathbf{k}$ and not simply on its magnitude as in Eq. (3.43), which applies to metals. Clearly, in the metamaterial arena, the debate is only in its starting stages.

The best example is given by one of the seemingly simplest metamaterials yet designed: the wire medium first studied by Pendry et al. (1996). This is the structure composed of thin, parallel wires disposed in a periodic lattice, generally square (see Fig. 3.7).

The debate surrounds the best ABCs to specify on the top and bottom interfaces, i.e., the ends of the wires. Arguments have been put forward supporting at least two different conditions: continuity of tangential and normal components of electric and magnetic fields (Silveirinha, 2006) and Neumann condition on the polarization current (Bouchitté and Felbacq, 2006). This problem is discussed in more detail in Section 8.2.

The effect of the interfaces can, in some cases, generate even further complexity, for example when the size of an object is on the same scale as l, the electromagnetic neighborhood. In fact, the scale of the object size s may be added as a *fourth* scale to the discussion of Section 3.1. When $s \approx l$, then, in a sense, the whole object is "close to the interface" and its dielectric properties become *shape dependent*. This is a common occurrence when studying micro- and nanoparticles and other fine structures, a venerable field tracing its roots back to the pioneering work of Mie and Maxwell Garnett at the beginning of the last century. As the size decreases and the behavior becomes increasingly complex, there is a progression of modeling approaches starting with the hydrodynamic approach, passing through random phase approximation and jellium approaches and finally time-dependent density functional approaches and Monte Carlo methods. Each particular problem must be matched with the most appropriate solution.

However, when shape dependence appears, one may question whether one may meaningfully speak of one and the same substance composing different objects, since for all practical purposes, each object is composed of a different, shape-dependent substance. We see, therefore, that the notion of substance itself is scale dependent; it is an intrinsically macroscopic notion, and once one is dealing with structures and features on the l scale or similar, that notion gradually starts to lose its meaning. Consequently, when dealing with metamaterial structures, it is important to identify the l scale right away, in order to determine to what extent the concept of "material" itself has meaning in the given context.

3.7 Summary

In this section, we take a step back and take a broader look at the theory developed in the previous sections. We have outlined a procedure whereby one starts with a set of microscopic quantities ($\mathbf{e}(\mathbf{x})$, $\mathbf{p}(\mathbf{x})$, $\gamma(\mathbf{x})$) and proceeds to obtain a set of macroscopic quantities ($\mathbf{E}(\mathbf{x})$, $\mathbf{P}(\mathbf{x})$, $\chi(\mathbf{x})$) via a series of steps involving, in particular, spatial averaging. The spatial averaging takes the form of a convolution by a smoothing function $f(\mathbf{x})$, and the macroscopic parameter of electric susceptibility is defined by the following relation between the macroscopic field and polarization

$$\mathbf{P}(\mathbf{k}) = \chi(\mathbf{k})\mathbf{E}(\mathbf{k}).$$

By writing out the above equation in detail, such that only microscopic quantities appear, we obtain the *master equation* of the

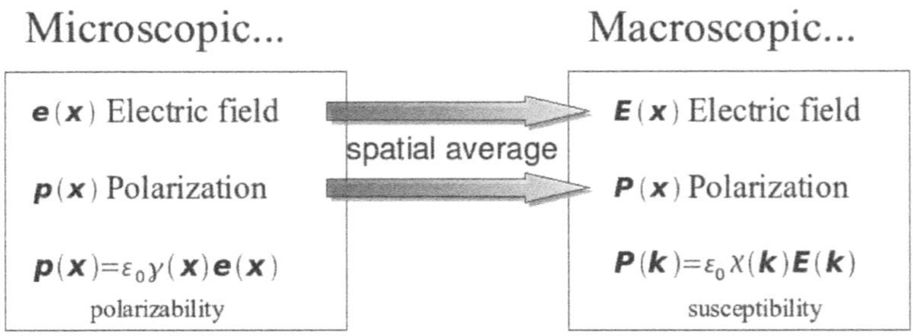

Figure 3.8 Schematic representation of the homogenization procedure.

effective medium model:

$$f(\mathbf{x}) \star (\gamma^{e}(\mathbf{x})\mathbf{e}(\mathbf{x})) = \chi^{e}(\mathbf{x}) \star (f(\mathbf{x}) \star \mathbf{e}(\mathbf{x})). \tag{3.44}$$

This equation is quite complex and non-intuitive due to the fact the the averaging function $f(\mathbf{x})$ has the role of erasing microscopic information. Consequently, the susceptibility defined by this equation is not unique. Any amount of microscopic spatial jitter can be added, and the equation will remain correct. A detailed discussion of this equation is beyond the scope of this review, but I must emphasize several aspects related to the role of the function $f(\mathbf{x})$, in particular in the context of periodic media.

In typical formulations of the derivation of the macroscopic Maxwell's equations (Jackson, 1999; Robinson, 1973; Russakoff, 1970), this function appears as a purely academic construct, and the sole concern of the authors is to state the required properties and to show that all of them can, in principle, be satisfied. These arguments were focused above all on the coherence and consistency of the theoretical construction, rather than on the notion of model-building.

The point of view I would like to emphasize here is different. In this approach, we do not know in advance the model that we must obtain. Instead, we make full use of the fact that $f(\mathbf{x})$ is a mathematical Construct, which we can choose as suits us. We treat the smoothing function $f(\mathbf{x})$ as a dial that we can tune to obtain different types of *custom-made effective medium models.*

The reason this is interesting and important is because it highlights that it is misleading to say that a medium has a "nonlocal response." What is local or nonlocal is not the *medium*, but the *model* we have made of it, which depends on our choice of $f(\mathbf{x})$. By tuning $f(\mathbf{x})$, we can choose to deal with a homogeneous model (though in some cases it may be nonlocal), or we can deal with a local model (though in some cases, it may be inhomogeneous). This can be a useful degree of freedom when trying to understand and design metamaterials.

In addition, one may argue that any nonlocal model is an incomplete description of the medium, in a sense, since it requires additional knowledge of the field (its phase and direction of propagation) in order to predict the response of the medium. One may explain this via the truncation argument: When the higher spatial harmonics are non-negligible, their removal via the truncation filter deprives the physicist of important information regarding the behavior of the structure. This lack of *direct space* information must then be compensated by a more complex, $\mathbf{k}$-dependent, *reciprocal space* description. From this point of view, the $f(\mathbf{x})$ adjustment is seen as a way to transition between a direct space and a reciprocal space description. Small averaging volumes favor a spatial picture, while large averaging volumes favor a $\mathbf{k}$ space picture.

References

Anderson, P. W. (1972). More is different, *Science* **177**, 4047, pp. 393–396, http://www.sciencemag.org.

Antsaklis, P. J. and Michel, A. N. (2006). *Linear Systems* (Springer).

Aspnes, D. E. (1986). Analysis of cermet films with large metal packing fractions, *Phys. Rev. B* **33**, 2, pp. 677–682.

Bergman, D. J. (1980). Exactly solvable microscopic geometries and rigorous bounds for the complex dielectric-constant of a 2-component composite-material, *Phys. Rev. Lett.* **44**, 19, pp. 1285–1287.

Bouchitté, G. and Felbacq, D. (2006). Homogenization of a wire photonic crystal: The case of small volume fraction, *SIAM J. Appl. Math.* **66**, 6, pp. 2061–2084.

Cabuz, A. I. (2007). Electromagnetic metamaterials: From photonic crystals to negative index composites, https://hal.archives-ouvertes.fr/tel-00161428.

Cabuz, A. I., Felbacq, D., and Cassagne, D. (2008). Spatial dispersion in negative-index composite metamaterials, *Phys. Rev. A* **77**, 1, p. 013807.

de Groot, S. R. (1969). *The Maxwell Equations, Studies in Statistical Mechanics*, Vol. IV (North-Holland, Amsterdam).

Delerue, C. and Lannoo, M. (2004). *Nanostructures: Theory and Modelling* (Springer).

Dendy, R. O. (1990). *Plasma Dynamics* (Oxford University Press).

Forstmann, F. and Gerhardts, R. R. (1986). *Metal Optics Near the Plasma Frequency* (Springer-Verlag).

Girard, C. (2005). Near fields in nanostructures, *Rep. Prog. Phys.* **68**, 8, pp. 1883–1933.

Haug, H. and Koch, S. W. (2004). *Quantum Theory of the Optical and Electronic Properties of Semiconductors* (World Scientific).

Hecht, B., Sick, B., Wild, U. P., Deckert, V., Zenobi, R., Martin, O. J. F., and Pohl, D. W. (2000). Scanning near-field optical microscopy with aperture probes: Fundamentals and applications, *J. Chem. Phys.* **112**, 18, pp. 7761–7774.

Jackson, J. D. (1999). *Classical Electrodynamics* (John Wiley and Sons).

Kittel, C. (1996). *Introduction to Solid State Physics* (John Wiley and Sons).

Kock, W. E. (1948). Metallic delay lenses, *Bell System Tech. J.* **27**, 1, pp. 58–82.

Kulzer, F. and Orrit, M. (2004). Single-molecule optics, *Annu. Rev. Phys. Chem.* **55**, pp. 585–611.

Laughlin, R. B. (2005). *A Different Universe: Reinventing Physics from the Bottom Down* (Basic Books).

Mahan, G. D. (2000). *Many Particle Physics* (Kluwer Academic).

Marton, L. and Leder, L. B. (1954). Energy loss of electrons in passage through thin films, *Phys. Rev.* **94**, 1, pp. 203–204, doi:10.1103/PhysRev.94.203.

Mazur, P. (1958). On statistical mechanics and electromagnetic properties of matter, *Adv. Chem. Phys.* **1**, pp. 309–360.

Milton, G. W. (1980). Bounds on the complex dielectric constant of a composite material, *Appl. Phys. Lett.* **37**, 3, pp. 300–302.

Novotny, L. and Stranick, S. J. (2006). Near-field optical microscopy and spectroscopy with pointed probes, *Annu. Rev. Phys. Chem.* **57**, pp. 303–331.

Pendry, J. B. (2000). Negative refraction makes a perfect lens, *Phys. Rev. Lett.* **85**, 18, pp. 3966–3969.

Pendry, J. B., Holden, A. J., Stewart, W. J., and Youngs, I. (1996). Extremely low frequency plasmons in metallic mesostructures, *Phys. Rev. Lett.* **76**, 25, pp. 4773–4776.

Rao, X. S. and Ong, C. K. (2003). Subwavelength imaging by a left-handed material superlens, *Phys. Rev. E* **68**, 6.

Robinson, F. N. H. (1973). *Macroscopic Electromagnetism* (Pergamon Press Ltd.).

Rovelli, C. (2004). *Quantum Gravity* (Cambridge University Press).

Russakoff, G. (1970). A derivation of the macroscopic Maxwell equations, *Am. J. Phys.* **38**, 10, pp. 1188–1195, http://link.aip.org/link/?AJP/38/1188/1.

Silveirinha, M. (2006). Additional boundary condition for the wire medium, arxiv.org.

van Vleck, J. H. (1932). *The Theory of Electric and Magnetic Susceptibilities* (Clarendon Press), http://www.archive.org/details/theoryofelectric031070mbp, http://www.archive.org/details/theoryofelectric031070mbp.

Veselago, V. G. (1968). Electrodynamics of substances with simultaneously negative values of sigma and mu, *Soviet Phys. USPEKHI-USSR* **10**, 4, p. 509.

Yablonovitch, E. (1987). Inhibited spontaneous emission in solid-state physics and electronics, *Phys. Rev. Lett.* **58**, 20, pp. 2059–2062.

Chapter 4

Transformation Optics in a Nutshell

André Nicolet
Institut FRESNEL, University of Aix-Marseille, Avenue Escadrille Normandie Niemen, 13013 Marseille, France
andre.nicolet@fresnel.fr

4.1 Transformation Optics

In recent years, transformation optics has become an active new field. It has been popularized through the idea of J. B. Pendry that an invisibility cloak can be designed by transforming space and considering the corresponding equivalent material properties (Pendry et al., 2006; Zolla et al., 2007). Indeed, a deep property of Maxwell's equations is that they are purely topological (when written in the proper formalism (Nicolet et al., 1994)) and that all the metric aspects can be encapsulated in the electromagnetic material properties. A direct consequence is that any continuous transformation of space can be encoded in an equivalent permittivitty and permeability. Extending this principle beyond continuous transformations allows to design exotic optical devices such as the invisibility cloak. Another example of transformation optics devices is a superlens (Pendry, 2000): Even if these devices were proposed a few years before the rise of transformation optics, they are nicely

Metamaterials Modeling and Design
Edited by Didier Felbacq and Guy Bouchitté

ISBN 978-981-4316-12-5 (Hardcover), 978-1-315-36500-8 (eBook)
www.panstanford.com

interpreted as corresponding to a folding of the space on itself. It has been suggested that such devices allow a kind of "remote action" of the scatterers making possible things such as immaterial waveguides called "invisible tunnels" (Zhang et al., 2007).

4.1.1 Geometrical Background

This section introduces some elementary notions of differential geometry, but the reader is advised to consult the relevant references (Burke, 1985; Deschamps, 1981).

Given an n-dimensional space with a (global) coordinate system $u_1, \cdots, u_n$ (not necessarily orthogonal), the exterior derivative d of a function $f(u_1, \cdots u_n)$ is its differential $df = \sum_i \frac{\partial f}{\partial u_i} du_i$. This is a 1-form. A general 1-form can be written as $\sum_i g_i(u_j)du_i$, where $g_i(u_j)$ are functions of the coordinates u_j. If a 1-form can be expressed as the differential of a function, it is an exact 1-form.

A curve γ is an application from an interval $I = [t_0, t_1]$ of $\mathbb{R}$ on the n-dimensional space: $\mathbf{r}(t) = (u_1(t), \cdots, u_n(t))$, where t is the parameter. The integral $\int_\gamma \alpha$ of a 1-form $\alpha = \sum_i g_i(u_j)du_i$ on the curve γ is defined by $\int_\gamma \alpha = \int_{t_0}^{t_1} (\sum_i g_i(u_j(t)) \frac{\partial u_i(t)}{\partial t})dt$. The value of the integral depends on γ but does not depend on the choice of the parameter.

The exterior product $\wedge$ is the skew-symmetric tensor product such that $du_i \wedge du_j = -du_j \wedge du_i = (du_i \otimes du_j - du_j \otimes du_i)$. A general 2-form is a linear combination $\sum_{i,j} g_{ij} du_i \wedge du_j$. The exterior derivative of the 1-form is $\sum_i g_i du_i$ is $d \sum_i g_i du_i = \sum_{i,j} \frac{\partial g_j}{\partial u_i} du_i \wedge du_j$.

A surface Σ is an application from a two-dimensional open domain $\Omega \subset \mathbb{R}^2$ on the n-dimensional space: $\mathbf{r}(s,t) = (u_1(s,t), \cdots, u_n(s,t))$, where s, t are the parameters. The integral $\int_\Sigma \beta$ of a 2-form $\beta = \sum_{i,j} g_{ij} du_i \wedge du_j$ on the surface Σ is defined by

$$\int_\Sigma \beta = \int \int_\Omega \sum_{i,j} \left(g_{ij} \frac{\partial(u_i, u_j)}{\partial(s,t)} \right) dsdt$$

where $\frac{\partial(u_i,u_j)}{\partial(s,t)}$ are the Jacobians. The value of the surface (flux) integral depends on Σ but does not depend on the way the parameters are chosen.

The Stokes theorem states that $\int_{\Sigma} d\alpha = \int_{\partial\Sigma} \alpha$, where $\partial\Sigma$ is the boundary (curve) of the surface Σ.

More generally, p-forms (with $0 \leq p \leq n$) are defined as totally skew-symmetric tensors and can be manipulated using the exterior derivative and the exterior product. Take $n = 3$.

Given a 1-form

$$\alpha = \alpha_1 du_1 + \alpha_2 du_2 + \alpha_3 du_3$$

and a 2-form

$$\beta = \beta_{23} du_2 \wedge du_3 + \beta_{31} du_3 \wedge du_1 + \beta_{12} du_1 \wedge du_2,$$

one has for instance:

$$\begin{aligned} d\alpha = &\left(\frac{\partial\alpha_2}{\partial u_1} - \frac{\partial\alpha_2}{\partial u_1}\right) du_1 \wedge du_2 + \left(\frac{\partial\alpha_3}{\partial u_2} - \frac{\partial\alpha_3}{\partial u_2}\right) du_2 \wedge du_3 \\ &+ \left(\frac{\partial\alpha_1}{\partial u_3} - \frac{\partial\alpha_1}{\partial u_3}\right) du_3 \wedge du_1, \\ d\beta = &\left(\frac{\partial\beta_{23}}{\partial u_1} + \frac{\partial\beta_{31}}{\partial u_2} + \frac{\partial\beta_{12}}{\partial u_3}\right) du_1 \wedge du_2 \wedge du_3, \end{aligned}$$

and

$$\alpha \wedge \beta = (\alpha_1\beta_{23} + \alpha_2\beta_{31} + \alpha_3\beta_{12}) du_1 \wedge du_2 \wedge du_3.$$

All the concepts here rely only on the topological and differential structure of the space.

The metric is a supplementary structure determined by a rank 2 covariant symmetric tensor $\mathbf{g}$ whose n^2 coefficients form a positive definite matrix. Given a metric, it is possible to introduce the concepts of scalar product, norm, distance, and angle. The metric allows the definition of a Hodge star operator $*$, which is a linear operator on differential forms mapping p-forms on $(n-p)$-forms.

Particular cases of spaces with a metric are the Euclidean spaces $\mathbb{E}^n$ where Cartesian coordinates can be chosen so that the coefficients of the metric form a unit matrix. For $\mathbb{E}^3$, Cartesian coordinates are denoted by $\{u_1 = x, u_2 = y, u_3 = z\}$ and the metric has the form: $\mathbf{g} = dx \otimes dx + dy \otimes dy + dz \otimes dz$. In these Cartesian coordinates, the Hodge operator has the following action:

$$\begin{aligned} &*dx = dy \wedge dz,\ *dy = dz \wedge dx,\ *dz = dx \wedge dy \\ &*(dx \wedge dy) = dz,\ *(dz \wedge dx) = dy,\ *(dy \wedge dz) = dx \\ &*1 = dx \wedge dy \wedge dz,\ *(dx \wedge dy \wedge dz) = 1. \end{aligned}$$

Exterior calculus is the most natural formalism to write Maxwell's equation (Burke, 1985; Deschamps, 1981) so that they have the following form (in the harmonic case with a pulsation ω and complex-valued fields):

$$\begin{cases} d\mathbf{H} = \mathbf{J} - i\omega\mathbf{D} \\ d\mathbf{E} = +i\omega\mathbf{B} \\ d\mathbf{D} = \rho \\ d\mathbf{B} = 0 \end{cases} \tag{4.1}$$

where d is the exterior derivative (d plays the role of **curl** in the first two equations and of div in the last two, see Appendix); the 1-forms **E**, **H** are the electric and magnetic fields, respectively; the 2-forms **D**, **B**, and **J** are the electric flux density or displacement, the magnetic flux density or induction, and the electrical current density, respectively; and the 3-form ρ is the electric charge density. The only operator involved is the exterior derivative that is completely independent of the metric.

The metric is involved in the Hodge star operator $*$ (see Appendix), which is necessary to introduce the constitutive laws of materials (including a void, where $\mathbf{D} = \varepsilon_0 * \mathbf{E}$ and $\mathbf{B} = \mu_0 * \mathbf{H}$). It can also be argued that it is, in fact, these electromagnetic properties of space that determine the metric (Guillemin and Sternberg, 1990). This formalism has proved to be very useful in the context of the numerical solution to Maxwell's equations (Henrotte et al., 1999; Nicolet et al., 1994). In this case, it has been shown that the topological structure of the equation can be preserved at the discrete level (for instance in Yee's FDTD algorithm or using Whitney discrete forms as finite elements), while the whole process of approximation is concentrated in the design of the discrete Hodge operator (Bossavit, 2001).

4.1.2 Change of Coordinates in Maxwell's Equations

In the exterior calculus formalism, the only task associated to changing a coordinate system is to determine an explicit expression for the Hodge star operator (Nicolet et al., 1994, 2010; Zolla et al., 2005). A very useful point of view is to consider weak formulations where integrals of volume forms (3-forms) are built with scalar products of forms, i.e., exterior products together with the Hodge operator acting on one of the factors.

For instance, the wave equation for the electric field (in homogeneous media):

$$d(\mu^{-1} * d\mathbf{E}) - \omega^2 \varepsilon * \mathbf{E} = 0, \tag{4.2}$$

has the following weak formulation: find $\mathbf{E} \in H(\mathrm{rot}, \Omega)$ such that

$$\begin{cases} \int_\Omega \mu^{-1} * d\mathbf{E} \wedge d\overline{\mathbf{E}}' d\mathbf{x} - \omega^2 \int_\Omega \varepsilon * \mathbf{E} \wedge \overline{\mathbf{E}}' d\mathbf{x} = 0 \ , \\ \qquad\qquad\qquad\qquad\qquad\qquad \forall \mathbf{E}' \in H_0(\mathrm{rot}, \Omega) \end{cases} \tag{4.3}$$

where $\wedge$ is the exterior product (see Appendix).

We can use the fact that we know how to write these expression components in the Cartesian coordinate system and that we also know how to transform the derivative and the multiple integrals to determine the action of the Hodge operator in other coordinate systems.

Considering a map from the coordinate system $\{u, v, w\}$ to the coordinate system $\{x, y, z\}$ given by the functions $x(u, v, w)$, $y(u, v, w)$, and $z(u, v, w)$, the transformation of the differentials is given by:

$$\begin{cases} dx = \dfrac{\partial x}{\partial u} du + \dfrac{\partial x}{\partial v} dv + \dfrac{\partial x}{\partial w} dw \\ dy = \dfrac{\partial y}{\partial u} du + \dfrac{\partial y}{\partial v} dv + \dfrac{\partial y}{\partial w} dw \\ dz = \dfrac{\partial z}{\partial u} du + \dfrac{\partial z}{\partial v} dv + \dfrac{\partial z}{\partial w} dw. \end{cases} \tag{4.4}$$

Given a p-form expressed in the $\{x, y, z\}$ coordinate system, it suffices to replace the dx, dy, dz by the corresponding 1-forms involving du, dv, dw in the basis exterior monomials to obtain the expression of the form in the new coordinate system. Note that the form travels naturally counter to the current with respect to the map, and this is why this transportation of the forms from x, y, z to u, v, w is called a pull-back.

This operation can be defined between not only two coordinate systems but also two different manifolds even if they do not have the same dimensions.

Consider two manifolds (or more simply, two open domains of $\mathbb{R}^m$ and $\mathbb{R}^n$, respectively) N and M and a (regular) map φ from N to M

such that (for simplicity) $\varphi(N) = M$. The example above shows that it is very easy to express the differentials of the coordinates on M in terms of the differentials of the coordinates on N and, therefore, to find the image on N of a 1-form on M given by the dual map φ^*, from M to N, also called, as indicated earlier, the pull-back. In fact, any covariant object such as a p-form or a metric can be pulled back by translating the differentials on M into the differentials on N. Defined in this way, the operation commutes, of course, with the exterior and tensor products but also with the exterior derivative and the Hodge star (defined with the pulled-back metric) (Nicolet et al., 1994).

As for contravariant objects such as vector fields, they travel forward just like the geometrical domains. Given a vector $\mathbf{v}$ at a point $\mathbf{p}$ on N, it suffices to choose a curve γ going through the point and such that the vector is the tangent vector to the curve at this point, to take the image of the curve $\varphi(\gamma)$ on M and the vector tangent to this curve at the point $\varphi(\mathbf{p})$ as the image of $\mathbf{v}$. Defined in this way, the map for vectors from N to M, denoted by $\varphi_*(\mathbf{v})$ or $d\varphi(\mathbf{v})$, is called the differential of φ or the push-forward, and it can be extended to any contravariant object.

Another fundamental property of the pull-back is its commutativity with integration in the sense that for any form α that is integrable on a subset $\varphi(\Omega)$ of M, which is the image of a subset Ω of N, one has:

$$\int_{\varphi(\Omega)} \alpha = \int_{\Omega} \varphi^*(\alpha). \tag{4.5}$$

All the information for the pull-back is, therefore, contained in the Jacobian matrix $\mathbf{J}$ (or maybe we should say matrix field since it depends on the point in space considered) in terms of which Eq. (4.4) can be written:

$$\begin{pmatrix} dx \\ dy \\ dz \end{pmatrix} = \mathbf{J} \begin{pmatrix} du \\ dv \\ dw \end{pmatrix} \tag{4.6}$$

with

$$\mathbf{J}(u, v, w) = \frac{\partial(x, y, z)}{\partial(u, v, w)} = \begin{pmatrix} \frac{\partial x}{\partial u} & \frac{\partial x}{\partial v} & \frac{\partial x}{\partial w} \\ \frac{\partial y}{\partial u} & \frac{\partial y}{\partial v} & \frac{\partial y}{\partial w} \\ \frac{\partial z}{\partial u} & \frac{\partial z}{\partial v} & \frac{\partial z}{\partial w} \end{pmatrix}.$$

Using matrix notation, the detailed computation of the relation between the coefficients of a 1-form α in $\{x, y, z\}$ and $\{u, v, w\}$ coordinates is performed as follows:

$$\alpha = \alpha_x dx + \alpha_y dy + \alpha_z dz = (\alpha_x\, \alpha_y\, \alpha_z) \begin{pmatrix} dx \\ dy \\ dz \end{pmatrix}$$

$$= (\alpha_x\, \alpha_y\, \alpha_z)\, \mathbf{J} \begin{pmatrix} du \\ dv \\ dw \end{pmatrix}$$

$$= \alpha_u du + \alpha_v dv + \alpha_w dw = (\alpha_u\, \alpha_v\, \alpha_w) \begin{pmatrix} du \\ dv \\ dw \end{pmatrix}$$

and the following relation is obtained:

$$(\alpha_x\, \alpha_y\, \alpha_z)\, \mathbf{J} = (\alpha_u\, \alpha_v\, \alpha_w). \tag{4.7}$$

Now the contributions to weak form integrals like Eq. (4.3) have the following generic form:

$$\int_\Omega \alpha \wedge *\alpha',$$

where α and α' are 1-forms The question is: How to deal with the Hodge operator? A direct attack would be to pull back the metric and use the explicit expression of the operator, but it is faster here to take advantage of the simple form of the scalar product in Cartesian coordinates that reduces to the dot product. Again using matrix notation (where $\mathbf{J}^{-T}$ is the inverse of $\mathbf{J}^T$):

$$\begin{aligned} \alpha \wedge *\alpha' &= (\alpha_x\, \alpha_y\, \alpha_z)(\alpha'_x\, \alpha'_y\, \alpha'_z)^T dx \wedge dy \wedge dz \\ &= (\alpha_u\, \alpha_v\, \alpha_w)\mathbf{J}^{-1}[(\alpha'_u\, \alpha'_v\, \alpha'_w)\mathbf{J}^{-1}]^T dx \wedge dy \wedge dz \\ &= (\alpha_u\, \alpha_v\, \alpha_w)\mathbf{J}^{-1}\mathbf{J}^{-T}(\alpha'_u\, \alpha'_v\, \alpha'_w)^T \det(\mathbf{J}) du \wedge dv \wedge dw. \end{aligned} \tag{4.8}$$

The first line is the definition of the scalar product of 1-forms equated to the scalar product in Cartesian coordinates.

The fact used here is that the transformation of 3-forms $dx \wedge dy \wedge dz = \det(\mathbf{J}) du \wedge dv \wedge dw$ only involves the Jacobian, i.e., the determinant of the Jacobian matrix. Hence, the only difference from the case of Cartesian coordinates is that one of the (column) vectors has to be multiplied on the left by a symmetric matrix $\mathbf{T}^{-1}$ before performing the dot product, where $\mathbf{T}$ is given by:

$$\mathbf{T} = \frac{\mathbf{J}^T\mathbf{J}}{\det(\mathbf{J})}. \tag{4.9}$$

It is now interesting to look at how a particular 2-form basis monomial transforms, for instance,

$$dx \wedge dy = \left[\frac{\partial x}{\partial u}du + \frac{\partial x}{\partial v}dv + \frac{\partial x}{\partial w}dw\right] \wedge \left[\frac{\partial y}{\partial u}du + \frac{\partial y}{\partial v}dv + \frac{\partial y}{\partial w}dw\right]$$
$$= \left(\frac{\partial x}{\partial u}\frac{\partial y}{\partial v} - \frac{\partial x}{\partial v}\frac{\partial y}{\partial u}\right) du \wedge dv + \left(\frac{\partial x}{\partial v}\frac{\partial y}{\partial w} - \frac{\partial x}{\partial w}\frac{\partial y}{\partial v}\right) dv \wedge dw$$
$$+ \left(\frac{\partial x}{\partial w}\frac{\partial y}{\partial u} - \frac{\partial x}{\partial u}\frac{\partial y}{\partial w}\right) dw \wedge du.$$

The cofactors of $\mathbf{J}$ are now involved in the transformation. (These are the elements of $\mathbf{J}^{-T}\det(\mathbf{J})$.)

Given a 2-form:

$$\begin{aligned} \beta &= \beta_x dy \wedge dz + \beta_y dz \wedge dx + \beta_z dx \wedge dy \\ &= \beta_u dv \wedge dw + \beta_v dw \wedge du + \beta_w du \wedge dv \end{aligned} \qquad (4.10)$$

the following relation is obtained:

$$(\beta_x\, \beta_y\, \beta_z)\, \mathbf{J}^{-T} \det(\mathbf{J}) = (\beta_u\, \beta_v\, \beta_w) \qquad (4.11)$$

and considering the scalar product $\beta \wedge *\beta'$ of two such 2-forms, it is straightforward to show that the matrix involved in the transformation of this scalar product is here $\mathbf{T}$ (instead of its inverse in the case of the scalar product of two 1-forms):

$$\beta \wedge *\beta' = (\beta_x\, \beta_y\, \beta_z)(\beta'_x\, \beta'_y\, \beta'_z)^T dx \wedge dy \wedge dz$$
$$= (\beta_u\, \beta_v\, \beta_w)\frac{\mathbf{J}^T}{\det(\mathbf{J})}\left[(\beta'_u\, \beta'_v\, \beta'_w)\frac{\mathbf{J}^T}{\det(\mathbf{J})}\right]^T \det(\mathbf{J}) du \wedge dv \wedge dw$$
$$= (\beta_u\, \beta_v\, \beta_w)\mathbf{J}^T\mathbf{J}\frac{1}{\det(\mathbf{J})}(\beta'_u\, \beta'_v\, \beta'_w)^T du \wedge dv \wedge dw. \qquad (4.12)$$

Everything can now be summarized in the following recipe that takes into account implicitly the Hodge star: Consider a 3-form γ to be integrated on a domain Ω in order to get $\int_\Omega \gamma$ to contribute to a weak form, then:

- If the integrand involves only scalars (0-forms or 3-forms, and it does not matter if the 3-forms are expressed as the divergence of a vector field) or if it is the exterior product of a 1-form and a 2-form (and it does not matter if they are, respectively, a gradient and a curl) looking superficially like a scalar product of vectors, only $\det(\mathbf{J})$ has to be introduced as a factor.

- If the integrand is the scalar product of two 1-forms (and it does not matter if one or both 1-forms are expressed as the gradient of a scalar field), multiply on the left one of the column vectors of coefficients by the matrix $\mathbf{T}^{-1}$.
- If the integrand is the scalar product of two 2-forms (and it does not matter if one or both 2-forms are expressed as the curl of a vector field), multiply on the left one of the column vectors of coefficients by the matrix $\mathbf{T}$.

The expression obtained for $\varphi^*(\gamma)$ depending on variables u, v, and w (the coordinates x, y, and z have been replaced by the functions $x(u, v, w)$, $y(u, v, w)$, and $z(u, v, w)$, respectively) is integrated on Ω to get the desired contribution to the volume integral of the weak formulation.

It can also be interesting to consider a compound transformation, i.e., the transformation of a transformation. Consider three systems of coordinates u_i, X_i, and x_i (possibly on different manifolds) and the maps $\varphi_{Xu} : u_i \to X_i$ given by functions $X_i(u_j)$ and $\varphi_{xX} : X_i \to x_i$ given by functions $x_i(X_j)$. The composition map $\varphi_{xX} \circ \varphi_{Xu} = \varphi_{xu} : u_i \to x_i$ is given by the functions: $x_i(X_j(u_k))$. If $\mathbf{J}_{xX}$ and $\mathbf{J}_{Xu}$ are the Jacobian matrices of the maps φ_{xX} and φ_{Xu}, respectively, the Jacobian matrix $\mathbf{J}_{xu}$ of the composition map φ_{xu} is simply the product of the Jacobian matrices:

$$\mathbf{J}_{xu} = \mathbf{J}_{xX}\mathbf{J}_{Xu}.$$

This rule naturally applies for an arbitrary number of maps.

It is also worth noting that the matrix $\mathbf{J}^T\mathbf{J}$ is nothing but the metric tensor whose coefficients are expressed in the local coordinates.

4.1.3 Geometric Transformation: Equivalent Material Principle

A very interesting interpretation of the preceding formulae is that matrix $\mathbf{T}$ and its inverse can be viewed as tensorial characteristics of equivalent materials.

By inspection of Eq. (4.3), it appears that μ^{-1} is present as a factor in the term involving the exterior derivatives, i.e., a scalar product of two 2-forms and that the $\mathbf{T}$ factor can be introduced by multiplying μ by $\mathbf{T}^{-1}$ (and, therefore, turning it to a tensor quantity). It appears also that ε is present as a factor in the term involving directly the electric field, i.e., a scalar product of two 1-forms and that the $\mathbf{T}^{-1}$ factor can be introduced by multiplying ε by $\mathbf{T}^{-1}$ (and, therefore, turning it also to a tensor quantity).

Therefore, the only thing to do to compute the integrals of the weak form in the transformed coordinates is to replace the materials (often homogeneous and isotropic) by equivalent ones that are inhomogeneous (their characteristics are no longer piecewise constant but merely depend on u, v, w coordinates) and anisotropic ones (tensorial nature) whose properties are given by

$$\underline{\underline{\varepsilon}}' = \varepsilon \mathbf{T}^{-1}\,, \quad \text{and} \quad \underline{\underline{\mu}}' = \mu \mathbf{T}^{-1}\,. \tag{4.13}$$

We note that there is no change in the impedance of the media since the permittivity and permeability undergo the same transformation. As for the vector analysis operator and product, everything works as if we were in Cartesian coordinates.

In electromagnetism, changing a material can thus be viewed as changing metric properties, and conversely a change of coordinates can be taken into account by introducing a fictitious equivalent material. For a general transformation, the equivalent material is inhomogeneous and anisotropic. It may be interesting in some cases to introduce non-orthogonal coordinate systems to facilitate the solution to particular problems, e.g., helicoidal geometries (Nicolet et al., 2004, 2007).

It is straightforward to generalize the present rules to initially anisotropic material properties. Also it has to be noted that they need not be initially homogeneous. Therefore, the basic principle of transformation optics can be stated in a very general setting: For all our practical purposes, M and N will be here the whole or parts of $\mathbb{R}^3$. Given a map φ from a space N to a space M determining a geometric transformation (i.e., given a Cartesian coordinate system $\mathbf{x}$ on M and an arbitrary coordinate system $\mathbf{x}'$ on N, $\varphi : N \to M$ is described by $\mathbf{x}(\mathbf{x}')$, i.e., $\mathbf{x}$ given as a function of $\mathbf{x}'$), when one has an electromagnetic system described by the tensor fields $\underline{\underline{\varepsilon}}(\mathbf{x})$ for the dielectric permittivity and $\underline{\underline{\mu}}(\mathbf{x})$ for the magnetic permeability in the space M, if one replaces the initial material properties by equivalent material properties given by the following rule (Milton et al., 2006;

Nicolet et al., 2004, 2007; Zolla et al., 2005):

$$\underline{\underline{\varepsilon}}'(\mathbf{x}') = \mathbf{J}^{-1}(\mathbf{x}')\underline{\underline{\varepsilon}}(\mathbf{x}(\mathbf{x}'))\mathbf{J}^{-T}(\mathbf{x}')\det(\mathbf{J}(\mathbf{x}')),$$
$$\underline{\underline{\mu}}'(\mathbf{x}') = \mathbf{J}^{-1}(\mathbf{x}')\underline{\underline{\mu}}(\mathbf{x}(\mathbf{x}'))\mathbf{J}^{-T}(\mathbf{x}')\det(\mathbf{J}(\mathbf{x}')), \qquad (4.14)$$

one gets an equivalent problem on N. Here, an equivalent problem means that the solution to the new problem on N, i.e., electromagnetic quantities described as differential forms, is the pulled back of the solution (Nicolet et al., 1994) to the original problem on M and that the same Maxwell's equations (i.e., as if we were in Cartesian coordinates or, more accurately, having the same form as (4.1) written with the exterior derivative) are still satisfied.

The transformation rules between M and N for components of the fields that are 1-forms, such as **E** and **H**, are given by Eq. (4.7) and those for the components of the fields that are 2-forms, such as **B**, **D**, and **J**, are given by Eq. (4.11). It clearly appears that the invariant quantities, according to Eq. (4.5), are the global quantities built as integral of p-forms on p-dimensional geometrical objects: the line

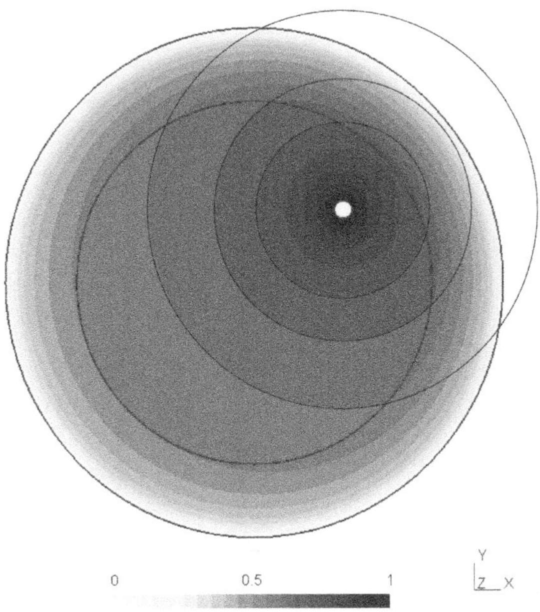

Figure 4.1 Electrostatic potential: circular cylinder $V = 1$, $V(r \to \infty) = 0$ $\Rightarrow$ circular equipotential lines.

integrals of the electric field and of the magnetic field along a curve, the fluxes across a surface of the electric displacement, the magnetic flux density, the current density, the Poynting vector, and so on.

4.1.4 Cylindrical Devices

A simple setting for transformation optics is to consider cylindrical devices where the transformation concerns only the radial distance. In order to obtain the components of the equivalent permeability and permittivity tensors in the Cartesian coordinates, it is necessary to perform first an initial transformation from Cartesian to cylindrical coordinates, then the radial transformation concerned, and finally the back transformation from cylindrical to Cartesian coordinates. This transformation from Cartesian coordinates $\{x, y, z\}$ to polar coordinates $\{r, \theta, z\}$ is introduced via the map from r, θ to x, y:

$$\begin{cases} x(r, \theta) = r\cos\theta \\ y(r, \theta) = r\sin\theta. \end{cases} \tag{4.15}$$

The associated Jacobian is

$$\mathbf{J}_{xr}(r, \theta) = \frac{\partial(x, y, z)}{\partial(r, \theta, z)} = \begin{pmatrix} \cos\theta & -r\sin\theta & 0 \\ \sin\theta & r\cos\theta & 0 \\ 0 & 0 & 1 \end{pmatrix} = \mathbf{R}(\theta)\ \mathbf{diag}(1, r, 1), \tag{4.16}$$

with

$$\mathbf{R}(\theta) = \begin{pmatrix} \cos\theta & -\sin\theta & 0 \\ \sin\theta & \cos\theta & 0 \\ 0 & 0 & 1 \end{pmatrix} \text{ and } \mathbf{diag}(1, r, 1) = \begin{pmatrix} 1 & 0 & 0 \\ 0 & r & 0 \\ 0 & 0 & 1 \end{pmatrix}.$$

$\mathbf{R}(\theta)$ has the well known properties: $\mathbf{R}(\theta)^{-1} = \mathbf{R}(\theta)^T = \mathbf{R}(-\theta)$.

Furthermore, the inverse transformation is given by the map:

$$\begin{cases} r(x, y) = \sqrt{x^2 + y^2} \\ \theta(x, y) = 2\arctan\left(\frac{y}{x+\sqrt{x^2+y^2}}\right), \end{cases} \tag{4.17}$$

and is associated with the Jacobian:

$$\mathbf{J}_{rx}(x, y) = \mathbf{J}_{xr}^{-1}(r(x, y), \theta(x, y)) = \mathbf{diag}\left(1, \frac{1}{r(x, y)}, 1\right)\mathbf{R}(-\theta(x, y)). \tag{4.18}$$

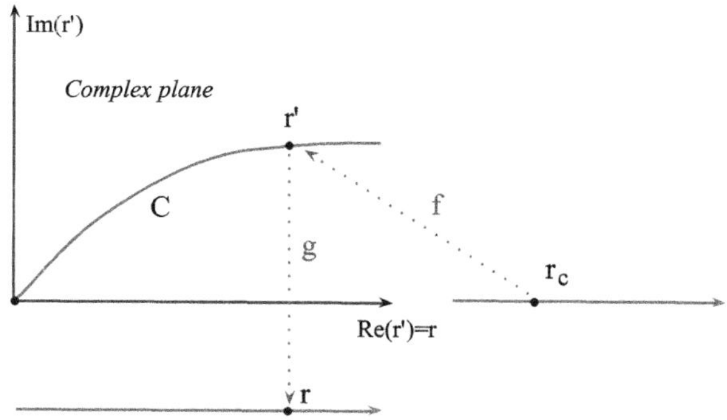

Figure 4.2 Complex-valued change of coordinates associated to the PML.

If a radial transformation is described by the function $r = f(r')$ expressing the original radial distance (in physical space) in terms of the transformed radial distance (transformed space), leaving $\theta = \theta'$ and $z = z'$, the Jacobian matrix is $\mathbf{J}_{rr'} = \mathbf{diag}(\frac{df(r')}{dr'}, 1, 1)$ and the Jacobian of the complete transformation (i.e., from Cartesian to transformed Cartesian coordinates) is

$$\mathbf{J}_{xx'} = \mathbf{J}_{xr}\mathbf{J}_{rr'}\mathbf{J}_{r'x'} = \mathbf{R}(\theta)\mathbf{diag}\left(\frac{df(r')}{dr'}, \frac{f(r')}{r'}, 1\right)\mathbf{R}(-\theta)$$

the transformation matrix of Eq. 4.13 is given by:

$$\begin{aligned}\mathbf{T}^{-1} &= \mathbf{J}_{xx'}^{-1}\mathbf{J}_{xx'}^{-T}\det(\mathbf{J}_{xx'}) \\ &= \mathbf{R}(\theta)\mathbf{diag}\left(\frac{f(r')}{\frac{df(r')}{dr'}r'}, \frac{\frac{df(r')}{dr'}r'}{f(r')}, \frac{\frac{df(r')}{dr'}f(r')}{r'}\right)\mathbf{R}(\theta)^T.\end{aligned}$$

This general formula has at least four very interesting special cases:

(1) **Open domain**: In potential theory, e.g., in electrostatics, when an unbounded problem is considered with the condition that the potential is decreasing to zero at infinity, the use of numerical methods such as the finite element method requires some special treatment in order to deal with a meshing limited to a finite domain. The radial transformation

$$r = f(r') = (R_1 - R_2)r'/(r' - R_2)$$

with $R_2 > R_1$ is such that

$$r' = R_1 \Rightarrow r = R_1 \text{ and } r' = R_2 \Rightarrow r \to \infty.$$

Surrounding the electrostatic problem with an annulus made of the equivalent permittivity allows a simple and rigorous solution to the problem (see Fig. 4.1). This technique is a precursor of transformation optics (Nicolet et al., 1994).

(2) **PML**: In this case, the PML corresponds to a complex stretch of the radial coordinate ρ; the *region of interest* is a disk defined by $\rho < R^*$, and the *PML region* is a circular annulus around the region of interest defined by $R^* < \rho < R^{\text{trunc}}$, where R^* and R^{trunc} are real constants. As the expressions of the material tensors in Cartesian coordinates are needed, the whole setting requires a transformation between Cartesian and cylindrical coordinates. The recipe involves a sequence of coordinate systems. We start here with the physical coordinates, and we finish with the modeling coordinates. The mapping will, therefore, be from the last system of the list to the first one, while the pull-back maps will be from the first system to the last one.

- (x, y, z) are real-valued classical Cartesian coordinates.
- $(\tilde{x}, \tilde{y}, \tilde{z})$ are a complex stretch of the previous Cartesian coordinates. They are complex valued and, it is fundamental to understand that this change is an active transformation rather than a mere change of coordinates in the sense that the ambient space is changed. (x, y, z) are a parametrization of $\mathbb{R}^3$, and the complex stretch corresponds to an extension of the problem to $\mathbb{C}^3$ and more precisely to a three-dimensional subspace Γ of $\mathbb{C}^3$ (in terms of real dimensions, $\mathbb{C}^3$ is six dimensional and $\mathbb{R}^3$ and Γ are three dimensional) (Lassas et al., 2001). The map from Γ to $\mathbb{R}^3$ is chosen in such way that the restriction of this map to the region of interest is the identity map. The solution to the original problem on $\mathbb{R}^3$ can be extended analytically to $\mathbb{C}^3$ and then restricted to Γ. If the complex stretch is correctly chosen, this "complexified" solution on Γ is evanescent where the physical solution involves outgoing or even exponentially diverging waves.
- $(\tilde{\rho}, \tilde{\theta}, \tilde{z})$ is a cylindrical representation of $(\tilde{x}, \tilde{y}, \tilde{z})$.
- (ρ_c, θ_c, z_c) are *real-valued* cylindrical coordinates on Γ. They are related to $(\tilde{\rho}, \tilde{\theta}, \tilde{z})$ via $\tilde{\theta} = \theta_c$, $\tilde{z} = z_c$, and a radial

complex stretch

$$\tilde{\rho} = \int_0^{\rho_c} s_\rho(\rho')\,d\rho' \tag{4.19}$$

where s_ρ is a *complex-valued function* of a *real variable*, i.e., $s_\rho = 1$ in the central region of interest defined by $\rho_c < R^*$ (the complex stretch corresponds to an identity map in this region) and s_ρ has a complex value in the PML defined by $R^* < \rho < R^{\mathrm{trunc}}$.

- (x_c, y_c, z_c) are the Cartesian representation of (ρ_c, θ_c, z_c) and are also *real-valued* coordinates that will be called *modeling coordinates*. This is the modeling space where the numerical approximations are written, where the finite element mesh is defined, and where all the outgoing waves are turned to evanescent ones so that the computation domain can be truncated.

In the end, only the real-valued coordinates x, y, z and x_c, y_c, z_c are involved, but the complex map corresponds to a complex-valued Jacobian. In the case of cylindrical coordinates, $\tilde{\rho}$ and ρ_c are just introduced to compute the radial stretch. Note also that $\theta_c = \tilde{\theta}$ and, therefore, will be simply denoted by θ. The transformation from Cartesian to cylindrical coordinates is just used to obtain the Cartesian expression of the corresponding metric tensor. The Jacobians associated to these changes of coordinates are: $\mathbf{J}_{\tilde{x}\tilde{\rho}} = \mathbf{J}_{x\rho}(\tilde{\rho}, \theta)$, $\mathbf{J}_{\tilde{\rho}\rho_c} = \mathbf{diag}(\frac{\partial\tilde{\rho}}{\partial\rho_c}, 1, 1) = \mathbf{diag}(s_\rho(\rho_c), 1, 1)$, $\mathbf{J}_{\rho_c x_c} = \mathbf{J}_{\rho x}(\rho_c(x_c, y_c), \theta(x_c, y_c))$. The global Jacobian $\mathbf{J}_s$ is the product of the individual Jacobians:

$$\mathbf{J}_s = \mathbf{J}_{\tilde{x}\tilde{\rho}}\mathbf{J}_{\tilde{\rho}\rho_c}\mathbf{J}_{\rho_c x_c} = \mathbf{R}(\theta)\mathbf{diag}\left(s_\rho, \frac{\tilde{\rho}}{\rho_c}, 1\right)\mathbf{R}(-\theta)\,, \tag{4.20}$$

where $\mathbf{R}(\theta)$ denotes the following matrix of rotation

$$\mathbf{R}(\theta) = \begin{pmatrix} \cos\theta & -\sin\theta & 0 \\ \sin\theta & \cos\theta & 0 \\ 0 & 0 & 1 \end{pmatrix}.$$

Note that, in fact, *we solve numerically the extended problem obtained by the complex stretch (4.19) and defined on* Γ *that have the very remarkable property to coincide with our original problem in the region of interest*. In order to comply with traditional notation in the PML context and to avoid cumbersome notations, we drop the c subscript associated to the modeling coordinates that will subsequently be denoted as ρ and (x, y, z) without any ambiguity. For isotropic uniform media outside the region of interest, the cylindrical PML characteristics are obtained by multiplying ε and μ by the following complex matrix:

$$\mathbf{T}_s^{-1} = \mathbf{J}_s^{-1}\mathbf{J}_s^{-T}\det(\mathbf{J}_s) = \mathbf{R}(\theta)\mathbf{diag}\left(\frac{\tilde{\rho}}{s_\rho\rho}, \frac{s_\rho\rho}{\tilde{\rho}}, \frac{s_\rho\tilde{\rho}}{\rho}\right)\mathbf{R}(-\theta)$$

$$= \begin{pmatrix} \dfrac{\rho s_\rho\sin(\theta)^2}{\tilde{\rho}} + \dfrac{\tilde{\rho}\cos(\theta)^2}{\rho s_\rho} & \sin(\theta)\cos(\theta)\left(\dfrac{\tilde{\rho}}{\rho s_\rho} - \dfrac{\rho s_\rho}{\tilde{\rho}}\right) & 0 \\ \sin(\theta)\cos(\theta)\left(\dfrac{\tilde{\rho}}{\rho s_\rho} - \dfrac{\rho s_\rho}{\tilde{\rho}}\right) & \dfrac{\rho s_\rho\cos(\theta)^2}{\tilde{\rho}} + \dfrac{\tilde{\rho}\sin(\theta)^2}{\rho s_\rho} & 0 \\ 0 & 0 & \dfrac{\tilde{\rho}s_\rho}{\rho} \end{pmatrix}.$$

This latest expression is the metric tensor in Cartesian coordinates (x, y, z) for the cylindrical PML and θ, ρ, $\tilde{\rho}$, and $s_\rho(\rho)$ are explicit functions of the variables x and y. Another remarkable property of the PML is that they provide the correct extension to non-Hermitian operators (since $\mathbf{T}_s$ is complex and symmetric) that allow the computation of the leaky modes, and this may be obtained via a correct choice of the PML parameters, namely, R^*, R^{trunc}, and $s_\rho(\rho)$.

(3) **Invisibility cloak**

Pendry's map for the invisibility cloak: Consider a geometric transformation that maps the field within the disk $r \leq R_2$ onto the annulus $R_1 \leq r \leq R_2$:

$$r = f(r') = (r' - R_1)R_2/(R_2 - R_1) \text{ for } R_1 \leq r' \leq R_2$$

and for $r \geq R_2$, it is the identity transformation. The material properties of the invisibility cloak are given by:

$$\mathbf{T}^{-1} = \mathbf{R}(\theta')\mathbf{diag}\left(\frac{r' - R_1}{r'}, \frac{r'}{r' - R_1}, (R_2/(R_2 - R_1))^2\frac{r' - R_1}{r'}\right)\mathbf{R}(\theta')^T \qquad (4.21)$$

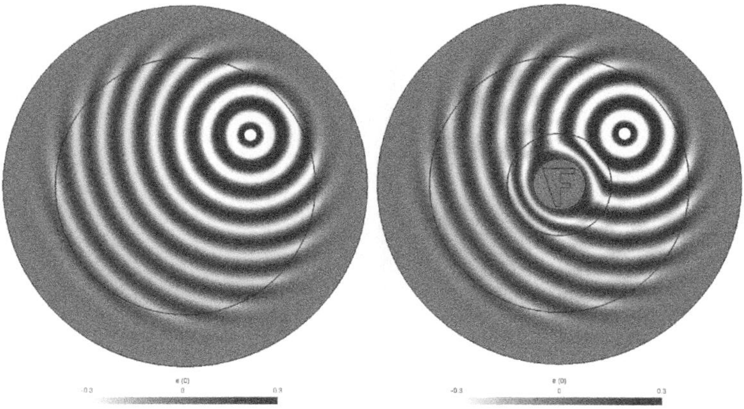

Figure 4.3 Invisibility cloaking.

Table 4.1 The cylindrical transformations

Type	$r = f(r')$	Domain
Open domain	$r = (R_1 - R_2)r'/(r' - R_2)$	$R_1 < r' < R_2$
PML	$r = \int_0^{r'} s_r(\rho)\,d\rho$	
Invisibility cloak	$r = (r' - R_1)R_2/(R_2 - R_1)$	$R_1 \leq r' \leq R_2$
Superlens	$r = r'(R_2 - \alpha R_1) + (\alpha - 1)\frac{R_1 R_2}{R_2 - R_1}$	$R_1 \leq r' \leq R_2$
	$r = r'\alpha R_1$	$r' \leq R_1$

(4) **Superlens**

$$r = f(r') = r'(R_2 - \alpha R_1) + (\alpha - 1)\frac{R_1 R_2}{R_2 - R_1}$$

$$r = f(r') = r'\alpha R_1$$

$$\alpha = \frac{R_2}{R_1} \text{ magnifying factor.}$$

4.2 Superlens Illusion

A dramatic example of transformation optics devices is the superlens (Pendry, 2000): Even if these devices were proposed a few years before the rise of transformation optics, they are nicely

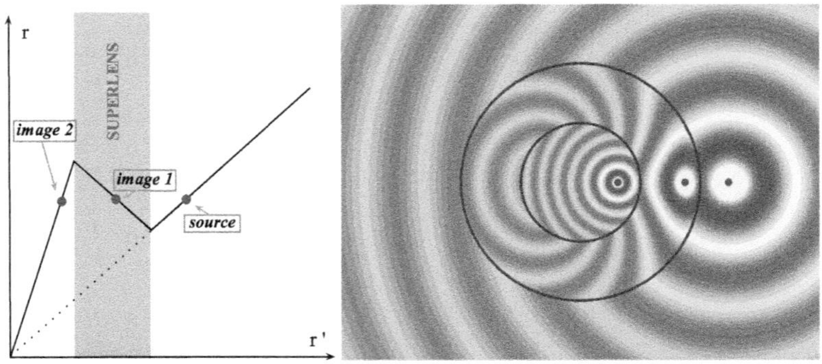

Figure 4.4 A superlens can be designed by folding the space on itself, here by transforming the radial distance (Yan et al., 2008). In this case, there is a part of the physical space that has a threefold image in the equivalent space.

interpreted as corresponding to a folding of the space on itself. It has been suggested that such devices allow a kind of "remote action" of the scatterers making possible things such as immaterial waveguides called "invisible tunnels" (Zhang et al., 2007). They can also be used to set up a new kind of invisibility devices (Lai et al., 2009a) and also illusion devices (Lai et al., 2009b), similar to the one obtained by generalized cloaking (Nicolet et al., 2010), but based on negative refraction index materials.

As an illustration of superlensing, consider the multiple-valued transformation of Fig. 4.4. The part with negative slope corresponds to a negative refraction index material ($\underline{\underline{\varepsilon}}'$ and $\underline{\underline{\mu}}'$ have negative eigenvalues, see Fig. 4.5) and acts as a superlens (Leonhardt and Philbin, 2006; Yan et al., 2008). As a transformation of empty space, it is supposed not to perturbate the cylindrical waves emitted by a wire antenna (in Fig. 4.6, the inner disk is the image of a four times larger disk with the same center), but the anomalous resonances (Milton and Nicorovici, 2006; Nicorovici et al., 2007) have a dazzling effect (see Fig. 4.6) that can be attenuated by introducing losses in the superlens material. For instance, in Fig. 4.7, permittivity of the ideal perfect lens $\underline{\underline{\varepsilon}}'$ has been multiplied by $1 - 0.001i$ and the attenuation is partial. Finally, one percent of losses have been introduced in the value of the permittivity of the ideal perfect lens ($\underline{\underline{\varepsilon}}'$ has been multiplied by $1 - 0.01i$) to get a suitable attenuation

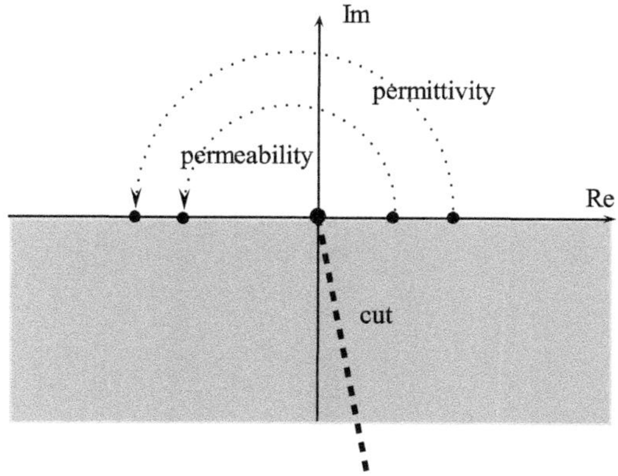

Figure 4.5 In the complex plane, the definition of the square root requires a cut. This explains why both negative permittivity and negative permeability lead to a negative refractive index that is the square root of their product (despite the fact that this product is positive).

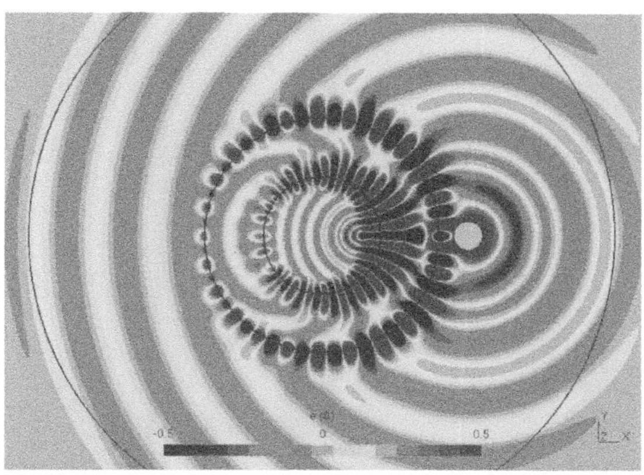

Figure 4.6 A superlens designed by folding the empty space on itself is supposed not to perturbate the cylindrical waves emitted by a wire antenna, but the anomalous resonances (Milton and Nicorovici, 2006; Nicorovici et al., 2007) have a dazzling effect.

(Figs. 4.8 and 4.9). The antenna on the right of the lens has two images, one inside the annular superlens and one inside the central part of the device so that we have three copies of the antenna. In Fig. 4.10, a small perfectly conducting deflector inside the region

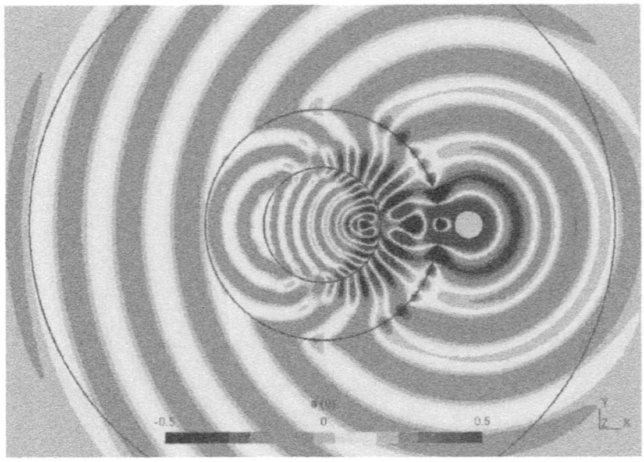

Figure 4.7 A superlens designed by folding the empty space on itself together with 1/1000 of losses to damp the anomalous resonances (Milton and Nicorovici, 2006; Nicorovici et al., 2007).

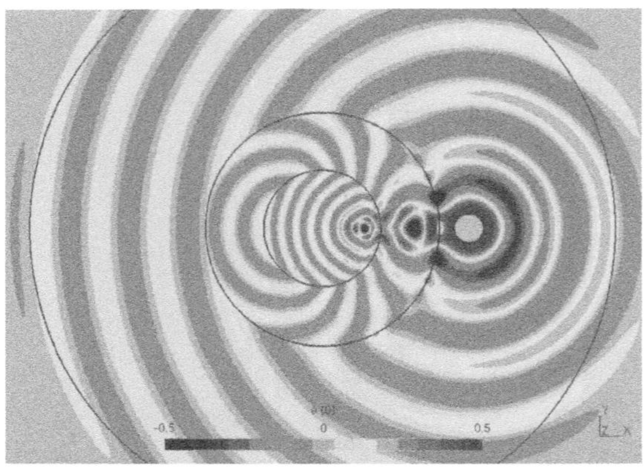

Figure 4.8 A superlens together with 1/100 of losses (in order to avoid the anomalous resonances) does not perturbate the cylindrical waves emitted by a wire antenna but for the attenuation due to the dissipation introduced in the superlens permittivity. The antenna has an image inside the superlens and inside the central part of the device.

surrounded by the perfect lens acts on the image of the antenna and forces the waves to propagate only to the right. This can also be interpreted as if the deflector had a four times larger image acting on the original antenna giving the illusion of a much larger

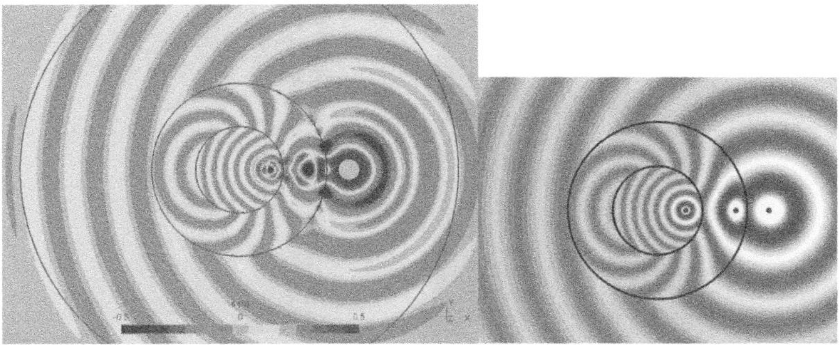

Figure 4.9 A superlens together with 1/100 of losses. Finite element numerical computation using GetDP (left) compared with analytical computation using Mathematica® (right).

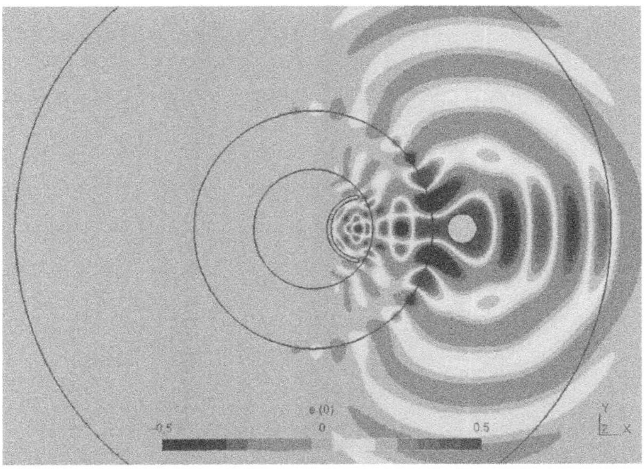

Figure 4.10 A small perfectly conducting deflector acts on the image of the antenna in the central part of the device and forces the waves to propagate only to the right. This can also be interpreted as if the deflector has a four times larger image acting on the original antenna.

object (this phenomenon is sometimes called superscattering). Now, if the image of the deflector is "perfect," one can wonder whether a deflector whose image is on the right of the source would have the effect of deflecting the waves to the left. Figure 4.12 shows that it is not the case: The small deflector inside the structure prevents the formation of the image of the source, and the global effect of

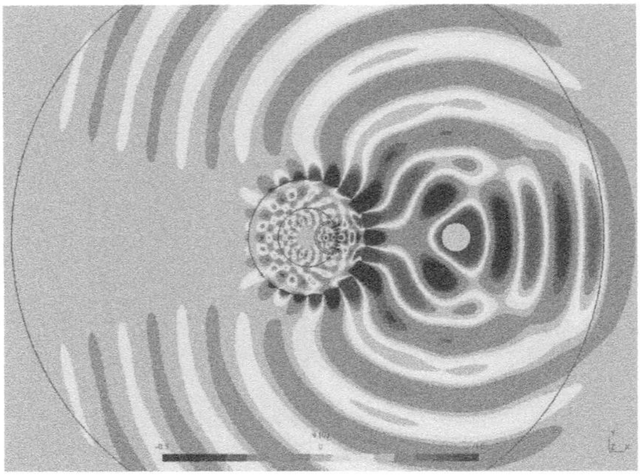

Figure 4.11 A small perfectly conducting deflector acts on the image of the antenna in the central part of the device and forces the waves to propagate only to the right. This can also be interpreted as if the deflector has a height time larger image acting on the original antenna (the structure is smaller and the magnification factor is 8, i.e., larger by a factor 2 with respect to Fig. 4.10.

the device is the one of a large object on the left of the original source. It should be remembered that the superlens comes from the folding of the space, but that the other elements, the source and the

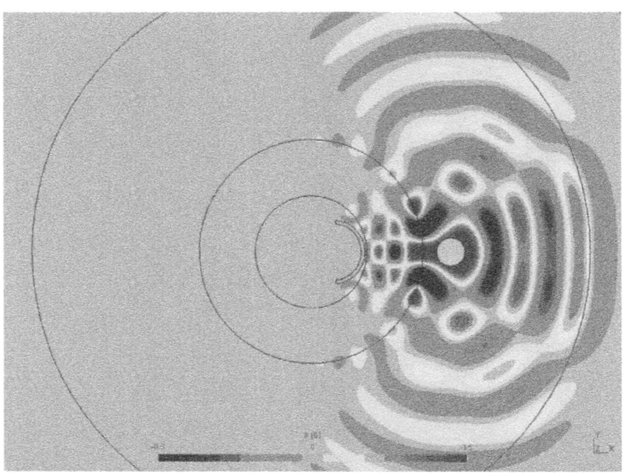

Figure 4.12 The scatterer is a perturbation of the folded geometry, and its presence prevents the correct formation of the image source. It does not have the effect of deflecting the waves to the left as a perfect image, i.e., the same deflector magnified by a factor 4 would do.

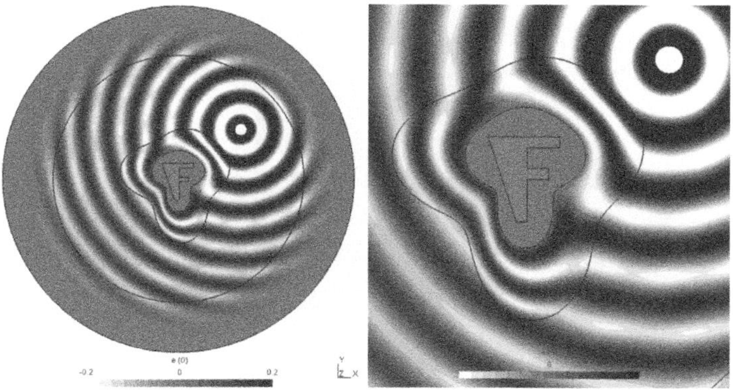

Figure 4.13 Cloak with a general shape given by Fourier series: $R_1(\theta)$ is with $a_0 = 1, b_1 = 0.1, a_2 = -0.15, b_3 = 0.2, a_4 = 0.1$; $R_2(\theta)$ is with $a_0 = 2, a_2 = -0.1, a_3 = -0.15, b_3 = 0.3, a_4 = 0.2$; all the other coefficients are equal to zero. The real part of the electric field E_z scattered by the cloak is represented here. Some residual interferences are due to numerical deviation mainly caused by the singular behavior of the equivalent material properties on the inner boundary of the cloak.

scatterer, are added afterward and are only a perturbation of the folded geometry. Therefore, the system may behave as if there were copies of these elements in the three regions but obviously not in all cases.

All the numerical computations were performed using GetDP (Dular et al., 1998).

4.3 Cylindrical Cloaks of Arbitrary Cross Section

The geometrical transformations can also be used in the reverse sense to design new materials. In this case, a geometrical transformation is applied to free space to guess interesting material properties given by the equivalence rule. A new device can be built if the new material properties may be approximated, e.g., using electromagnetic metamaterials (Ramakrishna, 2005). For instance, as proposed by Pendry (Pendry et al., 2006; Zolla et al., 2007), a convex domain is mapped on a holey domain with the same exterior boundary. The structure made of the transformed

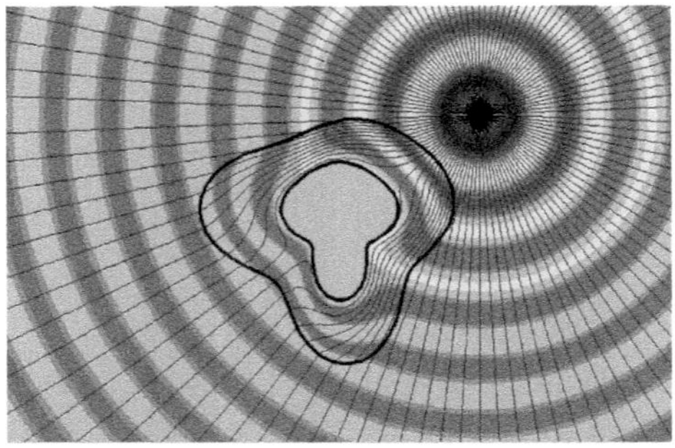

Figure 4.14 Analytical computation (with Mathematica®) of the field (and of associated rays) by directly applying the coordinate transformation to the source field: $E'_z(\rho', \theta') = E_z(\rho(\rho', \theta'), \theta = \theta'))$ with $\rho(\rho', \theta')$ obtained by inversion of the map defined by Eq. (4.22).

equivalent material is an invisibility cloak and any object can be perfectly hidden in the central hole.

A more general situation is now considered here, where the shape of the cloak is no more circular and even possibly non-convex but described by two arbitrary functions $R_1(\theta)$ and $R_2(\theta)$, giving an angle-dependent distance from the origin Corresponding, respectively, to the interior and exterior boundaries of the cloak (Nicolet et al., 2008c).

The geometric transformation that maps the field within the full domain $\rho \leq R_2(\theta)$ onto the hollow domain $R_1(\theta) \leq \rho \leq R_2(\theta)$ can be expressed as:

$$\rho'(\rho, \theta) = R_1(\theta) + \rho(R_2(\theta) - R_1(\theta))/R_2(\theta)\,,\ 0 \leq \rho \leq R_2(\theta) \quad (4.22)$$

with also $\theta' = \theta$, $0 < \theta \leq 2\pi$ and $z' = z$, $z \in \mathbb{R}$. Note that the transformation maps the field for $\rho \geq R_2(\theta)$ onto itself through the identity transformation. This leads to

$$\mathbf{J}_{\rho\rho'}(\rho', \theta') = \frac{\partial(\rho(\rho', \theta'), \theta, z)}{\partial(\rho', \theta', z')} = \begin{pmatrix} c_{11}(\theta') & c_{12}(\rho', \theta') & 0 \\ 0 & 1 & 0 \\ 0 & 0 & 1 \end{pmatrix}, \quad (4.23)$$

where

$$c_{11}(\theta') = R_2(\theta')/(R_2(\theta') - R_1(\theta')) \quad (4.24)$$

for $0 \leq \rho' \leq R_2(\theta')$
and $c_{11} = 1$ for $\rho' > R_2(\theta')$

and

$$c_{12}(\rho',\theta') = (\rho' - R_2(\theta'))R_2(\theta')\frac{dR_1(\theta')}{d\theta'} - \frac{(\rho' - R_1(\theta'))R_1(\theta')\frac{dR_2(\theta')}{d\theta'}}{(R_2(\theta') - R_1(\theta'))^2} \qquad (4.25)$$

for $0 \leq r' \leq R_2(\theta')$,
and $c_{12} = 0$ for $\rho' > R_2(\theta')$.

Finally, the properties of the cloak are given by:

$$\mathbf{T}^{-1} = \mathbf{R}(\theta') \begin{pmatrix} \dfrac{c_{12}^2 + f_\rho^2}{c_{11} f_\rho \rho'} & -\dfrac{c_{12}}{f_\rho} & 0 \\ -\dfrac{c_{12}}{f_\rho} & \dfrac{c_{11}\rho'}{f_\rho} & 0 \\ 0 & 0 & \dfrac{c_{11} f_\rho}{\rho'} \end{pmatrix} \mathbf{R}(\theta')^T, \qquad (4.26)$$

with

$$f_\rho = \frac{(\rho' - R_1)R_2}{(R_2 - R_1)}.$$

The parametric representation of the ellipse

$$\rho(\theta) = \frac{ab}{\sqrt{a^2\cos(\theta)^2 + b^2\sin(\theta)^2}}$$

corresponds to cloaks of elliptical cross section, and it has been checked that it provides exactly the same result as in Ref. (Nicolet et al., 2008b) where similar results have been obtained by a space dilatation.

To obtain general shapes, Fourier series

$$\rho(\theta) = a_0 + \sum_{k=1}^{n}(a_k\cos(k\theta) + b_k\sin(k\theta))$$

may be used. An example of such a general cloak is shown in Fig. 4.13: a source made of a wire of circular cross section (radius $= 0.25$) centered at point $\mathbf{r}_s = (2.5, 2)$ with a constant E_z imposed on its boundary, radiating in a vacuum with wavelength $\lambda = 1$ (Note that all lengths are given in arbitrary units, for instance μm for near infrared). The electric field E_z is, therefore, a cylindrical wave (Note that the electric field is given in arbitrary units, for instance V/m and $E_z = J_0(2\pi 0.25) - iY_0(2\pi 0.25) = 0.472001 - i0.410004$ on the boundary of the source wire) and is not perturbed at all

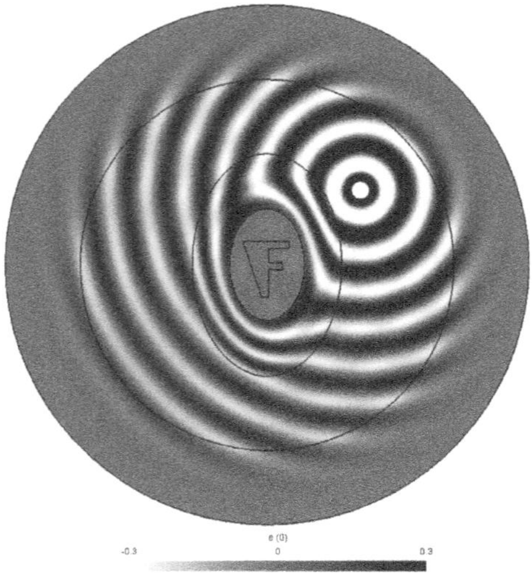

Figure 4.15 Elliptical cloak.

by a F-shaped scattering (lossy) obstacle of relative permittivity $1 + 4i$ placed near the origin (0, 0) and surrounded by the cloak. Note also that the unbounded space is simulated via a circular PML. Figure 4.14 shows the corresponding analytical model.

4.4 Generalized Cloaking

In this section, we present a generalization of cloaking able to arbitrarily transform the electromagnetic appearance of an object. The basic principle is to obtain the constitutive relations of the cloak by the application of a space transformation to a non-empty region. Invisibility can be considered a particular case that corresponds to choosing the empty space as the object to be faked.

In the case of the cylindrical Pendry's map (Nicolet et al., 2008a; Pendry et al., 2006; Zolla et al., 2007), described by the transformation of the two-dimensional cross section, the plane $\mathbb{R}^2$ minus a disk D_1 of radius R_1 is mapped on the whole plane $\mathbb{R}^2$ in such a way that a disk D_2 of radius $R_2 > R_1$, concentric with D_1, is the image of the annulus $D_2 \setminus D_1$ by a radial transformation (see

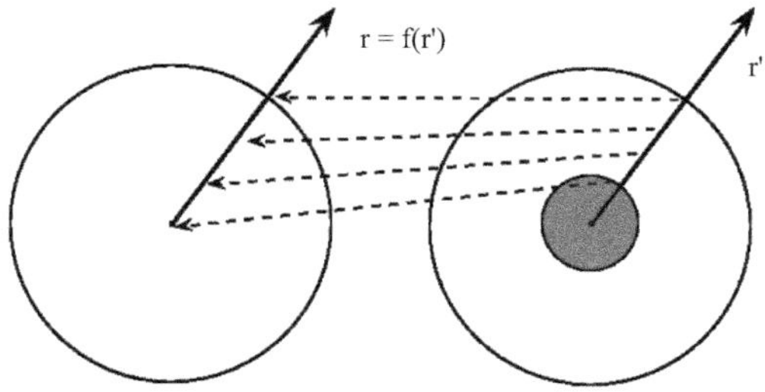

Figure 4.16 Pendry's map of an annulus to a disk used to determine the material properties of an invisibility cloak via the equivalence principles.

Fig. 4.16). In cylindrical coordinates, this transformation is given by:

$$\begin{cases} r = (r' - R_1)R_2/(R_2 - R_1) \text{ for } R_1 \leq r' \leq R_2, \\ \theta = \theta', \ z = z'. \end{cases} \tag{4.27}$$

As for the outside of the disk D_2, the map between the two copies of $\mathbb{R}^2 \backslash D_2$ is the identity map.

The material properties given by rule (4.14) corresponding to this transformation provide an ideal invisibility cloak: Outside D_2, everything behaves as if we were in free space, including the propagation of electromagnetic waves across the cloak and is completely independent of the content of D_1.

Now rule (4.14) may be applied to D_2 containing objects with arbitrary electromagnetic properties so that a region cloaked by this device is still completely hidden but has the appearance of the objects originally in D_2. We may call this optical effect masking (Teixeira, 2007) or "polyjuice" effect.

4.5 Numerical Modeling

Figures 4.19 and 4.20 show the effect of masking on a scattering structure. In Fig. 8.2, a cylindrical TM wave emitted by a circular cylindrical antenna is scattered by a conducting triangular cylinder (the longest side of the cross section is 1.62λ and $\varepsilon_r = 1 + 40i$). The field map represents the longitudinal electric field $E_z(x, y)$, and the outer boundary of the cloak is shown to ease the comparison

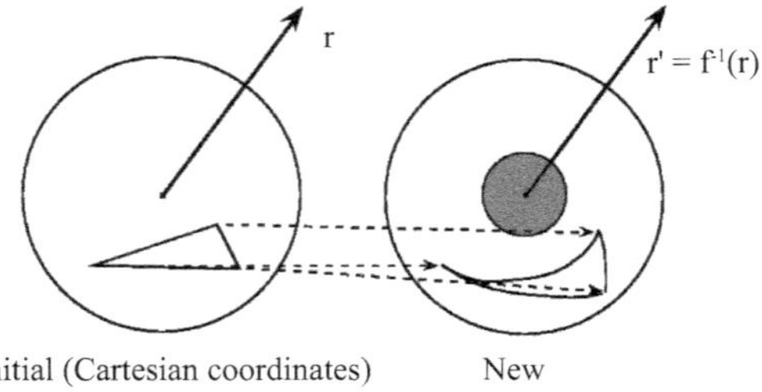

Figure 4.17 When the material properties are piecewise defined, a push forward of the geometry involving the inverse transformation is useful.

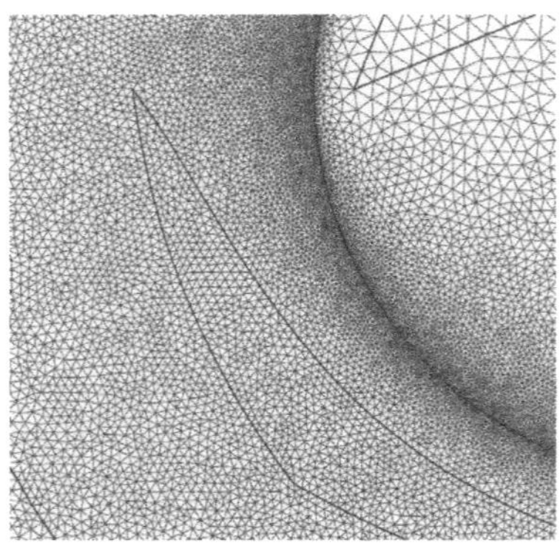

Figure 4.18 This figure shows a part of the triangular mesh used for the finite element modeling of the scattering problem of Fig. 4.20. The singular behavior of the permittivity and of the permeability requires a very fine mesh along the inner boundary of the cloak in order to achieve a satisfactory accuracy with the numerical model.

with the masked case. In Fig. 4.20, the same cylindrical TM wave is scattered by a masked triangular cylinder (but the scattering object inside the cloak may be arbitrarily shaped, as far as it is small enough to fit inside the cloak). This triangular cylinder is symmetric to the previous one with respect to the horizontal plane containing the central fiber of the cylindrical antenna. This bare scatterer would, therefore, give an inverted image to that shown in Fig. 4.19, but here this object is surrounded by a cloak in order to give the same scattering as before. Indeed, on both sides, the electric fields outside the cloak limit are alike.

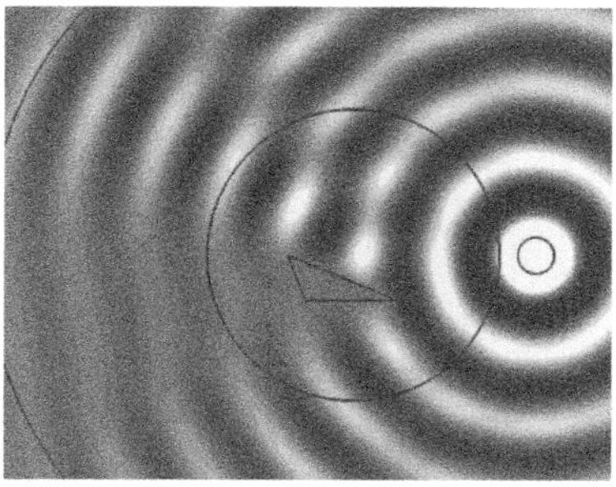

Figure 4.19 A conducting triangular cylinder is scattering cylindrical waves.

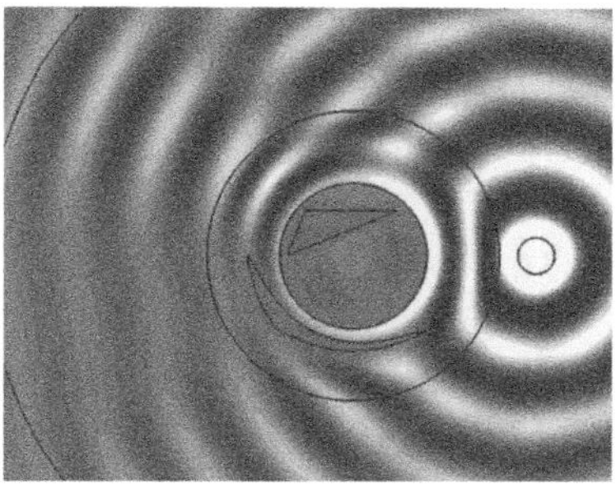

Figure 4.20 A triangular cylinder different from the one in Fig. 4.19 is surrounded by a cloak designed to reproduce the scattering pattern of the triangular cylinder in Fig. 4.19 in spite of the change of scattering object. Of course, the scattering object inside the cloak may be arbitrary as far as it is small enough to fit inside the cloak.

Figure 4.21 highlights the different scattering patterns by displaying the value of $\Re e(E_z)$ on a circle of radius 4λ located around the antenna-scatterer system in the three following cases: (1) the case of Fig. 4.19 (original) with the triangle alone, (2) the case of Fig. 4.20 (coated), and (3) the triangle of Fig. 4.20 without the coating (reversed). It is obvious that the coating restores the field distribution independently of the object present in the central hole.

The numerical computation is performed using the finite element method (via the free GetDP (Dular et al., 1998) and Gmsh (Geuzaine and Remacle, 2009) software tools). The mesh is made of 148,000

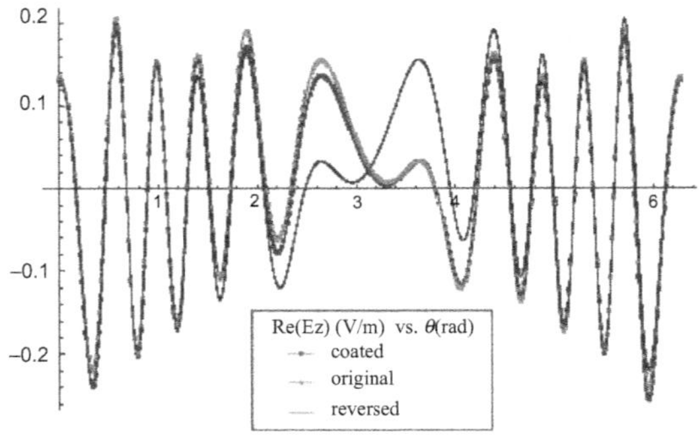

Figure 4.21 The value of the electric field (the real part of E_z) on a circle of radius 4λ concentric with the cloak is represented as a function of the position angle θ (increasing counterclockwise and with $\theta = 0$ corresponding to the point the most on the right). The three configurations considered here are the ones of Fig. 4.20 (coated), Fig. 4.19 (original), and the triangle of Fig. 4.20 without the coating (reversed).

second-order triangles, including the perfectly matched layers used to truncate the computation domain. The singularity of ε and μ requires a very fine mesh in the vicinity of the inner boundary of the cloak (see Fig. 4.18) and is also responsible for the small discrepancies between the numerical model and a perfect cloak (see Fig. 4.20), including the nonzero field in the hole of the cloak.

Note that a small technical problem arises in practice when rule (4.14) is applied: The material properties are defined piecewise on various domains, and it is very useful to know explicitly the boundaries of these domains, e.g., to build the finite element mesh (see Fig. 4.18). These boundaries are curves in the cross section and are thus contravariant objects. Therefore, their transformation requires the inverse map (see Fig. 4.17) φ^{-1} from M to N. Fortunately, map (4.27) is very simple to invert. More explicitly, for a given curve $\mathbf{x}(t)$ of parameter t in the initial Cartesian coordinates, its push forward by Pendry's map is:

$$\mathbf{x}'(t) = \varphi^{-1}(\mathbf{x}(t)) = \left(\frac{R_2 - R_1}{R_2} + \frac{R_1}{\|\mathbf{x}(t)\|} \right) \mathbf{x}(t), \tag{4.28}$$

with the same variation of the parameter t and $\|\mathbf{x}\| = r$. Note that the most common curves used in the design of devices, i.e., line segments and arc of circles, are transformed to less usual curves except for radial segments (with respect to the center of the cloak) and arc of circles concentric with the cloak.

In Fig. 4.20, the image by φ^{-1} of the triangle of Fig. 4.19 is the curvilinear triangle inside the coating region of the cloak. In practice, this anamorphosis of the triangle is described by three splines interpolating each 40 points that are images of points of the segments by φ^{-1}.

References

Bossavit, A. (2001). "Generalized finite differences" in computational electromagnetics, *Prog. Electromagn. Res., PIER 32 (Special Volume on Geometrical Methods for Comp. Electromagnetics)*, pp. 45–64 http://cetaweb.mit.edu/pier/pier32/02.bossavit.pdf.

Burke, W. L. (1985). *Applied Differential Geometry* (Cambridge University Press, Cambridge).

Deschamps, G. A. (1981). Electromagnetics and differential forms, *Proc. IEEE* **69**, p. 676.

Dular, P., Geuzaine, C., Henrotte, F., and Legros, W. (1998). A general environment for the treatment of discrete problems and its application to the finite element method, *IEEE Trans. Mag.* **34**, 5, pp. 3395–3398, see also the Internet address http://www.geuz.org/getdp/.

Geuzaine, C. and Remacle, J.-F. (2009). Gmsh: A three-dimensional finite element mesh generator with built-in pre- and post-processing facilities, *Int. J. Numer. Methods Eng.* **79**, p. 1309.

Guillemin, V. and Sternberg, S. (1990). *Symplectic Techniques in Physics* (Cambridge University Press, Cambridge).

Henrotte, F., Meys, B., Hedia, H., Dular, P., and Legros, W. (1999). Finite element modelling with transformation techniques, *IEEE Trans. Mag.* **35**, p. 1434.

Lai, Y., Chen, H., Zhang, Z.-Q., and Chan, C. T. (2009a). Complementary media invisibility cloak that cloaks objects at a distance outside the cloaking shell, *Phys. Rev. Lett.* **102**, p. 093901.

Lai, Y., J. Ng, H.-Y. C., Han, D.-Z., Xiao, J.-J., Zhang, Z.-Q., and Chan, C. T. (2009b). Illusion optics: The optical transformation of an object into another object, *Phys. Rev. Lett.* **102**, p. 253902.

Lassas, M., Liukkonen, J., and Somersalo, E. (2001). Complex riemannian metric and absorbing boundary condition, *J. Math. Pures Appl.* **80**, 9, pp. 739–768.

Leonhardt, U. and Philbin, T. G. (2006). General relativity in electrical engineering, *New J. Phys.* **8**, p. 247.

Milton, G. W., Briane, M., and Willis, J. R. (2006). On cloaking for elasticity and physical equations with a transformation invariant form, *New J. Phys.* **8**, p. 248.

Milton, G. W. and Nicorovici, N. A. (2006). On the cloaking effects associated with anomalous localized resonance, *Proc. R. Soc. Lond. A* **462**, p. 3027.

Nicolet, A., Remacle, J. F., Meys, B., Genon, A., and Legros, W. (1994). Transformation methods in computational electromagnetism, *J. Appl. Phys.* **75**, p. 6036–6038.

Nicolet, A., Zolla, F., and Guenneau, S. (2004). Modelling of twisted optical waveguides with edge elements, *Eur. Phys. J. Appl. Phys.* **28**, pp. 153–157.

Nicolet, A., Zolla, F., Agha, Y. O., and Guenneau, S. (2007). Leaky modes in twisted microstructured optical fibers, *Waves Random Complex Media* **17**, p. 559.

Nicolet, A., Zolla, F., Agha, Y. O., and Guenneau, S. (2008a). Geometrical transformations and equivalent materials in computational electromagnetism, *Int. J. Comput. Math. Electrical Electronic Eng.* **27**, p. 806.

Nicolet, A., Zolla, F., and Geuzaine, C. (2010). Transformation optics, generalized cloaking and superlenses, *IEEE Trans. Mag.* **46**, 8, pp. 2975–2981.

Nicolet, A., Zolla, F., and Guenneau, S. (2008b). Electromagnetic analysis of cylindrical cloaks of an arbitrary cross section, *Opt. Lett.* **33**, p. 1584.

Nicolet, A., Zolla, F., and Guenneau, S. (2008c). Finite element analysis of cylindrical invisibility cloaks of elliptical cross section, *IEEE Trans. Mag.* **44**, p. 1153.

Nicorovici, N. A., Milton, G. W., McPhedran, R. C., and Botten, L. C. (2007). Quasistatic cloaking of two-dimensional polarizable discrete systems by anomalous resonance, *Opt. Express* **15**, 10, pp. 6314–6323.

Pendry, J. B. (2000). Negative refraction makes a perfect lens, *Phys. Rev. Lett.* **85**, p. 3966.

Pendry, J. B., Shurig, D., and Smith, D. R. (2006). Controlling electromagnetic fields, *Science* **312**, p. 1780.

Ramakrishna, S. A. (2005). Physics of negative refractive index materials, *Rep. Prog. Phys.* **68**, p. 449.

Teixeira, F. L. (2007). Differential form approach to the analysis of electromagnetic cloaking and masking, *Microw. Opt. Technol. Lett.* **49**, pp. 2051–2053.

Yan, M., Yan, W., and Qiu, M. (2008). Cylindrical superlens by a coordinate transformation, *Phys. Rev. B* **78**, 12, 125113, doi:10.1103/PhysRevB.78.125113, http://link.aps.org/abstract/PRB/v78/e125113.

Zhang, J. J., Luo, Y., Chen, H. S., Huangfu, J., Wu, B.-I., Ran, L., and Kong, J. A. (2007). Guiding waves through an invisible tunnel, *Opt. Express* **17**, p. 6203.

Zolla, F., Guenneau, S., Nicolet, A., and Pendry, J. B. (2007). Electromagnetic analysis of cylindrical invisibility cloaks and the mirage effect, *Opt. Lett.* **32**, p. 1069.

Zolla, F., Renversez, G., Nicolet, A., Kuhlmey, B., Guenneau, S., and Felbacq, D. (2005). *Foundations of Photonic Crystal Fibres*, 1st edn. (Imperial College Press, London).

Section II

General methods: waves in periodic media

Chapter 5

Propagation in Periodic Media: Bloch Waves and Evanescent Waves

Didier Felbacq[a] **and Frédéric Zolla**[b]

[a]*Laboratory Charles Coulomb UMR CNRS-UM 5221, University of Montpellier, Place Bataillon, 34095 Montpellier Cedex 05, France*

[b]*Institut FRESNEL, University of Aix-Marseille, Avenue Escadrille Normandie Niemen, 13013 Marseille, France*

didier.felbacq@umontpellier.fr

5.1 Bloch Wave Theory

In this section, we study the fields that can exist in an infinite metamaterial. Considering the infinite structure allows us to obtain a very precise and elegant way of characterizing the (photonic) band structure and the dispersion curves of the medium.

5.1.1 The Periodic Structure

The structure is defined by repeating periodically an elementary cell Y along its basis vectors $\mathbf{a}_i$, where according to the dimension i belongs to $\{1\}$, $\{1, 2\}$, or $\{1, 2, 3\}$. The underlying structure is thus an integer lattice with basis $\mathbf{a}_i$. In one dimension, the metamaterial

Metamaterials Modeling and Design
Edited by Didier Felbacq and Guy Bouchitté

ISBN 978-981-4316-12-5 (Hardcover), 978-1-315-36500-8 (eBook)
www.panstanford.com

is simply characterized by its period $[0, d[$. In higher dimensions, the period is made of all the points M such that: $\mathbf{OM} = x^i\mathbf{a}_i$, $x^i \in [0, 1[$ (where a sum is implied over each pair of repeated index). Generically, a vector belonging to the lattice is denoted by $\mathbf{T} = n^i\mathbf{a}_i$ with integer coefficients n^i. We also define the so-called reciprocal lattice, which is a lattice whose basis vectors $\mathbf{a}^i$ are defined by: $\mathbf{a}^i \cdot \mathbf{a}_j = 2\pi\delta^i_j$. where δ^j_i is the Kronecker symbol. The Brillouin zone Y^* is defined as the set of points P such that $\mathbf{OP} = y_i\mathbf{a}^i$, $y_i \in [-1/2, 1/2[$. Generically, a vector belonging to the reciprocal lattice is denoted by $\mathbf{G} = n_i\mathbf{a}^i$ for some integers n_i.

5.1.2 Waves in a Homogeneous Space

We want to be able to characterize the waves that can exist in an infinite periodic medium. Let us first consider the case of a homogeneous medium and the scalar wave equation for harmonic waves: $\Delta u + k^2 u = 0$. We want to find all the bounded functions u satisfying this equation, for all values of k. Let us pretend for the moment that we do not know that a basis of solutions is the plane waves of the form $\exp(i\mathbf{k} \cdot \mathbf{y})$. If we rewrite the problem in the following form: Find a function u and a positive number E such that: $-\Delta u = Eu$, it appears as a spectral one: The point is now to determine the eigenvalues and eigenvectors of some linear operator (here the Laplacian). In order to do so, let us Fourier transform the function $u(\mathbf{y})$:

$$u(\mathbf{y}) = (2\pi)^{-N/2} \int_{\mathbb{R}^N} \widehat{u}(\mathbf{k}) e^{i\mathbf{k}\cdot\mathbf{y}} d\mathbf{k}. \tag{5.1}$$

This leads to the relation: $(\|\mathbf{k}^2\| - E)\widehat{u}(\mathbf{k}) = 0$. This shows that $\widehat{u}(\mathbf{k})$ is not a function but a Schwartz distribution, in fact: $\widehat{u} = A(\mathbf{k})\delta(\|\mathbf{k}^2\| - E)$, that is, it is proportional to the Dirac distribution whose support is a spherical shell of radius E.[a] We then obtain $u(\mathbf{y}) = (2\pi)^{-N/2} \int_{\mathbb{S}_E^{N-1}} A(\mathbf{k}) e^{i\mathbf{k}\cdot\mathbf{y}} d\mathbf{k}$[b]. A solution to the spectral

[a]Its action on a regular test function ϕ is $\langle \delta(\|\mathbf{k}\|^2 - E), \psi \rangle = \int_{\mathbb{S}_E^{N-1}} \phi(s) ds$, where $\mathbb{S}_E^{N-1}$ is the sphere of radius $\sqrt{E}$ in $\mathbb{R}^N$

[b]For $N = 2$, it reads as:

$$u(y_1, y_2) = (2\pi)^{-1} \int_{-\sqrt{E}}^{\sqrt{E}} e^{ik_1 y_1} \left(A^+(k_1) e^{i\sqrt{E - k_1^2} y_2} + A^-(k_1) e^{-i\sqrt{E-k_1^2} y_2} \right) dk_1.$$

In this formula, y_1 and y_2 can be exchanged.

problem (5.1) is, thus, a continuous sum of plane waves with some amplitude factor: The spectral problem is parametrized by plane waves. The very reason why this decomposition works is the fact that all translations of space $T_{\mathbf{u}}$, $\mathbf{u} \in \mathbb{R}^N$,[c] commute with the Laplacian; therefore, the translations and the Laplacian have a common basis of eigenvectors. This basis is formed with plane waves ($T_{\mathbf{u}}(e^{i\mathbf{k}\cdot\mathbf{y}}) = e^{-i\mathbf{k}\cdot\mathbf{u}}e^{i\mathbf{k}\cdot\mathbf{y}}$).

If the wavenumber $k = \|\mathbf{k}\|$ is given, the set of parameters is a spherical shell of dimension $N - 1$ (a shell of dimension 0 is just a pair of points symmetric with respect to the origin). The decomposition can also be taken in reverse order: We can begin by fixing the wavevector $\mathbf{k}$. Then for the plane wave with wavevector $\mathbf{k}$, the eigenvalue is k^2 and the associated frequency is $\omega = ck$. From this point of view, one frequency is associated with only one energy. Still, we can remark that it is possible to decompose any $\mathbf{k} \in \mathbb{R}^N$, in the following form:

$$\mathbf{k} = \mathbf{k}_b + 2\pi\mathbf{p},$$

where $\mathbf{p}$ is a vector with integer components (i.e., $\mathbf{p} \in \mathbb{Z}^N$) and $\mathbf{k}$ belongs to[d] $Y^* = [-\pi, \pi[^N$. If we use only Y^* and not the entire space $\mathbb{R}^N$ to parametrize the spectral problem, then with a wavevector $\mathbf{k}_b \in Y^*$ is now associated an infinite set of frequencies $\omega_p = c|\mathbf{k}_b + 2\pi\mathbf{p}|$ and an infinite set of eigenvectors, the so-called Bloch waves:

$$\psi_{\mathbf{p}}(\mathbf{k}_b, \mathbf{y}) = \exp(i\mathbf{k}_b \cdot \mathbf{y})\phi_{\mathbf{p}}(\mathbf{k}_b, \mathbf{y}), \tag{5.2}$$

where $\phi_{\mathbf{p}}(\mathbf{y}) = \exp(2i\pi\mathbf{p}\cdot\mathbf{y})$ (it is a Y-periodic function). Using this formulation, the Fourier integral of $u(\mathbf{y})$ can be written:

$$u(\mathbf{y}) = \int_{Y^*} \sum_{\mathbf{p}} u_{\mathbf{p}}(\mathbf{y}) e^{i\mathbf{k}_b\cdot\mathbf{y}}\phi_{\mathbf{p}}(\mathbf{y})d\mathbf{y}, \tag{5.3}$$

where $u_{\mathbf{p}}(\mathbf{y}) = (2\pi)^{-N/2}\widehat{u}(\mathbf{k} - 2\pi\mathbf{p})$. This expression can be extended so as to deal with non-homogeneous media.

---

[c]A translation acts on a function f of the variable $\mathbf{y} \in \mathbb{R}^N$ in the following way: $T_{\mathbf{u}}(f)(\mathbf{y}) = f(\mathbf{y} - \mathbf{u})$.

[d]The notation is not innocent. What we do amounts to decomposing the space $\mathbb{R}^N$ into cubic boxes of side 1, which endows it with a lattice structure of basic cell $Y = [0, 1[^N$, whose corresponding Brillouin zone is Y^*.

5.1.3 Bloch Modes

Let us now consider a metamaterial with basic cell Y and a partial differential operator with periodic coefficient $\mathcal{L}$ that describes wave propagation in the medium. For instance, $\mathcal{L}$ can be the Helmholtz-like operator: $-\varepsilon(\mathbf{y})^{-1}\Delta$ or $-\text{div}(\varepsilon(\mathbf{y})^{-1}\,\mathbf{grad}(\cdot))$, or else the full Maxwell system: $\mathbf{curl}(\varepsilon(\mathbf{y})^{-1}\,\mathbf{curl}(\cdot))$. In such a situation, the Fourier transform cannot lead easily to the solution because of the inhomogeneity of space. The idea behind Bloch waves is to find a way to generalize the Fourier transform to operators with periodic coefficients. The first point at issue is that the space is now really periodic and not homogeneous. It is no longer invariant under an arbitrary translation but only by those of the form $n^i\mathbf{a}_i$, $n^i \in \mathbb{Z}$. The consequence is that plane waves are no longer solutions to the propagation equation. However, the form of Bloch waves given in Eq. 5.2, where now $\phi_{\mathbf{p}}(\mathbf{k}, \mathbf{y})$ is an unknown Y-periodic function, shows that they transform in the following way under a translation of the direct lattice:

$$\psi_{\mathbf{p}}(\mathbf{k}_b, \mathbf{y} + \mathbf{T}) = \exp(i\mathbf{k}_b \cdot \mathbf{T})\psi_{\mathbf{p}}(\mathbf{k}_b, \mathbf{y}) \tag{5.4}$$

The function is then said to be pseudo-periodic. This suggests that we look for a decomposition such as that in Eq. 5.2, where the plane waves of the homogeneous space are now replaced by Bloch waves (i.e., the product of a plane wave by a Y-periodic function).

For such a decomposition to hold, we have to show that we can reduce the spectral problem by imposing the quasi-periodicity condition (5.4) and obtain an equivalent problem. In other words, we do not want to skip any solution by requesting that they be pseudo-periodic.

The first step consists in associating with any square integrable function u on $\mathbb{R}^N$ a family of pseudo-periodic functions indexed by $\mathbf{k}$. This is done by means of the Wannier transform:

$$\mathcal{W}(u)(\mathbf{k}, \mathbf{y}) = \sum_{\mathbf{T}} u(\mathbf{y} - \mathbf{T})e^{i\mathbf{k}\cdot\mathbf{T}}, \tag{5.5}$$

where the sum runs over all vectors of the direct lattice, and $\mathbf{k}$ belongs to the Brillouin zone Y^*. It is easy to check that the transformed function is quasi-periodic with respect to $\mathbf{y}$:

$$\mathcal{W}(\mathbf{k}, \mathbf{y} + \mathbf{T}') = \sum_{\mathbf{T}} u(\mathbf{y} + \mathbf{T}' - \mathbf{T}) e^{i\mathbf{k}\cdot\mathbf{T}} \tag{5.6}$$

$$= e^{i\mathbf{k}\cdot\mathbf{T}'} \sum_{\mathbf{T}''} u(\mathbf{y} - \mathbf{T}'') e^{i\mathbf{k}\cdot\mathbf{T}''} \tag{5.7}$$

$$= e^{i\mathbf{k}\cdot\mathbf{T}'} \mathcal{W}(\mathbf{k}, \mathbf{y} + \mathbf{T}'). \tag{5.8}$$

We can get back the original function by applying the inverse transform:

$$\mathcal{W}^*(\psi)(\mathbf{y}) = \frac{1}{|Y^*|} \int_{Y^*} \psi(\mathbf{k}, \mathbf{y}) d\mathbf{k}. \tag{5.9}$$

Let us apply the inverse Wannier transform to $\mathcal{W}(u)$:

$$\mathcal{W}^* \left(\mathcal{W}(u)(\mathbf{k}, \mathbf{y})\right) = \frac{1}{|Y^*|} \int_{Y^*} \mathcal{W}(u)(\mathbf{k}, \mathbf{y}) d\mathbf{k} \tag{5.10}$$

$$= \frac{1}{|Y^*|} \int_{Y^*} \sum_{\mathbf{T}} u(\mathbf{y} - \mathbf{T}) e^{i\mathbf{k}\cdot\mathbf{T}} d\mathbf{k} \tag{5.11}$$

$$= \sum_{\mathbf{T}} u(\mathbf{y} - \mathbf{T}) \frac{1}{|Y^*|} \int_{Y^*} e^{i\mathbf{k}\cdot\mathbf{T}} d\mathbf{k}. \tag{5.12}$$

The conclusion follows by the identity: $\frac{1}{|Y^*|} \int_{Y^*} e^{i\mathbf{k}\cdot\mathbf{T}} d\mathbf{k} = \delta_{\mathbf{0}}^{\mathbf{T}}$. Conversely, starting with a pseudo-periodic function $\psi(\mathbf{k}, \mathbf{y})$, we have:

$$\mathcal{W}\left(\mathcal{W}^*(\psi)\right)(\mathbf{k}', \mathbf{y}) = \frac{1}{|Y^*|} \sum_{\mathbf{T}} \int_{Y^*} \psi(\mathbf{k}, \mathbf{y} - \mathbf{T}) e^{i\mathbf{k}'\cdot\mathbf{T}} d\mathbf{k}$$

Using the pseudo-periodicity of ψ, we get

$$\mathcal{W}\left(\mathcal{W}^*(\psi)\right)(\mathbf{k}', \mathbf{y}) = \frac{1}{|Y^*|} \int_{Y^*} \psi(\mathbf{k}, \mathbf{y}) \sum_{\mathbf{T}} e^{i(\mathbf{k}'-\mathbf{k})\cdot\mathbf{T}} d\mathbf{k}$$

where the conclusion follows from the identity:

$$\frac{1}{|Y^*|} \sum_{\mathbf{T}} e^{i(\mathbf{k}'-\mathbf{k})\cdot\mathbf{T}} = \sum_{\mathbf{G}} \delta(\mathbf{k} - \mathbf{k}' - \mathbf{G}).$$

In order to be clearer, we state explicitly the spaces involved: The Wannier transform is defined on the space $\mathcal{H} = L^2(\mathbb{R}^N)$ and is into $\mathcal{V}$, which is the set of functions defined on $\mathbb{R}^N \times Y^*$ such that

$$\|\psi\|^2 = \int_{Y^* \times Y} |\psi(\mathbf{y}, \mathbf{k})|^2 d\mathbf{k} d\mathbf{y} < +\infty,$$

$\mathcal{V}$ is a Hilbert space for the scalar product:

$$(\psi_1, \psi_2) = \int_{Y^* \times Y} \psi_1(\mathbf{k}, \mathbf{y}) \overline{\psi_2(\mathbf{k}, \mathbf{y})}\, d\mathbf{k} d\mathbf{y}.$$

We have $\mathcal{W}\mathcal{W}^* = I_{\mathcal{V}}$ and $\mathcal{W}^*\mathcal{W} = I_{\mathcal{H}}$. We can obtain all functions[e] of $\mathcal{V}$ by fixing first the wavevector $\mathbf{k}$ and then by considering the functions $u_{\mathbf{k}}$ such that:

$$u_{\mathbf{k}}(\mathbf{y} + \mathbf{T}) = e^{i\mathbf{k}\cdot\mathbf{T}} u_{\mathbf{k}}(\mathbf{y}).$$

This set of functions is denoted by $\mathcal{H}_{\mathbf{k}}$:

$$\mathcal{H}_{\mathbf{k}} = \left\{ u,\, u(\mathbf{y} + \mathbf{T}) = e^{i\mathbf{k}\cdot\mathbf{T}} u(\mathbf{y}),\, \|u\|^2 = \frac{1}{|Y|} \int_Y |u(\mathbf{y})|^2 d\mathbf{y} < +\infty \right\}. \tag{5.13}$$

We now have a mathematical set-up that shows that any square integrable function can be considered a sum of quasi-periodic functions, by using the Wannier transform. The eigenvalues and eigenvectors of $\mathcal{L}$ can thus be obtained by solving the equation in $\mathcal{H}_{\mathbf{k}}$, then by varying $\mathbf{k}$ in Y^*. For each $\mathbf{k} \in Y^*$, we therefore look for functions $u \in \mathcal{H}_{\mathbf{k}}$ such that $\mathcal{L}(u) = E(\mathbf{k})u$. Once the eigenvalues $E(\mathbf{k})$ are obtained, the corresponding frequencies are $\omega/c = \sqrt{E(\mathbf{k})}$. In the Hilbert space $\mathcal{H}_{\mathbf{k}}$, the operator $\mathcal{L}$ has a set of quasi-periodic eigenfunctions that form a Hilbert-basis, the so-called Bloch waves. They are numbered by an integer[f] p and are of the form:

$$\psi_p(\mathbf{k}, \mathbf{y}) = e^{i\mathbf{k}\cdot\mathbf{y}} \phi_p(\mathbf{k}, \mathbf{y}), \tag{5.14}$$

where ϕ_p is a Y-periodic function. They are associated with a set of eigenvalues $E_p(\mathbf{k})$ that are ordered in ascending order: $E_1 < E_2 < \ldots < E_p < \ldots$ By varying $\mathbf{k}$ in Y^*, we obtain all the eigenvalues as a set of surfaces indexed by p.

We have, in fact, obtained a new way of decomposing a square integrable function[g] u. Indeed, for a given $\mathbf{k}$, $\mathcal{W}(u)(\cdot, \mathbf{k})$ belongs to

---

[e] Mathematically speaking, the space $\mathcal{V}$ can be identified with a direct Hilbertian integral: $\mathcal{V} = \int_{Y^*}^{\oplus} \mathcal{H}_{\mathbf{k}} d\mathbf{k}$, which corresponds to the notion of a "continuous" sum of Hilbert spaces.

[f] This is the band index, and it labels the various allowed frequencies for a given wavevector $\mathbf{k}$.

[g] The reader can remark that, in fact, the decomposition is valid in $\mathbb{R}^N$ for an arbitrary N, not necessarily for $N = 1, 2, 3$.

$\mathcal{H}_{\mathbf{k}}$; therefore, it can be expanded on the basis $\{\psi_p\}_p$:

$$\mathcal{W}(u)(\mathbf{k}, \mathbf{y}) = \sum_p W_p(\mathbf{k})\psi_p(\mathbf{k}, \mathbf{y})$$

where $W_p(\mathbf{k}) = \int_Y \mathcal{W}(u)(\mathbf{k}, \mathbf{y})\overline{\phi_p(\mathbf{k}, \mathbf{y})}e^{-i\mathbf{k}\cdot\mathbf{y}}d\mathbf{y}$, which can be written as:

$$W_p(\mathbf{k}) = \int_Y \sum_{\mathbf{T}} u(\mathbf{y}-\mathbf{T})e^{i\mathbf{k}\cdot\mathbf{T}}\overline{\phi_p(\mathbf{k}, \mathbf{y})}e^{-i\mathbf{k}\cdot\mathbf{y}}d\mathbf{y} \tag{5.15}$$

$$= \sum_{\mathbf{T}} \int_{Y+\mathbf{T}} u(\mathbf{y})\overline{\phi_p(\mathbf{k}, \mathbf{y})}e^{-i\mathbf{k}\cdot\mathbf{y}}d\mathbf{y} \tag{5.16}$$

$$= \int_{\mathbb{R}^N} u(\mathbf{y})\overline{\phi_p(\mathbf{k}, \mathbf{y})}e^{-i\mathbf{k}\cdot\mathbf{y}}d\mathbf{y} \tag{5.17}$$

By using the inverse transform $\mathcal{W}^*$, we get:

$$u(\mathbf{y}) = \frac{1}{|Y^*|}\int_{Y^*}\sum_p W_p(\mathbf{k})\phi_p(\mathbf{k}, \mathbf{y})e^{i\mathbf{k}\cdot\mathbf{y}}d\mathbf{k}.$$

Finally, we can state the Bloch decomposition theorem, which is a generalization of the Fourier transform:

Theorem 5.1. *Let u be a function of $L^2(\mathbb{R}^N)$, where its pth Bloch coefficient is defined by:*

$$\widehat{u}_p(\mathbf{k}) = \int_{\mathbb{R}^N} u(\mathbf{y})\overline{\phi_p(\mathbf{k}, \mathbf{y})}e^{-i\mathbf{k}\cdot\mathbf{y}}d\mathbf{y}\,.$$

We then have the Bloch decomposition formula:

$$u(\mathbf{y}) = \frac{1}{|Y^*|}\int_{Y^*}\sum_p \widehat{u}_p(\mathbf{k})\phi_p(\mathbf{k}, \mathbf{y})e^{i\mathbf{k}\cdot\mathbf{y}}d\mathbf{k}$$

and the Parseval identity:

$$\int_{\mathbb{R}^N} |u(\mathbf{y})|^2 d\mathbf{y} = \frac{1}{|Y^*|}\int_{Y^*}\sum_p |\widehat{u}_p(\mathbf{k})|^2 d\mathbf{k}.$$

5.2 Computation of Band Structures

5.2.1 Two-Dimensional Metamaterials

The previous section has shown that the bounded fields that can exist in an infinite crystal could be parametrized by the set Y^*. In

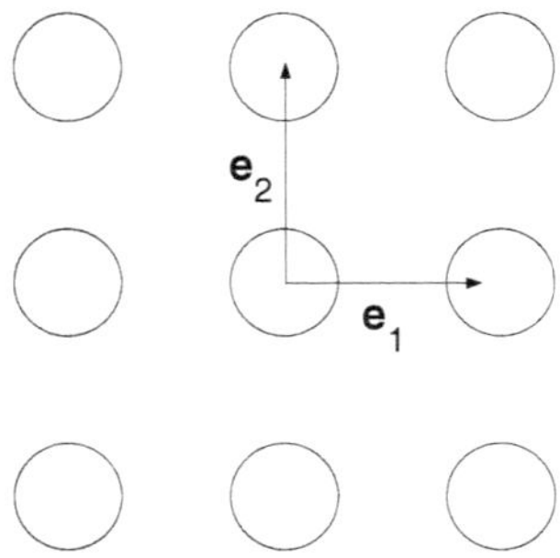

Figure 5.1 A few cells of a dielectric metamaterial with square lattice.

order to describe in a concise way these fields, only a reduced part of the Brillouin zone is used. Indeed, the crystal is invariant under some group of symmetries, and hence it is not necessary to use the entire Brillouin zone in order to compute the spectrum. For instance, let us consider a two-dimensional photonic crystal with a square lattice of side a. The crystal is made of rods of radius R and relative permittivity ε_2 embedded in a matrix of relative permittivity ε_1 (see Fig. 5.1). The Brillouin zone is a square of side $2\pi/a$ (see Fig. 5.2). It suffices to describe only $1/8^{\text{th}}$ of this square (namely, the triangle in bold lines in Fig. 5.2) in order to characterize the spectrum entirely. The description can be further reduced by restricting $\mathbf{k}$ to the lines connecting the points of higher symmetries: Γ, X, M (a more detailed treatment of the symmetries can be found in Ref. (Sakoda, 2005)).

Let us characterize the z-independent fields that can exist in such a structure. The Maxwell system can be reduced to two fundamental cases of polarization:

(1) The $E_{||}$ case in which the electric field is parallel to the z axis, and its z component E_z satisfies

$$-\varepsilon^{-1}(\mathbf{y})\Delta E_z = \left(\frac{\omega}{c}\right)^2 E_z \tag{5.18}$$

(2) The $H_{||}$ case in which the magnetic field is parallel to the z axis, and its z component H_z satisfies:

$$-\text{div}(\varepsilon^{-1}(\mathbf{y})\mathbf{grad}\, H_z) = \left(\frac{\omega}{c}\right)^2 H_z. \tag{5.19}$$

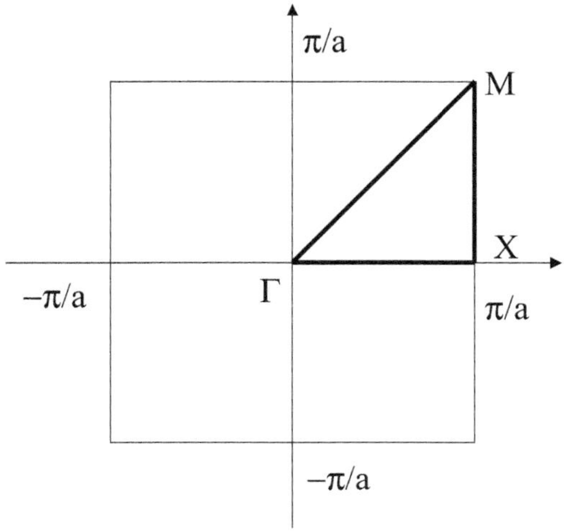

Figure 5.2 The Brillouin zone Y^*.

In both cases, we have to compute the Fourier series of $\varepsilon^{-1}(\mathbf{y})$. We write:

$$\varepsilon^{-1}(\mathbf{y}) = \sum_{\mathbf{G}} \widehat{\varepsilon^{-1}}(\mathbf{G}) e^{i\mathbf{G}\cdot\mathbf{y}}.$$

For a circular fiber, we have explicitly:

$$\widehat{\varepsilon^{-1}}(\mathbf{G}) = \begin{cases} \dfrac{1}{\varepsilon_1} f + \dfrac{1}{\varepsilon_2}(1-f), & \mathbf{G} = 0 \\ \left[\dfrac{1}{\varepsilon_1} - \dfrac{1}{\varepsilon_2}\right] f \dfrac{2J_1(\|\mathbf{G}\|R)}{\|\mathbf{G}\|R}, & \mathbf{G} \neq 0 \end{cases} \tag{5.20}$$

where $f = \frac{\pi R^2}{a^2}$ is the filling fraction and J_1 is the Bessel function of order 1.

We now choose a vector $\mathbf{k} \in Y^*$ and look for Bloch waves solving these equations. First, we expand any Bloch wave associated with E_z and H_z in Fourier series:

$$\begin{cases} E_z(\mathbf{k}, \mathbf{y}) = \sum_{\mathbf{G}} \widehat{E}(\mathbf{k}, \mathbf{G}) e^{i(\mathbf{G}+\mathbf{k})\cdot\mathbf{y}} \\ H_z(\mathbf{k}, \mathbf{y}) = \sum_{\mathbf{G}} \widehat{H}(\mathbf{k}, \mathbf{G}) e^{i(\mathbf{G}+\mathbf{k})\cdot\mathbf{y}}. \end{cases} \tag{5.21}$$

It suffices now to insert the expansions into Eqs. 5.18 and 5.19 to obtain two eigenvalues problems:

$$\begin{cases} \sum_{\mathbf{G}'} (\mathbf{k}+\mathbf{G}) \cdot (\mathbf{k}+\mathbf{G}') \widehat{\varepsilon^{-1}}(\mathbf{G}-\mathbf{G}') \widehat{H}(\mathbf{k}, \mathbf{G}') = \left(\frac{\omega}{c}\right)^2 \widehat{H}(\mathbf{k}, \mathbf{G}) \\ \sum_{\mathbf{G}'} (\mathbf{k}+\mathbf{G}')^2 \widehat{\varepsilon^{-1}}(\mathbf{G}-\mathbf{G}') \widehat{E}(\mathbf{k}, \mathbf{G}') = \left(\frac{\omega}{c}\right)^2 \widehat{E}(\mathbf{k}, \mathbf{G}) \end{cases} \tag{5.22}$$

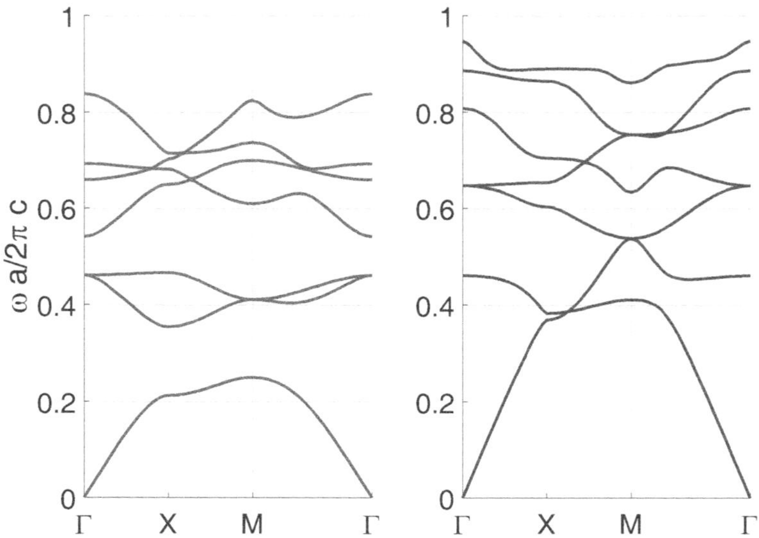

Figure 5.3 The band structure for the structure depicted in Fig. 5.1. The relative permittivity of the rods is 12, and the radius-to-period ratio is 1/4.

Solving these linear systems for a given value of **k** and keeping only the positive eigenvalues, we obtain the allowed frequencies $\frac{\omega_p}{c}$. By varying **k** along the lines connecting the points of high symmetry, we obtain the curves in Figs. 5.3 and 5.4 (a complete example of a triangular lattice is given in Ref. (Plihal and Maradudin, 1991)).

5.3 Periodic Waveguides

5.3.1 Bloch Modes

Although real structures are finite, and one is often interested in the study of defects, the determination of modes in ideal periodic structures is of foremost importance. The Floquet–Bloch theory reduces the problem to the study of a single cell (Langlet et al., 1995) as recalled in Section 5.1. The purpose of this section is to show how to combine this feature with finite element modeling in order to obtain numerical solutions to propagating modes in periodic structures. We consider a structure still invariant along the

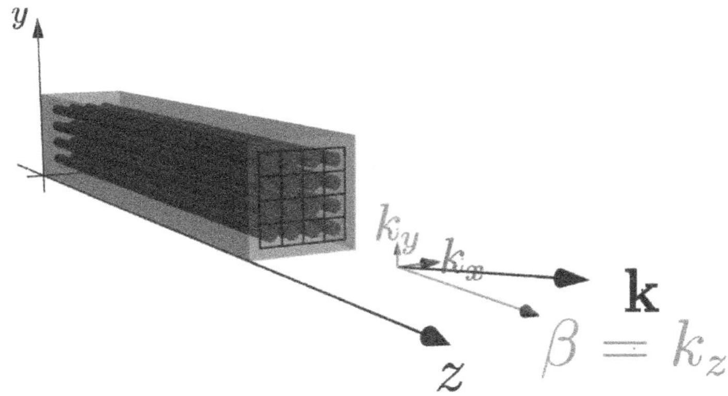

Figure 5.4 A system with a continuous translational invariance along the z axis together with a two-dimensional periodicity in the xy-plane and the general form of propagating modes $\mathbf{U_k}(x, y, z, t)$.

z axis but now also periodic in the xy-plane. Given two linearly independent vectors $\mathbf{a}$ and $\mathbf{b}$ in the xy-plane, the set of points $n\mathbf{a} + m\mathbf{b}$ is called the *lattice*. The *primitive cell* Y is a subset of $\mathbb{R}^2$ such that for any point $\mathbf{r}'$ of $\mathbb{R}^2$, there exist unique $\mathbf{r} = x\varepsilon_x + y\varepsilon_y \in Y$ and $n, m \in \mathbb{Z}$ such that $\mathbf{r}' = \mathbf{r} + n\mathbf{a} + m\mathbf{b}$. A function $U(\mathbf{r})$ is *Y-periodic* if $U(\mathbf{r} + n\mathbf{a} + m\mathbf{b}) = U(\mathbf{r})$ for any $n, m \in \mathbb{Z}$. The waveguide is Y-periodic if $\varepsilon_r(x, y)$ and $\mu_r(x, y)$ are Y-periodic functions. Possible PECs and PMWs have boundaries that form a Y-periodic pattern.

The problem reduces to looking for *Bloch wave* solutions $\mathbf{U_k}$ that have the form (Bloch theorem, see Section 5.1):

$$\mathbf{U_k}(\mathbf{r}) = e^{i\mathbf{k}\cdot\mathbf{r}}\mathbf{U}(\mathbf{r}) = e^{i(k_x x + k_y y)}\mathbf{U}(x, y)\,, \ \forall\,(x, y) \text{ in } \mathbb{R}^2 \qquad (5.23)$$

where $\mathbf{U}(x, y)$ is a Y-periodic function and $\mathbf{k} = k_x\varepsilon^x + k_y\varepsilon^y \in Y^* \subset \mathbb{R}^2$ is a parameter (the *Bloch vector* or quasi-momentum in solid state physics). $Y^* \subset \mathbb{R}^2$ is the *dual cell (first Brillouin zone)*, i.e., the primitive cell of the *reciprocal lattice* determined by the two vectors $\mathbf{a}^*$ and $\mathbf{b}^*$ such that $\mathbf{a}^* \cdot \mathbf{a} = 2\pi$, $\mathbf{a}^* \cdot \mathbf{b} = 0$, $\mathbf{b}^* \cdot \mathbf{a} = 0$, $\mathbf{b}^* \cdot \mathbf{b} = 2\pi$ (it is worth noting that this dot product is, in fact, a duality product: $\mathbf{k} \cdot \mathbf{r} = <\mathbf{k}, \mathbf{r}>$). Such solutions $\mathbf{U_k}$ are said to be $(\mathbf{k}, Y)$*-periodic* in the sequel (though they are not periodic but quasi-periodic).

To specify the class of solutions to our spectral problem, one introduces the Hilbert space

$$[L^2_\sharp(\mathbf{k}, Y)]^3 = \left\{ \mathbf{U_k}|_Y \in [L^2(Y)]^3 \,,\ \mathbf{U_k} \text{ is } (\mathbf{k}, Y)\text{-periodic} \right\} \quad (5.24)$$

of $(\mathbf{k}, Y)$-periodic square integrable functions with values in $\mathbb{C}^3$.

The pair $(\mathbf{E_k}, \mathbf{H_k})$ associated with the Bloch vector $\mathbf{k}$ is called an *electromagnetic propagating Bloch mode* if $\mathbf{E_k}$ and $\mathbf{H_k}$ are $(\mathbf{k}, Y)$-periodic fields satisfying the spectral problem:

$$\begin{cases} \mathbf{curl}_\beta\, \mathbf{H_k} = -i\omega\varepsilon_0\varepsilon_r(x, y)\mathbf{E_k} \\ \mathbf{curl}_\beta\, \mathbf{E_k} = i\omega\mu_0\mu_r(x, y)\mathbf{H_k} \end{cases} \quad (5.25)$$

with

$$\begin{cases} (\beta, \omega, \mathbf{k}) \in \mathbb{R}_+ \times \mathbb{R}_+ \times Y^* \\ (\mathbf{E_k}, \mathbf{H_k}) \neq (\mathbf{0}, \mathbf{0}) \\ \mathbf{E_k}, \mathbf{H_k} \in [L^2_\sharp(\mathbf{k}, Y)]^3 . \end{cases} \quad (5.26)$$

Looking for solutions that are Bloch functions in $[L^2_\sharp(\mathbf{k}, Y)]^3$ ensures the well-posedness of this spectral problem, as a replacement for the Sommerfeld radiation condition (or other decaying conditions for the far field), which is usually enforced in the presence of compact obstacles in the medium. The finite element formulation is completely identical to the non-periodic one. The only difference is that the study is now reduced to the primitive cell Y, which is meshed and in which the integrations are performed. Some technique must be found to ensure that the solution is a $(\mathbf{k}, Y)$-periodic Bloch mode. This can be enforced by using special boundary conditions as explained in the next section.

5.3.2 The Bloch Conditions

In order to find Bloch modes with the finite element method, some changes have to be made with respect to classical boundary value problems, which will be named *Bloch conditions* (Langlet et al., 1995; Nicolet et al., 2004). For the sake of simplicity, one considers first a square cell $Y\ =]0, 1[\times]0, 1[$ as an example. To avoid tedious notations, the case of a scalar field $U_\mathbf{k}(x, y)$ (time and z dependence are irrelevant here and it is no particular problem to extend this

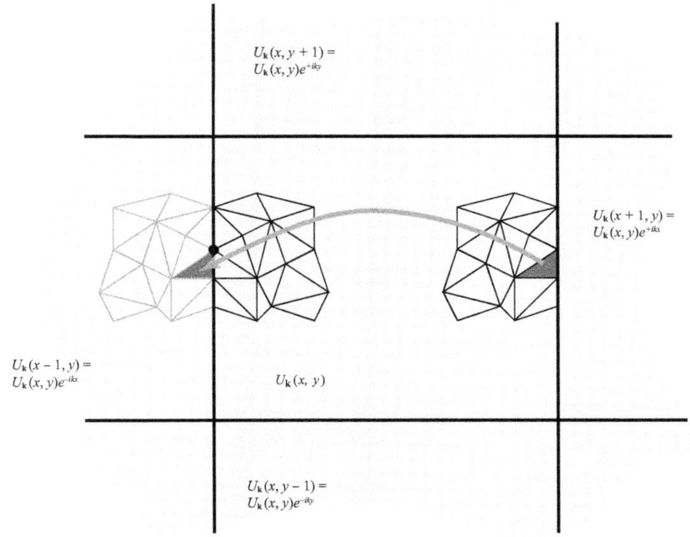

Figure 5.5 Bloch theorem and virtual periodic mesh.

method to vector quantities and edge elements) is considered on the square cell Y with Bloch conditions relating the left-hand and the right-hand sides (Fig. 5.5) since here, for simplification and clarity of the discussion, we take into account only the periodicity along the x coordinate. We consider the weak formulation for the scalar problem $\text{div}(\underline{\underline{\xi}}\ \mathbf{grad}\, U_{\mathbf{k}}) + k_0^2 \chi\, U_{\mathbf{k}} = 0$ (where $\underline{\underline{\xi}}$ and χ are periodic functions representing the material properties) along the classical approach. It is based on the construction of a weighted residual that is very classical but for the weight function $U'_{\mathbf{k}}(\mathbf{r})$ that is quasi-periodic (Demésy et al., 2007; 2009):

$$- \int_Y \left(\underline{\underline{\xi}}\ \mathbf{grad}\, U_{\mathbf{k}}\right) \cdot \mathbf{grad}\, \overline{U'_{\mathbf{k}}} + k_0^2 \chi\, U_{\mathbf{k}}\, \overline{U'_{\mathbf{k}}}\, d\Omega + \int_{\partial Y} \overline{U'_{\mathbf{k}}} \left(\underline{\underline{\xi}}\ \mathbf{grad}\, U_{\mathbf{k}}\right) \cdot \mathbf{n} dS = 0. \qquad (5.27)$$

The integrations are naturally performed on the basic cell Y and the only difference with respect to bounded problem in the treatment of the boundary conditions.

Denote by Γ_l and Γ_r the lines parallel to the y axis delimiting a cell of the lattice, respectively, from its left and right neighbor cells.

Considering explicitly that $U_{\mathbf{k}}(\mathbf{r}) = e^{ik_x x} U_{\#}(x, y)$ is quasi-periodic along x by introducing the x-periodic function $U_{\#}(x, y)$ and similarly that $U'_{\mathbf{k}}(\mathbf{r}) = e^{ik_x x} U'_{\#}(x, y)$, the boundary term for $\Gamma_l \cup \Gamma_r$ is

$$\int_{\Gamma_l \cup \Gamma_r} \overline{U'_{\mathbf{k}}} \left(\underline{\underline{\xi}} \ \mathbf{grad}\, U_{\mathbf{k}} \right) \cdot \mathbf{n} dS = \int_{\Gamma_l \cup \Gamma_r} \overline{U'_{\#}} e^{-ik_x x} \left(\underline{\underline{\xi}} \ \mathbf{grad}(U_{\#} e^{+ik_x x}) \right) \cdot \mathbf{n} dS$$

$$= \int_{\Gamma_l \cup \Gamma_r} \overline{U'_{\#}} \left(\underline{\underline{\xi}} \ (\mathbf{grad}\, U_{\#} + ik_x U_{\#} \mathbf{e}_x) \right) \cdot \mathbf{n} dS = 0$$

because the integrand $\overline{U'_{\#}} \left(\underline{\underline{\xi}} \ (\mathbf{grad}\, U_{\#} + ik_x U_{\#} \mathbf{e}_x) \right) \cdot \mathbf{n}$ is periodic along x and the normal $\mathbf{n}$ has opposite directions on Γ_l and Γ_r so that the contributions of these two boundaries have the same absolute value with opposite signs. *The contribution of the boundary terms vanishes, therefore, naturally in the case of quasi-periodicity.*

We consider now the problem of practically imposing the quasi-periodic Bloch conditions in the computing code. The set of nodes is separated into three subsets: the nodes on the left side, i.e., with $x = 0$, corresponding to the column array of unknowns $\mathbf{u_l}$; the nodes on the right side, i.e., with $x = 1$, corresponding to the column array of unknowns $\mathbf{u_r}$; and the internal nodes, i.e., with $x \in]0, 1[$, corresponding to the column array of unknowns $\mathbf{u}$. One has the following structure for the matrix problem (corresponding, in fact, to natural boundary conditions, i.e., Neumann homogeneous boundary conditions, as the degrees of freedom on the boundaries have to be kept as unknowns in the problem):

$$\mathbf{A} \begin{pmatrix} \mathbf{u} \\ \mathbf{u_l} \\ \mathbf{u_r} \end{pmatrix} = \mathbf{b} \tag{5.28}$$

where $\mathbf{A}$ is the (square Hermitian) matrix of the system and $\mathbf{b}$ is the right-hand side. The solution to be approximated by the numerical method is a Bloch function $U_{\mathbf{k}}(x, y) = U(x, y) e^{i(k_x x + k_y y)}$ with U being Y-periodic and in particular $U(x+1, y) = U(x, y)$. Therefore, the relation between the left-hand and the right-hand sides is:

$$U_{\mathbf{k}}(1, y) = U(1, y) e^{i(k_x + k_y y)} = U_{\mathbf{k}}(0, y) e^{ik_x} \Rightarrow \mathbf{u_r} = \mathbf{u_l} e^{ik_x}. \tag{5.29}$$

The set of unknowns can thus be expressed as a function of the reduced set $\mathbf{u}$ and $\mathbf{u_l}$ via:

$$\begin{pmatrix} \mathbf{u} \\ \mathbf{u_l} \\ \mathbf{u_r} \end{pmatrix} = \mathbf{P} \begin{pmatrix} \mathbf{u} \\ \mathbf{u_l} \end{pmatrix} \text{ with } \mathbf{P} = \begin{pmatrix} \mathbf{1} & \mathbf{0} \\ \mathbf{0} & \mathbf{1} \\ \mathbf{0} & \mathbf{1} e^{ik_x} \end{pmatrix} \tag{5.30}$$

where **1** and **0** are identity and null matrices, respectively, with suitable dimensions. The finite element equations related to the eliminated nodes have now to be taken into account. Due to the periodicity of the structure, the elements on the left of the right side correspond to the elements on the left of the left side (Fig. 5.5). Therefore, their contributions (i.e., the equations corresponding to $\mathbf{u_r}$) must be added to the equations corresponding to $\mathbf{u_l}$ with the right phase factor, i.e., e^{-ik_x}, which amounts to multiplying the system matrix by $\mathbf{P}^*$ (the Hermitian of $\mathbf{P}$). Finally, the linear system to be solved is:

$$\mathbf{P}^*\mathbf{A}\mathbf{P}\begin{pmatrix}\mathbf{u}\\ \mathbf{u_l}\end{pmatrix} = \mathbf{P}^*\mathbf{b} \tag{5.31}$$

where it is worth noting that the system matrix is still Hermitian, which is important for numerical computation. Now a generalized eigenvalue problem (with natural boundary conditions) $\mathbf{Au} = \lambda\mathbf{Bu}$ is transformed to a Bloch mode problem according to $\mathbf{P}^*\mathbf{APu}' = \lambda\mathbf{P}^*\mathbf{BPu}'$, which is still a large sparse Hermitian generalized eigenvalue problem.

5.3.3 A Numerical Example

As an illustration, the Bloch finite element method will be used to reproduce the results presented in (Maradudin and McGurn, 1994), where they were obtained using a plane wave method.

The basic cell is a rhombus made of two equilateral triangles: The lattice vectors are $\mathbf{a} = \Lambda\varepsilon_x$ and $\mathbf{b} = \frac{\Lambda}{2}\varepsilon_x + \frac{\Lambda\sqrt{3}}{2}\varepsilon_y$, where Λ is the nearest neighbor distance, i.e., the length of the sides of the cells. This cell contains a circular air inclusion (radius $R = 0.48\Lambda$, so that the filling fraction $f = 0.8358$ and $\varepsilon_r = 1.0$) surrounded by solid dielectric material ($\varepsilon_r = 13.0$). The vectors of the reciprocal lattice are $\mathbf{a}^* = \frac{2\pi}{\Lambda}\varepsilon^x - \frac{2\pi\sqrt{3}}{3\Lambda}\varepsilon^y$ and $\mathbf{b}^* = \frac{4\pi\sqrt{3}}{3\Lambda}\varepsilon^y$ and the first Brillouin zone is hexagonal. The irreducible part can be represented by the triangle with vertices $\Gamma = (0, 0)$, $M = (0, \frac{2\pi}{\sqrt{3}\Lambda})$, and $K = (\frac{2\pi}{3\Lambda}, \frac{2\pi}{\sqrt{3}\Lambda})$. The basic cell is meshed with 4628 triangles. All these data are summarized in Fig. 5.6. Note that the circular inclusion is too large to fit as a single piece inside the basic cell, hence the splitting into four parts in the corners.

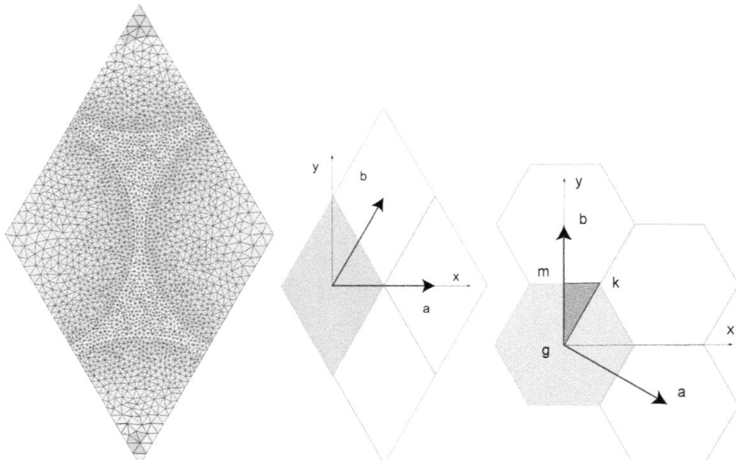

Figure 5.6 Two-dimensional periodic structure (the basic cell is rhombic with a side length Λ) with a circular air inclusion (radius $R = 0.48\Lambda$, $\varepsilon_r = 1.0$) surrounded by solid dielectric material ($\varepsilon_r = 13.0$): meshing of a basic rhombic cell with 4628 triangles (left), representation of some lattice cells with the lattice vectors $\mathbf{a} = \Lambda\varepsilon_x$ and $\mathbf{b} = \frac{\Lambda}{2}\varepsilon_x + \frac{\Lambda\sqrt{3}}{2}\varepsilon_y$ (centre), representation of some cells of the reciprocal lattice with the lattice vectors $\mathbf{a}^* = \frac{2\pi}{\Lambda}\varepsilon^x - \frac{2\pi\sqrt{3}}{3\Lambda}\varepsilon^y$ and $\mathbf{b}^* = \frac{4\pi\sqrt{3}}{3\Lambda}\varepsilon^y$ and the irreducible part of the first Brillouin zone represented by the triangle with vertices $\Gamma = (0, 0)$, $M = (0, \frac{2\pi}{\sqrt{3}\Lambda})$, and $K = (\frac{2\pi}{3\Lambda}, \frac{2\pi}{\sqrt{3}\Lambda})$ (right).

The Bloch boundary conditions connect the degrees of freedom on opposite sides of the rhombus: The degrees of freedom on the lower left-hand side are equal to the corresponding ones on the upper right-hand side multiplied by a phase factor equal to $e^{i(-k_x\frac{\Lambda}{2}-k_y\frac{\sqrt{3}\Lambda}{2})}$, and the degrees of freedom on the lower right-hand side are equal to the corresponding ones on the upper left-hand side multiplied by a phase factor equal to $e^{i(+k_x\frac{\Lambda}{2}-k_y\frac{\sqrt{3}\Lambda}{2})}$.

The dispersion curves shown in Figs. 5.7 and 5.8 correspond to pulsations ω (only the ω such that $\omega < \frac{2\pi c}{\Lambda}$ are represented here) of the propagation modes associated with a given value of the propagation constant β ($\beta\Lambda = 0.0, 2.0, 4.0, 2\pi$).

The value 1.0 is given to Λ for the numerical computations. The boundary of the irreducible Brillouin zone is sampled with 120

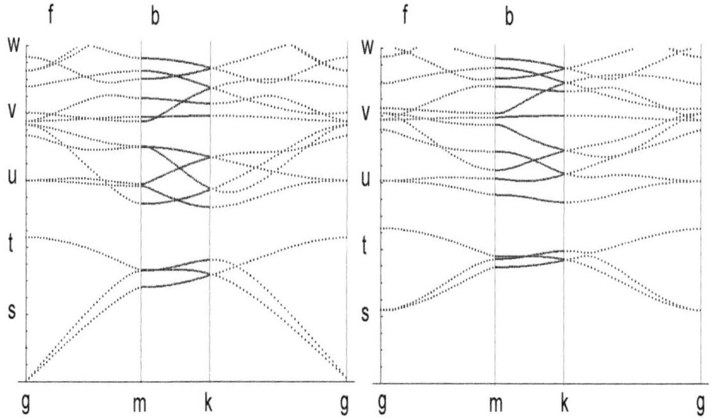

Figure 5.7 Dispersion curves corresponding to Bloch waves in conical mounting in the lattice of Fig. 5.6 for $\beta\Lambda = 0.0$ (left) and $\beta\Lambda = 2.0$ (right).

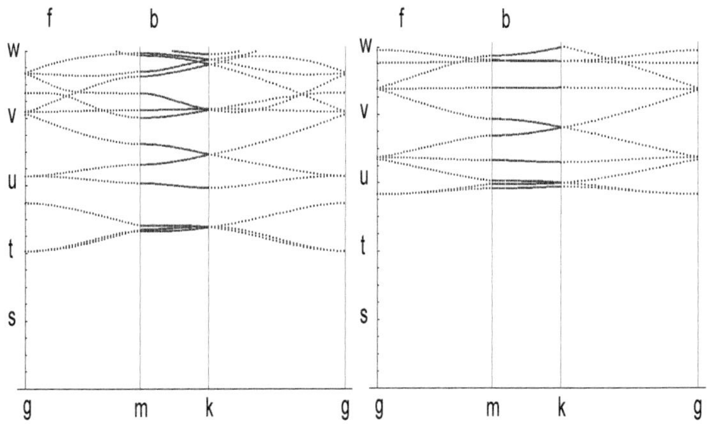

Figure 5.8 Dispersion curves corresponding to Bloch waves in conical mounting in the lattice of Fig. 5.6 for $\beta\Lambda = 4.0$ (left) and $\beta\Lambda = 2\pi$ (right).

points (40 on each side of the triangle). The computation of the eigenvalues associated with a particular Bloch vector takes a few minutes on a typical 2.6 GHz desktop microcomputer. The results are in good agreement with those of (Maradudin and McGurn, 1994).

5.3.4 Direct Determination of the Periodic Part

Another possible approach is to solve an equation with periodic boundary conditions, which gives directly the periodic vector field $\mathbf{U}(\mathbf{r})$ involved in the Bloch mode $\mathbf{U}(\mathbf{r})e^{i\mathbf{k}\cdot\mathbf{r}}$.

The equation $\mathbf{curl}_\beta(\mu_r^{-1}\,\mathbf{curl}_\beta\,\mathbf{E_k}) = k_0^2\varepsilon_r\mathbf{E_k}$ with $\mathbf{E_k}(\mathbf{r}) = \mathbf{E}(\mathbf{r})e^{i\mathbf{k}\cdot\mathbf{r}}$ gives for $\mathbf{E}$:

$$\mathbf{curl}_{\beta,\mathbf{k}}(\mu_r^{-1}\,\mathbf{curl}_{\beta,\mathbf{k}}\,\mathbf{E}) = k_0^2\varepsilon_r\mathbf{E}$$

where $\mathbf{curl}_{\beta,\mathbf{k}}\,\mathbf{U} = \mathbf{curl}_\beta(\mathbf{U}(\mathbf{r})e^{i\mathbf{k}\cdot\mathbf{r}})e^{-i\mathbf{k}\cdot\mathbf{r}}$. The following transverse operators are defined for a scalar function $\varphi(x, y)$ and a transverse field $\mathbf{v} = v_x(x, y)\mathbf{e}^x + v_y(x, y)\mathbf{e}^y$:

$$\begin{aligned}\mathbf{grad}_{t,\mathbf{k}}\,\varphi &= \mathbf{grad}_t\,\varphi + i\mathbf{k}\varphi\\ \mathrm{rot}_{t,\mathbf{k}}\mathbf{v} &= \mathrm{rot}_t\mathbf{v} + i\mathbf{k}\times\mathbf{v}\\ \mathrm{div}_{t,\mathbf{k}}\,\mathbf{v} &= \mathrm{div}_t\,\mathbf{v} + i\mathbf{k}\cdot\mathbf{v}\end{aligned}$$

and one has:

$$\begin{aligned}\mathbf{curl}_{\beta,\mathbf{k}}\,(\mathbf{v}+\varphi\mathbf{e}^z) &= \mathrm{rot}_{t,\mathbf{k}}\mathbf{v} + (\mathbf{grad}_{t,\mathbf{k}}\,\varphi - i\,\beta\mathbf{v})\times\mathbf{e}^z\\ &= \mathrm{rot}_t\mathbf{v} + i\mathbf{k}\times\mathbf{v} + (\mathbf{grad}_t\,\varphi + i\mathbf{k}\varphi - i\,\beta\mathbf{v})\times\mathbf{e}^z\,.\end{aligned}$$

The weak formulation is now:

$$\begin{aligned}\mathcal{R}(\beta;\mathbf{E},\mathbf{E}') = \int_\Omega \mu_r^{-1}\,\mathbf{curl}_{\beta,\mathbf{k}}\,\mathbf{E}\cdot\overline{\mathbf{curl}_{\beta,\mathbf{k}}\,\mathbf{E}'}\,dxdy\\ -k_0^2\int_\Omega \mathbf{E}\cdot\overline{\mathbf{E}'}\,\varepsilon_r dxdy = 0\quad,\forall\,\mathbf{E}'\in H(\mathbf{curl}_{\beta,\mathbf{k}},\Omega)\end{aligned}$$

with periodic boundary conditions on $\mathbf{E}$ (Boffi et al., 2002) (which can be seen as a particular case of the Bloch boundary conditions with $k_x = k_y = 0$).

5.4 Evanescent Waves

5.4.1 Introduction

The preceding section was devoted to the study of the bounded solutions to the Maxwell system in a periodic medium. This restriction to waves that do not grow at infinity has led to the onset of forbidden bands. In fact, solutions to the Maxwell system do exist in the forbidden bands: they are exponentially growing.

For a perfect infinite periodic structure, these solutions are not relevant from a physical point of view. They become, however, of prime importance whenever the translational invariance of the medium is broken, for instance by the insertion of defects, sources, interfaces, and so on. Let us give a very simple example in case of a vacuum. By using the Fourier transform, any field can be written as an integral of plane waves: $U(x) = \int u(\mathbf{k})e^{i\mathbf{k}\cdot\mathbf{x}}d^3k$, which are the Bloch waves of a homogeneous medium. If now there is a localized source, say a source for a cylindrical wave $H_0(k_0 PM)$ at some point P, then the radiated field contains dissociated plane waves (i.e., propagative in direction x and evanescent in direction y) as it is shown by considering the plane wave expansion of the source: $H_0(\sqrt{x^2+y^2}) = \frac{1}{\pi}\int \frac{e^{i(\alpha x+\beta|y|)}}{\beta}d\alpha$. All the plane waves with $\alpha > (\omega/c)$ are obviously evanescent in the y direction, away from the source.

In any physical situation, the structures used are always finite and, therefore, present one or more interfaces with the open space, from which evanescent waves are radiated. Bloch waves are no longer sufficient to describe the electromagnetic field and should be complemented by evanescent waves.

A theoretical approach only involving equivalent medium theories, group velocity, and more generally quantities only derived from the band structure is certainly incomplete. The goal of this section is to analyze the relative importance of the evanescent waves (the near-field) with respect to Bloch modes for describing the propagation of the electromagnetic field inside a metamaterial.

5.4.2 Propagating and Non-Propagating Modes

We consider a structure that is periodic in the x direction with period d, finite in the y direction with height h, and invariant in the z direction. It can be a structured membrane or a grating or a metasurface.

With these hypotheses, the Maxwell system can be reduced to two scalar equations corresponding to the two usual polarized cases: s-polarized (electric field linearly polarized along z) or p-polarized (magnetic field linearly polarized along z). According to the polarization, $u(x, y)$ will represent the z component of the electric field (s-polarization) or of the magnetic field (p-polarization).

Our aim is to characterize the field by taking specifically into account the existence of an interface.

We denote

$$U = \begin{cases} (u, \partial_y u) : s\text{-polarization} \\ (u, \varepsilon^{-1}\partial_y u) : p\text{-polarization} \end{cases}$$

where ε is the relative permittivity. We start by rewriting the Helmholtz equation in the form of an evolution equation with respect to the variable y:

$$\partial_y U = M(x, y)U. \tag{5.32}$$

where the matrix M is defined by:

$$M = \begin{pmatrix} 0 & I_d \\ B & 0 \end{pmatrix} \tag{5.33}$$

and

$$B = \begin{cases} -\partial_x^2 - k^2\varepsilon(x, y) : s\text{-polarization} \\ -\partial_x\left(\varepsilon^{-1}(x, y)\,\partial_x\right) - k^2 : p\text{-polarization} \end{cases} \tag{5.34}$$

For an infinite medium, the spectral analysis of these operators may be performed by means of Bloch waves theory exposed in the preceding chapter. However, the medium that is considered here is finite in the y direction; therefore, a bi-dimensional Bloch analysis is not sufficient. Because, for simplicity reasons, it was assumed that the medium is infinite in the x direction, only a Bloch analysis in the x direction is performed: The field can be decomposed as the sum of pseudo-periodic fields in the x direction: $u(x + d_x, y) = \exp(i\alpha d_x)\, u(x, y)$, where α belongs to the interval $Y = [-\pi/d, \pi/d[$.[h]

We obtain a family of operators B_α which are defined by (5.34) and with domain $D_\alpha = \{u \in L^2([0, d]), Bu \in L^2([0, d]),\ u(x + d_x, y) = e^{i\alpha d_x} u(x, y)\}$. B_α can be put in diagonal form (see the appendix at the end of this chapter). Let us denote by $\{\lambda_{j,\alpha}\}_j$ its eigenvalues and $\{|U_{j,\alpha}\rangle\}$ the corresponding eigenvectors. This

[h] In a diffraction problem, where the structure is illuminated by a plane wave under the incidence θ, it holds $\alpha \equiv k \sin\theta \bmod (\pi/d)$.

diagonal form allows to consider the square root $\sqrt{B_\alpha}$ of B_α. It is the operator whose matrix elements in the basis $\{|U_{j,\alpha}\rangle\}$ are simply $\langle U_{i,\alpha}|\sqrt{B_\alpha}|U_{j,\alpha}\rangle = \sqrt{\lambda_{j,\alpha}}$.

Proposition 1: B_α *is self-adjoint with compact resolvent and is bounded below.*

Proof: That B is self-adjoint is a standard result (Reed and Simon, 1978). Let us prove that B is bounded below:

$$(Bu, u) = \int |\partial_x u|^2\, dx - k^2 \int \varepsilon(x, y)\, |u|^2\, dx,$$

denoting by ε_m the essential sup of $\varepsilon(x)$, we have:

$$(Bu, u) + k^2 \varepsilon_m (u, u) = \int |\partial_x u|^2\, dx + k^2 \int (\varepsilon_m - \varepsilon(x, y))\, |u|^2\, dx > 0$$

showing that $B + k^2\varepsilon_m$ is positive and hence B is bounded below. Finally, let us consider the equation $Bu_n + \alpha u_n = f_n$, with (f_n) a bounded sequence of functions of $L^2([0, d])$ and γ such that $B + \gamma$ is strictly positive. Then we have: $\|f_n\|_{L^2} > C\,\|u\|_{H^1}$, for some constant C. Hence, by Rellich–Kondrachov theorem, $(B + \alpha)^{-1}$ is compact, and therefore B has compact resolvent. □

We denote by $\{\lambda_{j,\alpha}\}_j$ the spectrum of B_α. As a corollary to proposition 1, we may define the square root of B_α (with a cut along $i\mathbb{R}^-$ and with $\sqrt{1} = 1$).

We can now define a propagator $R(y, y_0)$, which is an operator depending on two variables (y, y_0) satisfying:

$$U(y) = \mathcal{R}(y, y_0)U(y_0) \tag{5.35}$$

The propagator allows to compute the solution at y when it is known at y_0, hence the word "propagator," because it propagates the solution from one point to another. Let us denote $\mathcal{T}$ the so-called transfer matrix that relates $U(0)$ to $U(h)$. Evidently, we have $\mathcal{T} = \mathcal{R}_s(0, h)$.

Proposition: *The spectrum of* $\mathcal{T}$ *is invariant under the transformation* $\tau \to 1/\tau$.

Proof: Let us denote: $\mathbb{U} = \begin{pmatrix} \not{\Bbb 1} & 0 \\ 0 & -\not{\Bbb 1} \end{pmatrix}$ Then $\mathcal{T}^{-1} = \mathbb{U}\mathcal{T}\mathbb{U}$ implying that $\mathcal{T}$ and $\mathcal{T}^{-1}$ are isospectral. In the case of a product of such operators, it holds:

$$\Pi_n T_n = \Pi_n \mathbb{U} T_n^{-1} \mathbb{U} = \mathbb{U} \Pi_n T_n^{-1} \mathbb{U}$$

Now it remains to show that T_1T_2 and T_2T_1 are isospectral. Consider a spectral couple (X, μ) of T_1T_2:

$$T_1T_2X = \mu X$$

keeping in mind that T is invertible, it comes:

$$T_2T_1(T_1^{-1}X) = \mu(T_1^{-1}X)$$

and the result. □

The propagative modes correspond to the eigenvalues of $\mathcal{T}$ of modulus 1 and thus to the negative eigenvalues of B_α. They are finitely many.

Theorem: *$\mathcal{T}$ can be put in diagonal form.*
This is a very important result, because it shows that propagating modes should be supplemented by evanescent ones.

The transfer matrix $\mathcal{T}$ of the basic layer links $\left(u(x, h), \partial_y u\,(x, h)\right)$ to $\left(u\,(x, 0), \partial_y u\,(x, 0)\right)$, but as such it is not very easy to handle. A simpler representation can be obtained by considering the Rayleigh expansions of the field at the boundaries of the structure.

Above or under the structure, the field can be expanded on a Rayleigh basis:

$$\begin{aligned} u(x, 0) &= \sum_n \left(A_n^+ e^{i\beta_n y} + A_n^- e^{-i\beta_n y}\right) e^{i\alpha_n x},\ u(x, h) \\ &= \sum_n \left(B_n^+ e^{i\beta_n (y-h)} + B_n^- e^{-i\beta_n (y-h)}\right) e^{i\alpha_n x}. \end{aligned}$$

where $\alpha_n = \alpha + nK$, $K = \frac{2\pi}{d}$ and $\beta_n^2 = k^2 - \alpha_n^2$. By considering the field directly on the boundary, we have the following expansions for the field and its normal derivative:

$$u(x, 0) = \sum_n \left(A_n + A_n^-\right) e^{i\alpha_n x},\ u(x, h) = \sum_n \left(B_n^+ + B_n^-\right) e^{i\alpha_n x} \tag{5.36}$$

$$\begin{aligned} \partial_y u(x, 0) &= \sum_n i\beta_n \left(A_n^+ - A_n^-\right) e^{i\alpha_n x},\ \partial_y u(x, h) \\ &= \sum_n i\beta_n \left(B_n^+ - B_n^-\right) e^{i\alpha_n x}. \end{aligned} \tag{5.37}$$

The knowledge of $\widehat{A^\pm} = \left\{A_n^\pm\right\}_n$ (resp. $\widehat{B^\pm} = \left\{B_n^\pm\right\}$) gives the value of the derivatives. Therefore, rather than computing the monodromy matrix as defined above, we compute the matrix $\mathcal{T}(\alpha, \lambda)$ such that

$$\mathcal{T}(\alpha, \lambda)\begin{pmatrix}\widehat{A^+}\\ \widehat{A^-}\end{pmatrix} = \begin{pmatrix}\widehat{B^+}\\ \widehat{B^-}\end{pmatrix} \tag{5.38}$$

Let us denote $e^{ik_y h}$, $k_y \in \left[-\frac{\pi}{h}, \frac{\pi}{h}\right]$, an eigenvalue of $\mathcal{T}(\alpha, \lambda)$ of modulus one, and $\Psi = \left(\widehat{\psi^+}, \widehat{\psi^-}\right)$ an associated eigenvector, $\mathcal{T}\left(\widehat{\psi^+}, \widehat{\psi^-}\right) = e^{ik_y h}\left(\widehat{\psi^+}, \widehat{\psi^-}\right)$ and therefore, $\Psi(x, y+h) = e^{ik_y h}\psi(x, y)$. Consequently, $(x, y) \to e^{i\alpha x}\psi(x, y)$ is a Bloch wave associated to the Bloch vector (α, β). The non-propagating modes inside the crystal correspond to eigenvectors associated with eigenvalues that are not of modulus one. We have thus obtained a decomposition of the modes by means of a family of monodromy operators parametrized by $\alpha \in Y$.

5.4.3 Analysis of the Spectrum

At a given wavelength, Bloch waves are not sufficient to compute the scattering properties of the metamaterial; the evanescent waves should be added. In order to quantify the relative importance of the evanescent waves, a decomposition of the field is needed.

5.4.3.1 Decomposition of the field

Suppose that the electromagnetic field is known on the upper face of the structure (by means of the coefficients $\widehat{A^\pm}$); it is then possible to expand it on the various modes that exist in the structure. More precisely, the matrix $\mathcal{T}(\alpha, \lambda)$ can be put in the diagonal form:

$$\mathcal{T}(\alpha, \lambda) = \mathcal{T}_p \oplus \mathcal{T}_e \oplus \mathcal{T}_a$$

where $\mathcal{T}_p$ is a finite rank operator corresponding to propagative waves and $\mathcal{T}_e$, $\mathcal{T}_a$ correspond to the evanescent and anti-evanescent modes.

With this decomposition, the vector $\Psi = \left(\widehat{\psi}^+, \widehat{\psi}^-\right)$ can be put in the form $\Psi = \left(\Psi_p, \Psi_e, \Psi_a\right)$. Whence we define the branching ratios π_p (resp. π_e, π_a) of the field on the propagating (resp. evanescent, anti-evanescent) modes by:

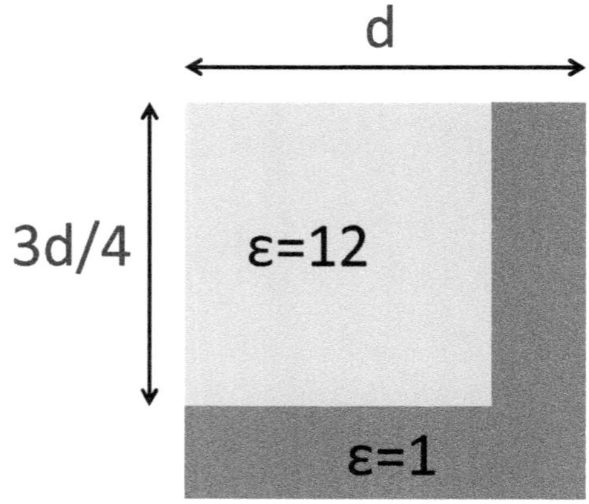

Figure 5.9 Basic cell of the lamellar grating.

$$\pi_p = \frac{\|\Psi_p\|^2}{N(\Psi)}, \pi_e = \frac{\|\Psi_e\|^2}{N(\Psi)}, \pi_a = \frac{\|\Psi_a\|^2}{N(\Psi)} \tag{5.39}$$

where $N(\Psi) = \|\Psi_p\|^2 + \|\Psi_e\|^2 + \|\Psi_a\|^2$.

The point of the above decomposition is to quantify the relative importance of the various modes in the total field existing in the crystal, in order to understand to what extend the field is not solely described by Bloch waves.

5.4.3.2 Cut wavelengths and classification of the conduction bands

Let us now turn to some numerical examples. The structure is a lamellar grating made of square rods of permittivity 12 embedded in vacuum (see Fig. 5.9). In Fig. 5.10, we give the absolute values of the eigenvalues of $\mathcal{T}(\theta, \lambda)$ versus the wavelength. The conduction bands are the regions with a horizontal straight line ($|\mu| = 1$). For each wavelength λ, there is a finite, possibly empty, set of eigenvalues of modulus one $\{e^{i\beta_n(\lambda)}\}_n$ and an infinite set of eigenvalues that do not belong to $\mathbb{U} = \{z \in \mathbb{C}, |z| = 1\}$.

Within a given conduction band, hence locally, it is possible to define continuous sections $\lambda \to \mu_n(\lambda) \in sp(\mathbf{T}_{(\theta,\lambda)})$ representing the evolution of the eigenvalues of the monodromy operator with respect to the wavelength. At some values of the wavelength,

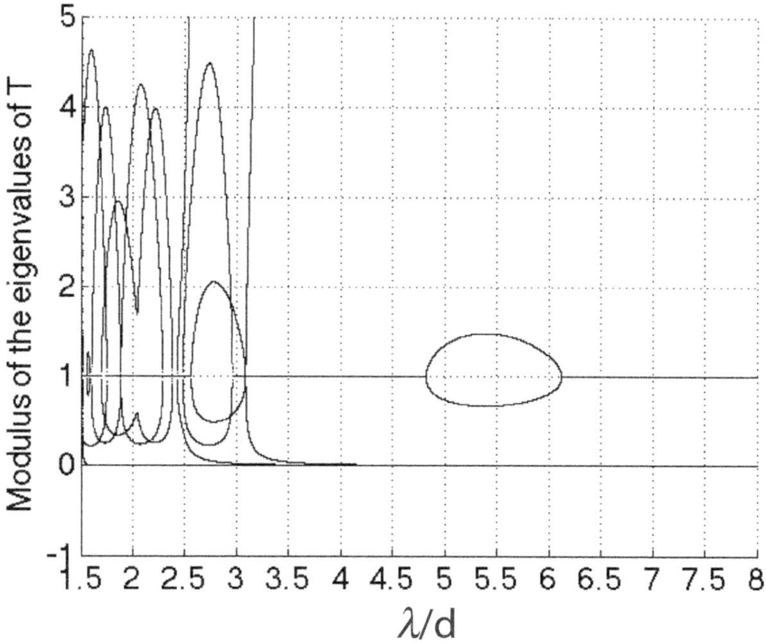

Figure 5.10 (a) Absolute value of the eigenvalues of $\mathbf{T}_\theta(\lambda)$.

however, these sections may encounter a bifurcation, or cut-off: The eigenvalue leaves $\mathbb{U}$ and the associated modes give rise to an evanescent mode and an anti-evanescent mode. At such a branch point, the section $\lambda \rightarrow \mu_n(\lambda)$ is not differentiable and may cross other sections. As a consequence, a global description of the sections is not possible in that case.

A very simple example of branch point is the extinction of a diffracted order of a diffraction grating. The diffracted order becomes then evanescent. Another elementary situation is that of a stratified medium (a Bragg mirror, for instance) in which case there are only two propagative modes in the conduction bands and one evanescent and one anti-evanescent mode inside the gaps.

For a given incidence, a gap is then an interval of wavelengths over which all the propagative eigenvalues have encountered a bifurcation. In Fig. 5.10, we have this situation, for instance, within the intervals (2.56, 2.95) and (4.8, 6.12).

When the wavelength tends to infinity, it is known that the device finally behaves as a homogeneous slab (Felbacq and Bouchitté, 1997; Felbacq et al., 1998; Jikov et al., 1994), and then there are only two propagative modes (up and down), which means that all "sections" finally bifurcate definitely, except the one corresponding to the homogenization regime. In that case, there are still evanescent (and anti-evanescent) waves, but with a very huge damping exponent so that π_e and π_a are small.

However, before that regime, eigenvalues may experience a local bifurcation: that is, they leave $\mathbb{U}$ over a finite interval but finally come back on it (in Fig. 5.10, this situation happens over the interval (2.48, 2.95)). What is important to note is that such a local bifurcation may affect only one eigenvalue, so that within a conduction band, there may be evanescent field coming from such a bifurcation, hence with a small damping exponent.

We give in Fig. 5.11 the absolute values of the eigenvalues of matrix $\mathbf{T}_{(0,\lambda)}$ (normal incidence) as well as the projection ratio $\pi_{p,e,a}$.

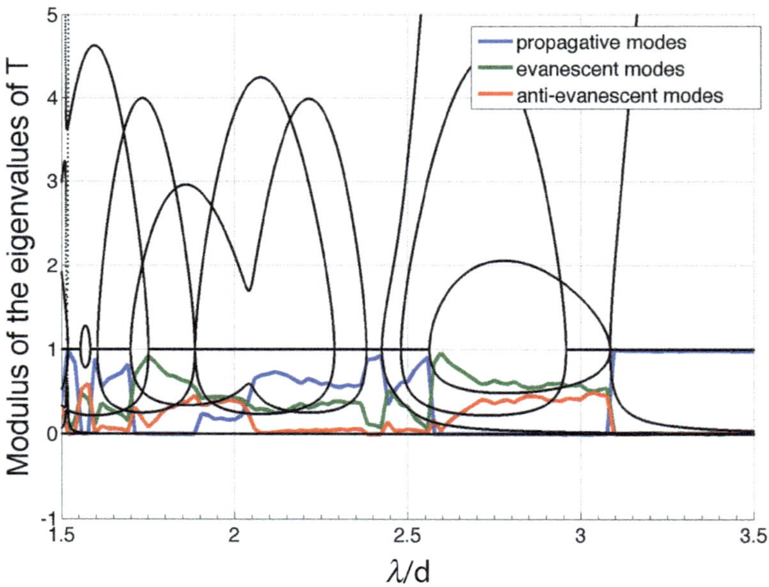

Figure 5.11 Branching ratios for $\varepsilon_{\text{ext}} = 2.26$, $\varepsilon_2 = 4$, $\varepsilon_1 = 1$, $h/d = 1$, $d_1/d = 0.5$, s-polarized waves.

The region (1.75, 2.5) corresponds to a local conduction band, i.e., in which there is a local bifurcation of an eigenvalue. We see that the part of the field on the non-propagating modes is not at all negligible so that the field cannot be described solely by Bloch modes. It is, therefore, important to distinguish between various kinds of conduction bands: An important part of the field inside the structure can be made of non-propagative modes. Especially near a band edge, one should be very careful before deriving the behavior of the field solely by looking at the dispersion diagram: It does not take into account the evanescent waves.

References

Boffi, D., Gastaldi, L., and Naldi, G. (2002). Application of Maxwell equations, *Proc. SIMAI*.

Demésy, G., Zolla, F., Nicolet, A., Commandré, M., and Fossati, C. (2007). The finite element method as applied to the diffraction by an anisotropic grating, *Opt. Express* **15**, 26, pp. 18089–18102.

Demésy, G., Zolla, F., Nicolet, A., and Commandré, M. (2009). Versatile full-vectorial finite element model for crossed gratings, *Opt. Lett.* **34**, 14, pp. 2216–2218.

Felbacq, D. and Bouchitté, G. (1997). Homogenization of a set of parallel fibers, *Waves Random Media* **7**, p. 245.

Felbacq, D., Guizal, B., and Zolla, F. (1998). Wave propagation in one-dimensional photonic crystals, *Opt. Comm.* **152**, pp. 119–126.

Jikov, V. V., Kozlov, S. M., and Oleinik, O. A. (1994). *Homogenization of Differential Operators and Integral Functionals* (Springer, New York).

Langlet, P., Hladky-Hennion, A.-C., and Decarpigny, J.-N. (1995). Analysis of the propagation of plane acoustic waves in passive periodic materials using the finite element method, *J. Acoust. Soc. Am.* **98**, 5, pp. 2792–2800.

Maradudin, A. A. and McGurn, A. R. (1994). Out of plane propagation of electromagnetic waves in a two-dimensional periodic dielectric medium, *J. Modern Opt.* **41**, 2, pp. 275–284.

Nicolet, A., Guenneau, S., Geuzaine, C., and Zolla, F. (2004). Modeling of electromagnetic waves in periodic media with finite elements, *J. Comput. Appl. Math.* **168**, pp. 321–329.

Plihal, M. and Maradudin, A. A. (1991). Photonic band structure of two-dimensional systems: The triangular lattice, *Phys. Rev. B* **44**, p. 8565.

Reed, M. and Simon, B. (1978). *Methods of Modern Mathematical Physics vol. IV: Analysis of Operators* (Academic Press Inc., San Diego).

Sakoda, K. (2005). *Optical Properties of Photonic Crystals* (Springer, Berlin).

Chapter 6

Scattering Problems: Numerical Methods (FEM, Multiple Scattering)

Didier Felbacq,[a] Frédéric Zolla,[b] and André Nicolet[b]
[a]*Laboratory Charles Coulomb UMR CNRS-UM 5221, University of Montpellier, Place Bataillon, 34095 Montpellier Cedex 05, France*
[b]*Institut FRESNEL, University of Aix-Marseille, Avenue Escadrille Normandie Niemen, 13013 Marseille, France*
didier.felbacq@umontpellier.fr

6.1 Finite Element Method

6.1.1 Introduction

The problem of diffraction of electromagnetic waves by a monodimensional grating is abundantly treated in scalar and conical cases. Much less numerical methods allow the calculation of the vector field diffracted by a bidimensional grating (also called bigrating or crossed-grating). We can refer to the work of Moharam et al. (Moharam et al., 1995) for a description of the so-called *rigorous coupled wave method* (RCWA), also known as *Fourier modal method* (see, for instance, Li, 1997 or Noponen and Turunen, 1994). The recent work of Schuster et al. (2007) combines the approaches

Metamaterials Modeling and Design
Edited by Didier Felbacq and Guy Bouchitté

ISBN 978-981-4316-12-5 (Hardcover), 978-1-315-36500-8 (eBook)
www.panstanford.com

of Moharam et al. (2015) and Popov and Nevière (2001) to improve the convergence of the *differential method* (DM) introduced in 1978 (Maystre and Nevière, 1978; Vincent, 1978). These two close methods are generally considered to be low memory consuming and consequently mainly employed for the electromagnetic modeling of crossed-gratings. We can also cite the *method of transformation of coordinates* (or *C method* (Derrick et al., 1979; Granet, 1998; McPhedran et al., 1982), whose most recent developments were made by Harris et al. (1996)), the *Rayleigh method* (RM) (Greffet et al., 1992), and the *method of variation of boundaries* (Bruno and Reitich, 1993, 1998). Finally, the *finite-difference time-domain* (FDTD) method (Yee, 1966; Yee and Chen, 1997) allows the computation of electromagnetic vector fields. Its principle relies on the numerical propagation of a pulse along a temporal grid together with a spatial one. Therefore, this method is not well adapted to the harmonic domain that we are addressing here.

The finite element method (FEM), a very general method dedicated to solving partial differential equations, is massively used in mechanics, fluid mechanics for instance, but not much in electromagnetism at visible frequencies. Volakis et al. (1994) wrote a formulation adapted to tridimensional diffusion problems. Wei et al. (2007) described another formulation adapted to diffusion problems as well and suggested it could be applied to periodic problems without giving any numerical illustrations.

In this chapter, we propose a new formulation of the FEM dedicated to the modeling of vector diffraction by crossed-gratings and entirely based on the use of second-order edge elements. The main advantage of this method is its complete independence toward the shape of the diffractive element, whereas methods listed above require time- or memory-consuming adjustments depending on whether the geometry of the groove region presents oblique edges (e.g., RCWA, (Popov et al., 2002)), high permittivity contrasts or inappropriate height-to-period ratios (e.g., RM, (Greffet and Maassarani, 1990)). Its principle relies on a rigorous treatment of the plane wave sources through an equivalence of the diffraction problem with a radiation one, whose sources are localized inside the diffractive element itself, as proposed in the scalar case for monodimensional gratings (Demésy et al., 2007, 2009).

This approach combined with the use of second-order edge elements allowed us to retrieve with a good accuracy the few numerical academic examples found in the literature. Furthermore, we provide a new reference case combining major difficulties such as a nontrivial toric geometry together with strong losses and a high permittivity contrast. Finally, we discuss computation time and convergence as a function of the mesh refinement as well as the choice of the direct solver.

6.1.2 Theoretical Developments

6.1.2.1 Set up of the problem and notations

We denote by **x**, **y**, and **z** the unit vectors of the axes of an orthogonal coordinate system $Oxyz$. We only deal with time-harmonic fields; consequently, electric and magnetic fields are represented by the complex vector fields $\boldsymbol{E}$ and $\boldsymbol{H}$, with a time dependance in $\exp(-i\,\omega\,t)$.

Besides, in this chapter, for the sake of simplicity, the materials are assumed to be isotropic and, therefore, are optically characterized by their relative permittivity ε and relative permeability μ (note that the inverse of relative permeabilities in this chapter are denoted by ν). It is important to note that lossy materials can be studied, the relative permittivity and relative permeability being represented by complex-valued functions. The crossed-gratings that we are addressing in this chapter can be split into the following regions as suggested in Fig. 6.1:

- *The superstrate* ($z > z_0$) is supposed to be homogeneous, isotropic, and lossless and, therefore, characterized by its relative permittivity ε^+ and its relative permeability $\mu^+(= 1/\nu^+)$ and we denote $k^+ := k_0\,\sqrt{\varepsilon^+\mu^+}$, where $k_0 := \omega/c$.
- *The multilayered stack* ($z_N < z < z_0$) is made of N layers, which are supposed to be homogeneous and isotropic and, therefore, characterized by their relative permittivity ε^n, their relative permeability $\mu^n(= 1/\nu^n)$, and their thickness e_n. We denote $k_n := k_0\,\sqrt{\varepsilon^n\,\mu^n}$ for n integer between 1 and N.
- *The groove region* ($z_g < z < z_{g-1}$), which is embedded in the layer indexed g (ε^g, μ^g) of the previously described domain,

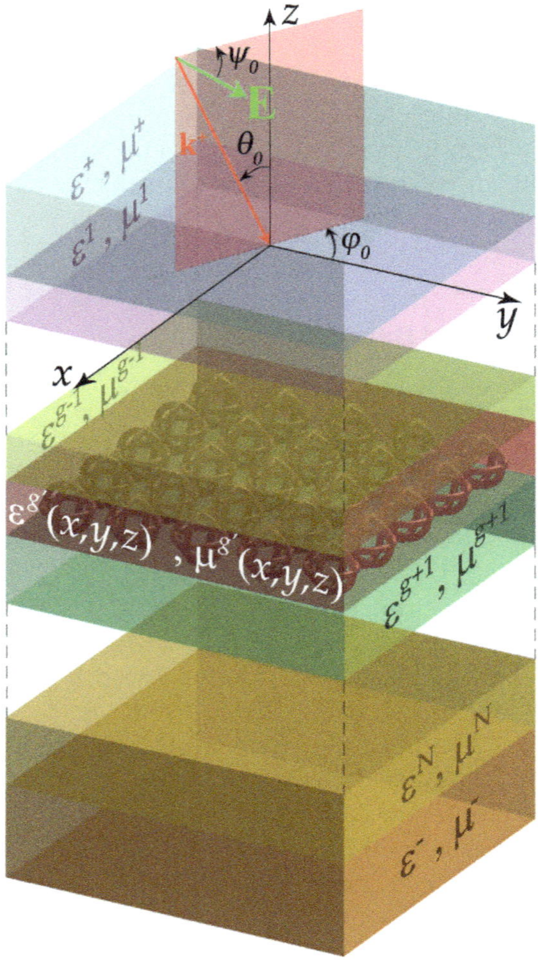

Figure 6.1 Scheme and notations of the studied bi-gratings.

is heterogeneous. Moreover, the method used in this chapter works irrespective of whether the diffractive elements are homogeneous: The permittivity and permeability can vary continuously (gradient index gratings) or discontinuously (step index gratings). This region is, thus, characterized by the scalar fields $\varepsilon^{g'}(x, y, z)$ and $\mu^{g'}(x, y, z)(= 1/\nu^{g'}(x, y, z))$.

The groove periodicities along the x axis and y axis are denoted by d_x and d_y, respectively, in the sequel.

- *The substrate* $(z < z_N)$ is supposed to be homogeneous and isotropic and, therefore, characterized by its relative permittivity ε^- and its relative permeability $\mu^-(= 1/\nu^-)$ and we denote $k^- := k_0\sqrt{\varepsilon^-\mu^-}$.

Let us emphasize the fact that the method principles remain unchanged in the case of several diffractive patterns made of distinct geometry and/or material.

The incident field on this structure is denoted by:

$$\boldsymbol{E}^{\text{inc}} = \mathbf{A}_0^e \exp(i\,\mathbf{k}^+ \cdot \mathbf{r}) \tag{6.1}$$

with

$$\mathbf{k}^+ = \begin{bmatrix} \alpha_0 \\ \beta_0 \\ \gamma_0 \end{bmatrix} = k^+ \begin{bmatrix} -\sin\theta_0\,\cos\varphi_0 \\ -\sin\theta_0\,\sin\varphi_0 \\ -\cos\theta_0 \end{bmatrix} \tag{6.2}$$

and

$$\mathbf{A}_0^e = \begin{bmatrix} E_x^0 \\ E_y^0 \\ E_z^0 \end{bmatrix} = A^e \begin{bmatrix} \cos\psi_0\,\cos\theta_0\,\cos\varphi_0 - \sin\psi_0\,\sin\varphi_0 \\ \cos\psi_0\,\cos\theta_0\,\sin\varphi_0 + \sin\psi_0\,\cos\varphi_0 \\ -\cos\psi_0\,\sin\theta_0 \end{bmatrix}, \tag{6.3}$$

where $\varphi_0 \in [0, 2\pi]$, $\theta_0 \in [0, \pi/2]$ and $\psi_0 \in [0, \pi]$ (polarization angle).

The problem of diffraction that we address in this chapter is, therefore, to find the solution to Maxwell's equations in harmonic regime, i.e., the unique solution $(\boldsymbol{E}, \boldsymbol{H})$ of:

$$\begin{cases} \mathbf{curl}\ \boldsymbol{E} = i\,\omega\,\mu_0\,\mu\,\boldsymbol{H} & (6.4a) \\ \mathbf{curl}\ \boldsymbol{H} = -i\,\omega\,\varepsilon_0\,\varepsilon\,\boldsymbol{E} & (6.4b) \end{cases}$$

such that the diffracted field satisfies the so-called *outgoing waves condition* (OWC (Zolla and Petit, 1996)) and where $\boldsymbol{E}$ and $\boldsymbol{H}$ are quasi-biperiodic functions with respect to the x and y coordinates.

One can choose arbitrarily to calculate $\boldsymbol{E}$, since $\boldsymbol{H}$ can be deduced from Eq. (6.4a). The diffraction problem amounts to looking for the unique solution $\boldsymbol{E}$ to the so-called vectorial Helmholtz propagation equation, deduced from Eqs. (6.4a,b):

$$\mathscr{M}_{\varepsilon,\nu} := -\text{rot}\,(\nu\,\text{rot}\boldsymbol{E}) + k_0^2\,\varepsilon\,\boldsymbol{E} = \mathbf{0} \tag{6.5}$$

such that the diffracted field satisfies an OWC and where $\boldsymbol{E}$ is a quasi-biperiodic function with respect to the x and y coordinates.

6.1.2.2 From a diffraction problem to a radiative one with localized sources

According to Fig. 6.1, the scalar relative permittivity ε and inverse permeability ν fields associated to the studied diffractive structure can be written using complex-valued functions defined by part and taking into account the notations adopted in Section 6.1.2.1:

$$\upsilon(x, y, z) := \begin{cases} \upsilon^{+} & \text{for} \quad z > z_0 \\ \upsilon^{n} & \text{for } z_{n-1} > z > z_n \text{ with } 1 \leq n < g \\ \upsilon^{g'}(x, y, z) & \text{for } z_{g-1} > z > z_g \\ \upsilon^{n} & \text{for } z_{n-1} > z > z_n \text{ with } g < n \leq N \\ \upsilon^{-} & \text{for} \quad z < z_N \end{cases} \tag{6.6}$$

with $\upsilon = \{\varepsilon, \nu\}$, $z_0 = 0$ and $z_n = -\sum_{l=1}^{n} e_l$ for $1 \leq n \leq N$.

It is now convenient to introduce two functions ε_1 and ν_1 corresponding to the associated multilayered case (i.e., the same stack without any diffractive element) constant over Ox and Oy:

$$\upsilon_1(x, y, z) := \begin{cases} \upsilon^{+} \text{ for} \quad z > 0 \\ \upsilon^{n} \text{ for } z_{n-1} > z > z_n \text{ with } 1 \leq n \leq N \\ \upsilon^{-} \text{ for} \quad z < z_N \end{cases} \tag{6.7}$$

with $\upsilon = \{\varepsilon, \nu\}$.

We denote by $\boldsymbol{E}_0$ the restriction of $\boldsymbol{E}^{\text{inc}}$ to the superstrate region:

$$\boldsymbol{E}_0 := \begin{cases} \boldsymbol{E}^{\text{inc}} \text{ for } z > z_0 \\ \boldsymbol{0} \quad \text{for } z \leq z_0 \end{cases} \tag{6.8}$$

We are now in a position to define more explicitly the vectorial diffraction problem that we are dealing with in this chapter. It amounts to looking for the unique vector field $\boldsymbol{E}$ solution to:

$$\mathcal{M}_{\varepsilon,\nu}(\boldsymbol{E}) = \boldsymbol{0} \quad \text{such that } \boldsymbol{E}^{d} := \boldsymbol{E} - \boldsymbol{E}_0 \text{ satisfies an OWC.} \tag{6.9}$$

In order to reduce this diffraction problem to a radiation one, an intermediary vector field denoted by $\boldsymbol{E}_1$ is necessary and is defined as the unique solution to:

$$\mathcal{M}_{\varepsilon_1,\nu_1}(\boldsymbol{E}_1) = \boldsymbol{0} \quad \text{such that } \boldsymbol{E}_1^{d} := \boldsymbol{E}_1 - \boldsymbol{E}_0 \text{ satisfies an OWC.} \tag{6.10}$$

The vector field $\boldsymbol{E}_1$ corresponds to an *ancillary problem* associated to the *general vectorial case of a multilayered stack*, which can be calculated *independently*. This general calculation is seldom treated in the literature; we present a development in Appendix. Thus, from now on, $\boldsymbol{E}_1$ is considered a *known vector field*. It is now relevant to introduce the unknown vector field $\boldsymbol{E}_2^d$, simply defined as the difference between $\boldsymbol{E}$ and $\boldsymbol{E}_1$, which can finally be calculated thanks to the FEM and:

$$\boldsymbol{E}_2^d := \boldsymbol{E} - \boldsymbol{E}_1 = \boldsymbol{E}^d - \boldsymbol{E}_1^d\,. \tag{6.11}$$

It is important to note that the presence of the superscript d is not fortuitous: As a difference between two diffracted fields Eq. (6.11), $\boldsymbol{E}_2^d$ satisfies an OWC, which is of prime importance in our formulation. By taking into account these new definitions, Eq. (6.9) can be written as:

$$\mathscr{M}_{\varepsilon,\nu}(\boldsymbol{E}_2^d) = -\mathscr{M}_{\varepsilon,\nu}(\boldsymbol{E}_1)\,, \tag{6.12}$$

where the right-hand member is a vector field, which can be interpreted as *a known vectorial source term* $-\mathscr{S}_1(x, y, z)$ *whose support is localized inside the diffractive element itself*. To prove it, let us introduce the null term defined in Eq. (6.10) and use the linearity of $\mathscr{M}$, which leads to:

$$\mathscr{S}_1 := \mathscr{M}_{\varepsilon,\nu}(\boldsymbol{E}_1) = \mathscr{M}_{\varepsilon,\nu}(\boldsymbol{E}_1) - \underbrace{\mathscr{M}_{\varepsilon_1,\nu_1}(\boldsymbol{E}_1)}_{=\mathbf{0}} = \mathscr{M}_{\varepsilon-\varepsilon_1,\nu-\nu_1}(\boldsymbol{E}_1)\,. \tag{6.13}$$

6.1.2.3 Quasi-periodicity and weak formulation

The weak form is obtained by multiplying scalarly Eq. (6.9) by weighted vectors $\boldsymbol{E}'$ chosen among the ensemble of quasi-biperiodic vector fields of $L^2(\text{rot})$ (denoted by $L^2\left(\text{rot}, (d_x, d_y), \mathbf{k}\right)$) in Ω:

$$\mathscr{R}_{\varepsilon,\nu}(\boldsymbol{E}, \boldsymbol{E}') = \int_\Omega -\text{rot}\,(\nu\,\text{rot}\boldsymbol{E})\cdot\overline{\boldsymbol{E}'} + k_0^2\,\varepsilon\,\boldsymbol{E}\cdot\overline{\boldsymbol{E}'}\,\mathrm{d}\Omega \tag{6.14}$$

Integrating Eq. (6.14) by part and using the Green–Ostrogradsky theorem lead to:

$$\mathscr{R}_{\varepsilon,\nu}(\boldsymbol{E}, \boldsymbol{E}') = \int_\Omega -\nu\,\text{rot}\boldsymbol{E}\cdot\text{rot}\overline{\boldsymbol{E}'} + k_0^2\,\varepsilon\,\boldsymbol{E}\cdot\overline{\boldsymbol{E}'}\,\mathrm{d}\Omega - \int_{\partial\Omega}(\mathbf{n}\times(\nu\,\text{rot}\boldsymbol{E}))\cdot\overline{\boldsymbol{E}'}\,\mathrm{d}S \tag{6.15}$$

where $\mathbf{n}$ refers to the exterior unit vector normal to the surface $\partial\Omega$ enclosing Ω.

The first term of this sum concerns the volume behavior of the unknown vector field, whereas the right-hand term can be used to set boundary conditions (Dirichlet, Neumann, or so-called quasi-periodic Bloch–Floquet conditions).

The solution $\boldsymbol{E}_2^d$ to the *weak form associated to the diffraction problem*, expressed in its previously defined *equivalent radiative form* in Eq. (6.12), is the element of $L^2\left(\text{rot}, (d_x, d_y), \mathbf{k}\right)$ such that:

$$\forall \boldsymbol{E}' \in L^2(\mathbf{curl}, d_x, d_y, \mathbf{k}), \quad \begin{cases} \forall (x, y, z) \in \Omega_s, \\ \mathscr{R}_{\varepsilon^g, \nu^g}(\boldsymbol{E}_2^d, \boldsymbol{E}') = -\mathscr{R}_{\varepsilon^g, \nu^g}(\mathscr{S}_1, \boldsymbol{E}') \\ \forall (x, y, z) \in \Omega \backslash \Omega_s, \ \mathscr{R}_{\varepsilon, \nu}(\boldsymbol{E}_2^d, \boldsymbol{E}') = 0 \end{cases} . \tag{6.16}$$

In order to rigorously truncate the computation, a set of Bloch boundary conditions are imposed on the pair of planes defined by $(y = -d_y/2, y = d_y/2)$ and $(x = -d_x/2, x = d_x/2)$. One can refer to (Nicolet et al., 2004) for a detailed implementation of Bloch conditions adapted to the FEM. A set of perfectly matched layers is used to truncate the substrate and the superstrate along the z axis (see (Agha et al., 2008) for practical implementation of PML adapted to the FEM). Since the proposed unknown $\boldsymbol{E}_2^d$ is quasi-biperiodic and satisfies an OWC, this set of boundary conditions is perfectly reasonable: $\boldsymbol{E}_2^d$ is radiated from the diffractive element toward the infinite regions of the problem and decays exponentially inside the PMLs along the z axis. The total field associated to the diffraction problem $\boldsymbol{E}$ is deduced at once from Eq. (6.11).

6.1.2.4 Edge or Whitney 1-form second-order elements

In the vectorial case, edge elements (or Whitney forms) make a more relevant choice (Dular et al., 1995) than nodal elements. Note that a lot of work (see for instance (Ingelstrom, 2006)) has been done on higher-order edge elements since their introduction by Bossavit and Mayergoyz (1989). These elements are suitable to the representation of vector fields such as $\boldsymbol{E}_2^d$, by letting their normal component to be discontinuous and imposing the continuity of their tangential components. Instead of linking the degrees of freedom (DOF) of the final algebraic system to the nodes of the mesh, the

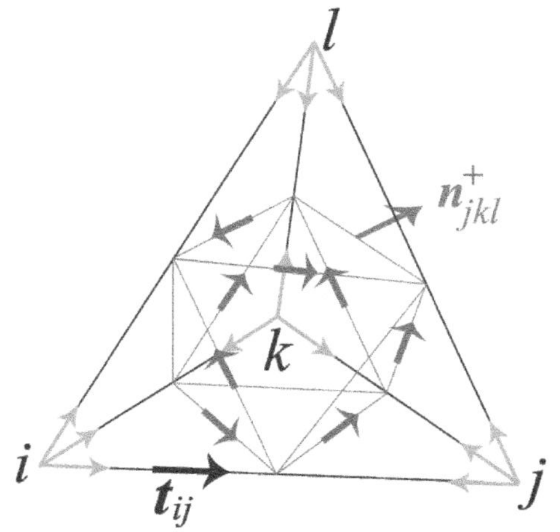

Figure 6.2 Degrees of freedom of a second-order tetrahedral element.

DOFs associated to edge (face) elements are the *circulations* (*flux*) of the unknown vector field along (across) its *edges* (*faces*).

Let us consider the computation cell Ω together with its exterior boundary $\partial\Omega$. This volume is sampled in a finite number of tetrahedron according to the following rules: Two distinct tetrahedrons have to either share a node, an edge, or a face or have no contact. Let us denote by $\mathcal{T}$ the set of tetrahedrons, $\mathcal{F}$ the set of faces, $\mathcal{E}$ the set of edges, and $\mathcal{N}$ the set of nodes. In the sequel, one will refer to the node $n = \{i\}$, the edge $e = \{i, j\}$, the face $f = \{i, j, k\}$, and the tetrahedron $t = \{i, j, k, l\}$.

Twelve DOFs (two for each of the six edges of a tetrahedron) are classically derived from line integral of weighted projection of the field $\boldsymbol{E}_2^d$ on each oriented edge $e = \{i, j\}$:

$$\begin{cases} \vartheta_{ij} = \displaystyle\int_i^j \boldsymbol{E}_2^d \cdot \mathbf{t}_{ij}\, \lambda_i \,\mathrm{d}l \\ \vartheta_{ji} = \displaystyle\int_j^i \boldsymbol{E}_2^d \cdot \mathbf{t}_{ji}\, \lambda_j \,\mathrm{d}l \end{cases}, \tag{6.17}$$

where $\mathbf{t}_{ij}$ is the unit vector and λ_i, the barycentric coordinate of node i, is the chosen weight function.

According to Yioultsis and Tsiboukis (1996b), a judicious choice for the remaining DOFs is to make use of a tangential projection of the 1-form $\boldsymbol{E}_2^d$ on the face $f = \{i, j, k\}$.

$$\begin{cases} \vartheta_{ijk} = \iint_f \left(\boldsymbol{E}_2^d \times \mathbf{n}_{ijk}^{+}\right) \cdot \mathbf{grad}\,\lambda_j \,\mathrm{d}s \\ \vartheta_{ikj} = \iint_f \left(\boldsymbol{E}_2^d \times \mathbf{n}_{ijk}^{-}\right) \cdot \mathbf{grad}\,\lambda_k \,\mathrm{d}s \end{cases}. \tag{6.18}$$

The expressions for the shape functions, or basis vectors, of the second-order 1-form Whitney element are given by:

$$\begin{cases} \mathbf{w}_{ij} = (8\,\lambda_i^2 - 4\,\lambda_i)\ \mathbf{grad}\,\lambda_j + (-8\,\lambda_i\,\lambda_j + 2\,\lambda_j)\ \mathbf{grad}\,\lambda_i \\ \mathbf{w}_{ijk} = 16\,\lambda_i\,\lambda_j\ \mathbf{grad}\,\lambda_k - 8\,\lambda_j\,\lambda_k\ \mathbf{grad}\,\lambda_i - 8\,\lambda_k\,\lambda_i\ \mathbf{grad}\,\lambda_j \end{cases}. \tag{6.19}$$

This choice of shape function ensures (Yioultsis and Tsiboukis, 1996a) the following fundamental properties: every degree of freedom associated with a shape function should be zero for any other shape function. Finally, an approximation of the unknown $\boldsymbol{E}_2^d$ projected on the shape functions of the mesh m ($\boldsymbol{E}_2^{d,m}$) can be derived by:

$$\boldsymbol{E}_2^{d,m} = \sum_{e \in \mathcal{E}} \vartheta_e\,\mathbf{w}_e + \sum_{f \in \mathcal{F}} \vartheta_f\,\mathbf{w}_f. \tag{6.20}$$

Weight functions $\boldsymbol{E}'$ (c.f. Eq. (6.16) are chosen in the same space as the unknown $\boldsymbol{E}_2^d$, $L^2(\mathrm{rot}, (d_x, d_y), \mathbf{k})$. According to the Galerkin formulation, this choice is made so that their restriction to one biperiod belongs to the set of shape functions mentioned above. Inserting the decomposition of $\boldsymbol{E}_2^d$ of Eq. (6.20) in Eq. (6.16) leads to the final algebraic system, which is solved in the following numerical examples, thanks to direct solvers.

6.1.3 Energetic Considerations: Diffraction Efficiencies and Losses

Contrary to modal methods based on the determination of Rayleigh coefficients, the rough results of the FEM are three complex components of the vector field $\boldsymbol{E}^d$ interpolated over the mesh of the computation cell. Diffraction efficiencies are deduced from this field maps as follows.

As a difference between two quasi-periodic vector fields (see Eq. (6.9)), $\boldsymbol{E}^d$ is quasi-biperiodic and its components can be expanded as a double Rayleigh sum:

$$E_x^d(x, y, z) = \sum_{(n,m)\in\mathbb{Z}^2} u_{n,m}^{d,x}(z)\, e^{i\,(\alpha_n\, x+\beta_m\, y)}, \tag{6.21}$$

with $\alpha_n = \alpha_0 + \frac{2\pi}{d_x}\, n$, $\beta_m = \beta_0 + \frac{2\pi}{d_y}\, m$ and

$$u_{n,m}^{d,x}(z) = \frac{1}{d_x\, d_y} \int_{-d_x/2}^{d_x/2} \int_{-d_y/2}^{d_y/2} E_x^d(x, y, z)\, e^{-i\,(\alpha_n\, x+\beta_m\, y)}\, \mathrm{d}x\, \mathrm{d}y\,. \tag{6.22}$$

By inserting the decomposition of Eq. (6.21), which is satisfied by E_x^d everywhere but in the groove region, into the Helmholtz propagation equation, one can express Rayleigh coefficients in the substrate and the superstrate as follows:

$$u_{n,m}^{d,x}(z) = e_{n,m}^{x,p}\, e^{-i\,\gamma_{n,m}^{+}\, z} + e_{n,m}^{x,c}\, e^{i\,\gamma_{n,m}^{+}\, z} \tag{6.23}$$

with $\gamma_{n,m}^{\pm^2} = k^{\pm^2} - \alpha_n^2 - \beta_m^2$, where $\gamma_{n,m}$ (or $-i\,\gamma_{n,m}$) is positive. The quantity $u_{n,m}^{d,x}$ is the sum of a propagative plane wave (which propagates toward decreasing values of z, superscript p) and a counterpropagative wave (superscript c). The OWC verified by $\boldsymbol{E}^d$ imposes:

$$\forall(n, m) \in \mathbb{Z}^2 \begin{cases} e_{n,m}^{x,p} = 0 & \text{for} \quad z > z_0 \\ e_{n,m}^{x,c} = 0 & \text{for} \quad z < z_N \end{cases} \tag{6.24}$$

Equation (6.22) allows to evaluate numerically $e_{n,m}^{x,c}$ ($e_{n,m}^{x,p}$) by double trapezoidal integration of a slice of the complex component E_x^d at an altitude z_c fixed in the superstrate (substrate). It is well known that the mere trapezoidal integration method is very efficient for smooth and periodic functions (integration on one period). The same holds for E_y^d and E_z^d components as well as their coefficients $e_{n,m}^{y,\{c,p\}}$ and $e_{n,m}^{z,\{c,p\}}$.

The dimensionless expression of the efficiency of each reflected and transmitted (n, m) order (Noponen and Turunen, 1994) is deduced from Eqs. (6.23) and (6.24):

$$\begin{cases} R_{n,m} = \frac{1}{A_e^2}\, \frac{\gamma_{n,m}^{+}}{\gamma_0}\, \mathbf{e}_{n,m}^{c}(z_c) \cdot \overline{\mathbf{e}_{n,m}^{c}(z_c)} & \text{for} \quad z_c > z_0 \\ T_{n,m} = \frac{1}{A_e^2}\, \frac{\gamma_{n,m}^{-}}{\gamma_0}\, \mathbf{e}_{n,m}^{p}(z_c) \cdot \overline{\mathbf{e}_{n,m}^{p}(z_c)} & \text{for} \quad z_c < z_N \end{cases}, \tag{6.25}$$

with $\mathbf{e}_{n,m}^{\{c,p\}} = e_{n,m}^{x,\{c,p\}}\, \mathbf{x} + e_{n,m}^{y,\{c,p\}}\, \mathbf{y} + e_{n,m}^{z,\{c,p\}}\, \mathbf{z}$.

Furthermore, normalized losses Q can be obtained through the computation of the following ratio:

$$Q = \frac{\displaystyle\int_V \frac{1}{2}\,\omega\,\varepsilon_0\,\Im\mathrm{m}(\varepsilon^{g'})\,\boldsymbol{E}\cdot\overline{\boldsymbol{E}}\,\mathrm{d}V}{\displaystyle\int_S \frac{1}{2}\Re\mathrm{e}\{\boldsymbol{E}_0 \times \overline{\boldsymbol{H}_0}\}\cdot\mathbf{n}\,\mathrm{d}S}\,. \tag{6.26}$$

The numerator in Eq. (6.26) clarifies losses in watts by biperiod of the considered crossed-grating and is computed by integrating the Joule effect losses density over the volume V of the lossy element. The denominator normalizes these losses to the incident power, i.e., the time-averaged incident Poynting vector flux across one biperiod (a rectangular surface S of area $d_x\, d_y$ in the superstrate parallel to Oxy, whose normal oriented along decreasing values of z is denoted by $\mathbf{n}$). Since $\boldsymbol{E}_0$ is nothing but the plane wave defined at Eqs. (6.2) and (6.3), this last term is equal to $(A_e^2\,\sqrt{\varepsilon_0/\mu_0}\,d_x\,d_y)/(2\cos(\theta_0))$. Volumes and normal to surfaces being explicitly defined, normalized losses (Q) are quickly computed once $\boldsymbol{E}$ is determined and interpolated between mesh nodes.

Finally, the accuracy and self-consistency of the whole calculation can be evaluated by summing the real parts of transmitted and reflected efficiencies (n, m) to normalized losses:

$$Q + \sum_{(n,m)\in\mathbb{Z}^2} \Re\mathrm{e}\{R_{n,m}\} + \sum_{(n,m)\in\mathbb{Z}^2} \Re\mathrm{e}\{T_{n,m}\}\,,$$

quantity to be compared to 1. The sole diffraction orders taken into account in this conservation criterium correspond to propagative orders whose efficiencies have a non-null real part. Indeed, diffraction efficiencies of evanescent orders, corresponding to pure imaginary values of $\gamma_{n,m}^{\pm}$ for higher values of (n, m) (see Eq. (6.23)) are also pure imaginary values as it appears clearly in Eq. (6.25). Numerical illustrations of such global energy balances are presented in the next section.

6.1.4 Accuracy and Convergence

6.1.4.1 Classical crossed gratings

There are only a few references in the literature containing numerical examples. For each of them, the problem only consists of

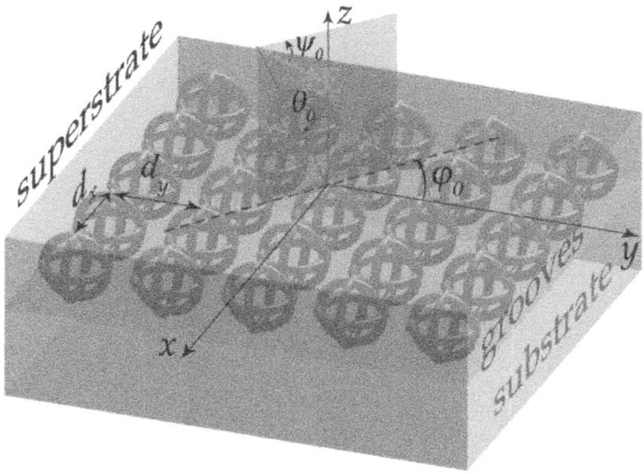

Figure 6.3 Configuration of the studied cases.

three regions (superstrate, grooves, and substrate) as summed up in Fig. 6.3. For the four selected cases (among the six found in the literature) published, results are compared to the ones given by our formulation of the FEM. Moreover, in each case, a satisfying global energy balance is detailed. Finally, a new validation case combining all the difficulties encountered when modeling crossed-gratings is proposed: a nontrivial geometry for the diffractive pattern (a torus), made of an arbitrary lossy material leading to a large step of index and illuminated by a plane wave with an oblique incidence. Convergence of the FEM calculation as well as computation time will be discussed in Section 6.1.4.2.

Checkerboard grating: In this example worked out by Li (1997), the diffractive element is a rectangular parallelepiped as shown in Fig. 6.4 and the grating parameters highlighted in Fig. 6.3 are as follows: $\varphi_0 = \theta_0 = 0°$, $\psi_0 = 45°$, $d_x = d_y = 5\,\lambda_0\,\sqrt{2}/4$, $h = \lambda_0$, $\varepsilon^+ = \varepsilon^{g'} = 2.25$, and $\varepsilon^- = \varepsilon^g = 1$.

Our formulation of the FEM shows good agreement with the Fourier modal method developed by Li (1997), since the maximal relative difference between the array of values presented in Table 6.1 remains lower than 10^{-3}. Moreover, the sum of the efficiencies of propagative orders given by the FEM is very close to 1 in spite of the addition of all errors of determination upon the efficiencies.

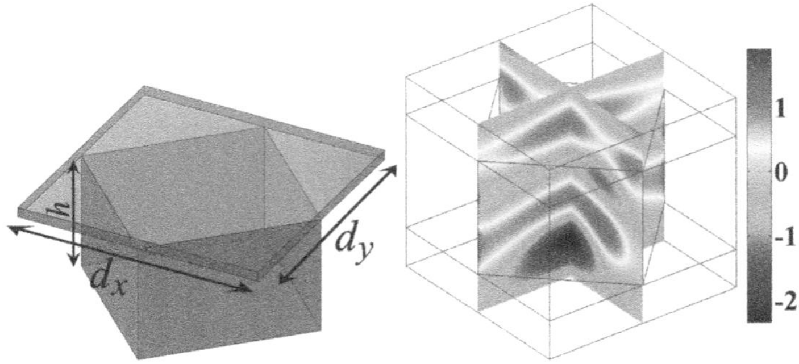

Figure 6.4 Diffractive element with vertical edges (a). $\Re e\{E_x\}$ in V/m (b).

Table 6.1 Energy balance

	FMM (Li, 1997)	FEM
$T_{-1,-1}$	0.04308	0.04333
$T_{-1,0}$	0.12860	0.12845
$T_{-1,+1}$	0.06196	0.06176
$T_{0,-1}$	0.12860	0.12838
$T_{0,0}$	0.17486	0.17577
$T_{0,+1}$	0.12860	0.12839
$T_{+1,-1}$	0.06196	0.06177
$T_{+1,0}$	0.12860	0.12843
$T_{+1,+1}$	0.04308	0.04332
$\sum_{(n,m)\in\mathbb{Z}} \Re e\{R_{n,m}\}$	-	0.10040
TOTAL	-	1.00000

Source: Ref. (Li, 1997).

Pyramidal crossed grating: In this example first worked out by Derrick et al. (1979), the diffractive element is a pyramid with rectangular basis as shown Fig. 6.5, and the grating parameters highlighted in Fig. 6.3 are the following: $\lambda_0 = 1.533$, $\varphi_0 = 45°$, $\theta_0 = 30°$, $\psi_0 = 0°$, $d_x = 1.5$, $d_y = 1$, $h = 0.25$, $\varepsilon^+ = \varepsilon^g = 1$, and $\varepsilon^- = \varepsilon^{g'} = 2.25$.

Results given by the FEM show good agreement with the ones of the C-method (Derrick et al., 1979; Granet, 1998), the Rayleigh method (Greffet et al., 1992), and the RCWA (Bräuer and Bryngdahl,

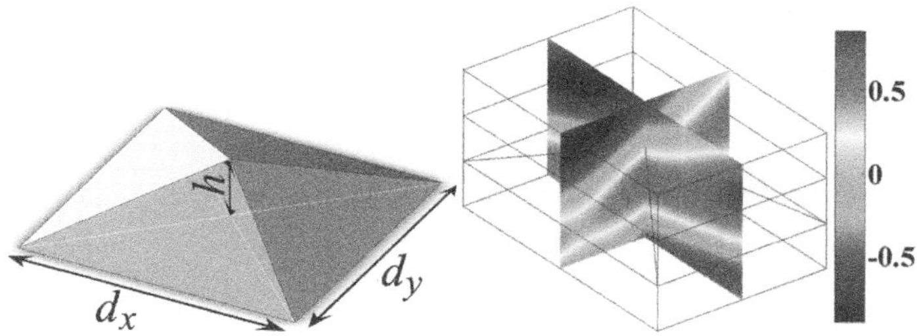

Figure 6.5 Diffractive element with oblique edges (a). $\Re\{E_y\}$ in V/m (b).

Table 6.2 Comparison of results from various studies

	Derrick et al. 1979	Greffet et al. 1992	Bräuer and Bryngdahl 1993	Granet 1998	FEM
$R_{-1,0}$	0.00254	0.00207	0.00246	0.00249	0.00251
$R_{0,0}$	0.01984	0.01928	0.01951	0.01963	0.01938
$T_{-1,-1}$	0.00092	0.00081	0.00086	0.00086	0.00087
$T_{0,-1}$	0.00704	0.00767	0.00679	0.00677	0.00692
$T_{-1,0}$	0.00303	0.00370	0.00294	0.00294	0.00299
$T_{0,0}$	0.96219	0.96316	0.96472	0.96448	0.96447
$T_{1,0}$	0.00299	0.00332	0.00280	0.00282	0.00290
TOTAL	0.99855	1.00001	1.00008	0.99999	1.00004

1993). Note that, in this case, some edges of the diffractive element are oblique.

***Bi-sinusoidal grating*:** In this example worked out by Bruno and Reitich (1993), the surface of the grating is bi-sinusoidal (see Fig. 6.6) and described by the function f defined by:

$$f(x, y) = \frac{h}{4}\left[\cos\left(\frac{2\pi x}{d}\right) + \cos\left(\frac{2\pi y}{d}\right)\right] \tag{6.27}$$

The grating parameters highlighted in Fig. 6.3 are the following: $\lambda_0 = 0.83$, $\varphi_0 = \theta_0 = \psi_0 = 0°$, $d_x = d_y = 1$, $h = 0.2$, $\varepsilon^+ = \varepsilon^g = 1$, and $\varepsilon^- = \varepsilon^{g'} = 4$.

Note that to define this surface, the bi-sinusoid was first sampled (15×15 points) and then converted to a 3D file format. This

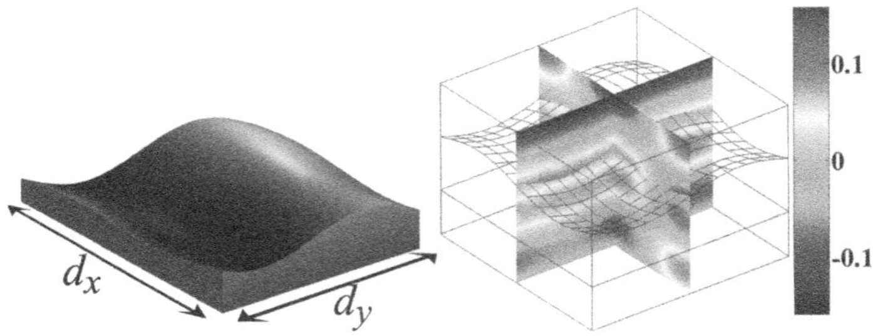

Figure 6.6 Diffractive element with oblique edges (a). $\Re e\{E_z\}$ in V/m (b).

Table 6.3 Energy balance

	(Bruno and Reitich, 1993)	FEM
$R_{-1,0}$	0.01044	0.01164
$R_{0,-1}$	0.01183	0.01165
$T_{-1,-1}$	0.06175	0.06299
$\sum_{(n,m)\in\mathbb{Z}} \Re e\{R_{n,m}\}$	-	0.10685
$\sum_{(n,m)\in\mathbb{Z}} \Re e\{T_{n,m}\}$	-	0.89121
TOTAL	-	0.99806

Source: Ref. (Bruno and Reitich, 1993).

sampling can account for the slight differences with the results obtained using the method of variation of boundaries developed by Bruno and Reitich (1993).

Circular apertures in a lossy layer: In this example worked out by Schuster et al. (2007), the diffractive element is a circular aperture in a lossy layer, as shown in Fig. 6.7, and the grating parameters highlighted in Fig. 6.3 are the following: $\lambda_0 = 500$ nm, $\varphi_0 = \theta_0 = 0\,^\circ$, $\psi_0 = 45\,^\circ$, $d_x = d_y = 1\ \mu$m, $h = 500$ nm, $\varepsilon^+ = \varepsilon^g = 1$, $\varepsilon^{g'} = 0.8125 + 5.2500\,i$, and $\varepsilon^- = 2.25$.

In this lossy case, results obtained with the FEM show good agreement with the ones obtained with the FMM (Li, 1997), the differential method (Arnaud, 2008; Schuster et al., 2007), and the RCWA (Moharam et al., 1995). Joule losses inside the diffractive element can be easily calculated, which allows to provide a global

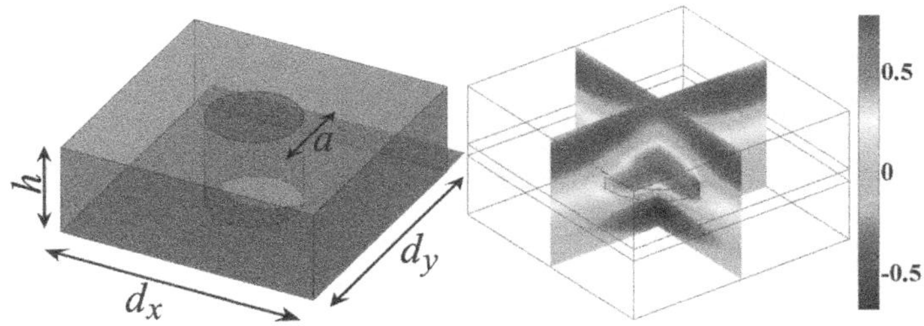

Figure 6.7 Lossy diffractive element with vertical edges (a). $\Re e\{E_y\}$ in V/m (b).

Table 6.4 Comparison of various studies with energy balance

	(Moharam et al. 1995)	(Li 1997)	(Schuster et al. 2007)	FEM
$R_{0,0}$	0.24657	0.24339	0.24420	0.24415
$\sum_{(n,m)\in\mathbb{Z}} \Re e\{T_{n,m}\}$	–	–	–	0.29110
$\sum_{(n,m)\in\mathbb{Z}} \Re e\{R_{n,m}\}$	–	–	–	0.26761
Q	–	–	–	0.44148
TOTAL	–	–	–	1.00019

energy balance for this configuration. Finally, the convergence of the value $R_{0,0}$ as a function of the mesh refinement will be examined.

Lossy torus grating: We finally propose a new test case for crossed-grating numerical methods. The major difficulty of this case lies both in the nontrivial geometry (see Fig. 6.8) of the diffractive object and in the fact that it is made of a material chosen so that losses are optimal inside it. The grating parameters highlighted in Fig. 6.3 and Fig. 6.8 are the following: $\lambda_0 = 1$, $\varphi_0 = \psi_0 = 0^\circ$, $d_x = d_y = 0.3$, $a = 0.1$, $b = 0.05$, $R = 0.15$, $h = 500\,\text{nm}$, $\varepsilon^+ = \varepsilon^g = 1$, $\varepsilon^{g'} = -21 + 20\,i$, and $\varepsilon^- = 2.25$.

Table 6.5 illustrates the independence of our method toward the geometry of the diffractive element. $\varepsilon^{g'}$ is chosen so that the skin depth has the same order of magnitude as b, which maximizes losses. Note that energy balances remain very accurate at normal and

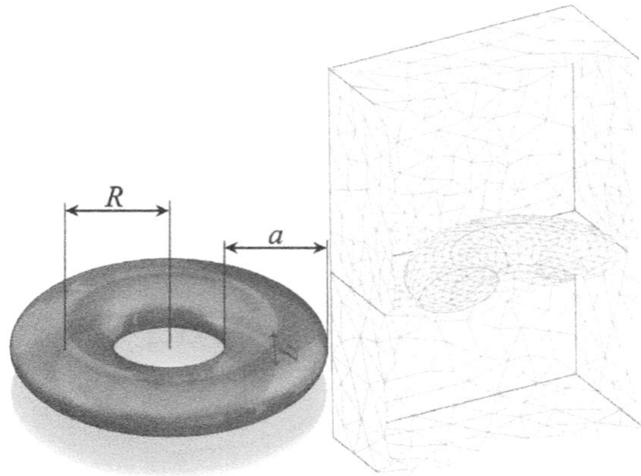

Figure 6.8 Torus parameters (a); coarse mesh of the computational domain (b).

Table 6.5 Energy balances at normal and oblique incidence

FEM 3D	$\theta = 0^\circ$	$\theta = 40^\circ$
$R_{0,0}$	0.36376	0.27331
$T_{0,0}$	0.32992	0.38191
Q	0.30639	0.34476
TOTAL	1.00007	0.99998

oblique incidence, in spite of both the nontriviality of the geometry and the strong losses.

6.1.4.2 Convergence and computation time

***Convergence as a function of mesh refinement*:** When using modal methods such as the RCWA or the differential method, based on the calculation of Rayleigh coefficients, a number proportional to N_R has to be determined *a priori*. The unknown diffracted field is then expanded as a Fourier series, injected under this form in Maxwell's equations, which annihilates the x- and y-dependencies. This leads to a system of coupled partial differential equations whose coefficients can be structured in a matrix formalism. The resulting matrix is sometimes directly invertible (RCWA) depending on whether the geometry allows to suppress the z-dependence, which makes this method adapted to diffractive elements with vertically (or decomposed in staircase functions) shaped edge. In some other cases, one has to use integral methods to solve the system, as in the pyramidal case for instance, which leads to the so-called differential method. The diffracted field map can be deduced from these coefficients. If the grating configuration only calls for a few propagative orders and if the field inside the groove region is not the main information sought for, these two close methods allow to determine the repartition of the incident energy quickly. However, if the field inside the groove region is the main piece of information, it is advisable to calculate many Rayleigh coefficients corresponding to evanescent waves, which increases the computation time as $(N_R)^3$ or even $(N_R)^4$.

FEM relies on the direct calculation of the vectorial components of the complex field. Rayleigh coefficients are determined *a posteriori*. The parameter limiting the computation time is the number of tetrahedral elements along which the computational domain is split up. We suppose that it is necessary to calculate at least two or three points (or mesh nodes) per period of the field $(\lambda_0/\sqrt{\Re e\{\varepsilon\}})$. Figure 6.9 shows the convergence of the efficiency $R_{0,0}$ (circular apertures case, see Fig. 6.7) as a function of the mesh refinement characterized by the parameter N_M: The maximum size of each element is set to $\lambda_0/(N_M\sqrt{\Re e\{\varepsilon\}})$.

It is interesting to note that even if $N_M < 3$, the FEM still gives pertinent diffraction efficiencies: $R_{0,0} = 0.2334$ for $N_M = 1$ and $R_{0,0} = 0.2331$ for $N_M = 2$. The Galerkin method (see Eq. (6.15)) corresponds to a minimization of the error (between the exact solution and the approximation) with respect to a norm that can be physically interpreted in terms of energy-related quantities. Therefore, the finite element methods usually provide energy-related quantities that are more accurate than the local values of the fields themselves.

Computation time: All the calculations were performed on a server equipped with 8 dual core Itanium1 processors and 256Go of RAM. Tetrahedral quadratic edge elements were used together with the direct solver PARDISO. Among different direct solvers adapted

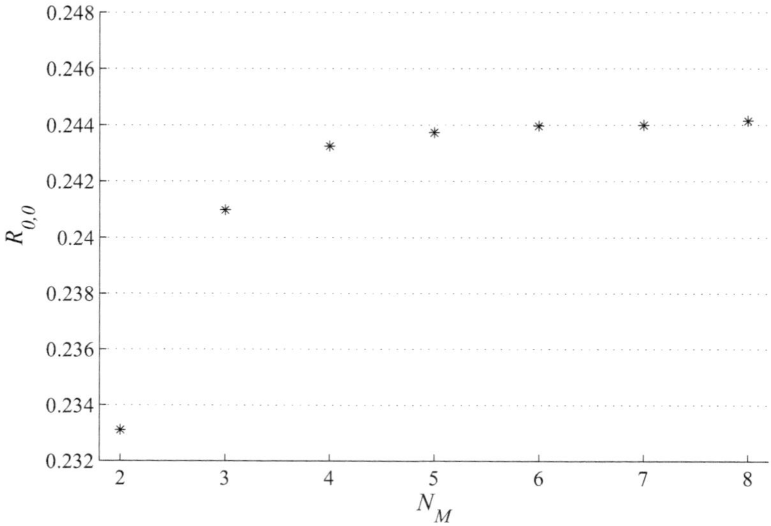

Figure 6.9 Convergence of $R_{0,0}$ in function of N_m (circular apertures crossed grating).

Table 6.6 Computation time variations from various solvers

Solver	Computation time for 41720 DOF	Computation time for 205198 DOF
SPOOLES	15 min 32 s	14 h 44 min
UMFPACK	2 min 07 s	1 h 12 min
PARDISO	57 s	16 min

to sparse matrix algebra (UMFPACK, SPOOLES, and PARDISO), PARDISO turned out to be the less time-consuming one as shown in Table 6.6.

Figure 6.10 shows the computation time required to perform the whole FEM computational process for a system made of a number of DOF indicated on the right-hand ordinate. It is important to note that for values of N_M lower than 3, the problem can be solved in less than a minute on a standard laptop (4Go RAM, 2×2 GHz) with three significant digits on the diffraction efficiencies. This accuracy is more than sufficient in numerous experimental cases. Furthermore, as far as integrated values are at stake, relatively coarse meshes

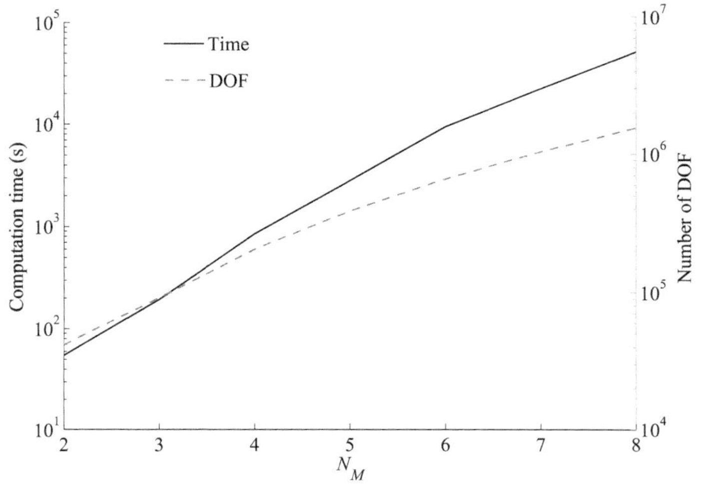

Figure 6.10 Computation time and number of DOF as a function of N_M.

($N_M \approx 1$) can be used trustfully, authorizing fast geometric, spectral, or polarization studies.

6.1.5 Conclusion

We have established a new vectorial formulation of the FEM, allowing to calculate the diffraction efficiencies from the electric field maps of an arbitrarily shaped grating embedded in multilayered stack lightened by a plane wave of arbitrary incidence and polarization angle. It relies on a rigorous treatment of the plane wave sources problem through an equivalent radiation problem with localized sources. Bloch conditions and PML have been implemented to rigorously truncate the computational domain. Nowadays, the efficiency of the numerical algorithms for sparse matrix algebra, together with the available power of computers and the fact that the problem reduces to a basic cell with a size of a small number of wavelengths, makes the 3D problem very tractable as proved here.

The main advantage of this formulation is its complete generality with respect to the studied geometries and the material properties, as illustrated with the lossy tori grating nontrivial case. Its principle remains independent of both the number of diffractive elements by period and the number of stack layers. Finally, choosing fully anisotropic materials for the groove region or stack layers (see Appendix for the principle of calculation of the ancillary problem in this case) is possible. The weak form associated to the problem would involve more terms, but it would not add any degree of freedom to the final algebraic system. Its flexibility allowed us to retrieve with accuracy the few numerical academic examples found in the literature and established with independent methods.

Its remarkable accuracy observed in the case of coarse meshes makes it a fast tool for the design and optimization of diffractive optical components (e.g., reflection and transmission filters, polarizers, beam shapers, pulse compression gratings). Finally, its complete independence toward both the geometry and the isotropic constituent materials of the diffractive elements makes it a useful tool for the study of metamaterials, finite-size photonic crystals, periodic plasmonic structures.

6.1.6 Electric Vector Field in Multilayered Stack Illuminated by a Plane Wave of Arbitrary Incidence and Polarization

This appendix is dedicated to the determination of the vectorial electric field in a dielectric stack illuminated by a plane wave of arbitrary polarization and incidence angle. This calculation, abundantly treated in the 2D scalar case, is generally not presented in the literature since, as far as isotropic cases are concerned, it is possible to project the general vectorial case on the two reference TE and TM cases. However, the presented formulation can be extended to a fully anisotropic case for which this TE/TM decoupling is no longer valid and the three components of the field have to be calculated as follows.

Let us consider the *ancillary problem* mentioned in Section 6.1.2.2, i.e., a dielectric stack made of N homogeneous, isotropic, lossy layers characterized by their relative permittivity denoted by ε^j and their thickness e_j. This stack is deposited on a homogeneous, isotropic, possibly lossy substrate characterized by its relative permittivity denoted by $\varepsilon^{N+1} = \varepsilon^-$. The superstrate is air, and its relative permittivity is denoted by $\varepsilon^+ = 1$. Finally, we denote by

z_j the altitude of the interface between the jth and $j+1^{th}$ layers. The restriction of the incident field $\boldsymbol{E}^{\mathrm{inc}}$ to the superstrate region is denoted by $\boldsymbol{E}_0$. The problem amounts to looking for $(\boldsymbol{E}_1, \boldsymbol{H}_1)$ satisfying Maxwell's equations in harmonic regime (see Eqs. (6.4a) and (6.4b)).

Across the interface $z = z_j$

By projection on the main axis of the vectorial Helmholtz propagation equation (Eq. (6.5)), the total electric field inside the jth layer can be written as the sum of a propagative and a counterpropagative plane waves:

$$\boldsymbol{E}_1(x, y, z) = \begin{bmatrix} E_1^{x,j,+} \\ E_1^{y,j,+} \\ E_1^{z,j,+} \end{bmatrix} \exp\left(j\,(\alpha_0\, x + \beta_0\, y + \gamma_j\, z)\right) + \begin{bmatrix} E_1^{x,j,-} \\ E_1^{y,j,-} \\ E_1^{z,j,-} \end{bmatrix} \exp\left(j\,(\alpha_0\, x + \beta_0\, y - \gamma_j\, z)\right) \quad (6.28)$$

where

$$\gamma_j^2 = k_j^2 - \alpha_0^2 - \beta_0^2 \quad (6.29)$$

What follows consists in writing the continuity of the tangential components of $(\boldsymbol{E}_1, \boldsymbol{H}_1)$ across the interface $z = z_j$, i.e., the continuity of the vector field Ψ defined by:

$$\Psi = \begin{bmatrix} E_1^x \\ E_1^y \\ i\, H_1^x \\ i\, H_1^y \end{bmatrix}. \quad (6.30)$$

The continuity of Ψ along Oz together with its analytical expression inside the jth and $j+1^{th}$ layers allows to establish a recurrence relation for the interface $z = z_j$.

Then, by projection of Eqs. (6.4a) and (6.4b) on Ox, Oy, and Oz:

$$\begin{bmatrix} i\,\beta_0\, H_1^z - \frac{\partial H_1^y}{dz} \\ \frac{\partial H_1^x}{dz} - i\,\alpha_0\, H_1^z \\ i\,\alpha_0\, H_1^y - i\,\beta_0\, H_1^x \end{bmatrix} = -i\,\omega\,\varepsilon \begin{bmatrix} E_1^x \\ E_1^y \\ E_1^z \end{bmatrix} \quad (6.31)$$

and

$$\begin{bmatrix} i\,\beta_0\,E_1^z - \frac{\partial E_1^y}{\partial z} \\ \frac{\partial E_1^x}{\partial z} - i\,\alpha_0\,E_1^z \\ i\,\alpha_0\,E_1^y - i\,\beta_0\,E_1^x \end{bmatrix} = i\,\omega\,\mu \begin{bmatrix} H_1^x \\ H_1^y \\ H_1^z \end{bmatrix}. \tag{6.32}$$

Consequently, the tangential components of $\boldsymbol{H}_1$ can be expressed in function of tangential components of $\boldsymbol{E}_1$:

$$\underbrace{\begin{bmatrix} \omega\,\mu & 0 & \beta_0 \\ 0 & \omega\,\mu & -\alpha_0 \\ -\beta_0 & \alpha_0 & -\omega\,\varepsilon \end{bmatrix}}_{B} \begin{bmatrix} i\,H_1^x \\ i\,H_1^y \\ i\,H_1^z \end{bmatrix} = \begin{bmatrix} \frac{\partial E_1^y}{dz} \\ -\frac{\partial E_1^x}{dz} \\ 0 \end{bmatrix}. \tag{6.33}$$

By noticing the invariance and linearity of the problem along Ox and Oy, the following notations are adopted:

$$\begin{cases} U_x^{j,\pm} = E_1^{x,j,\pm}\,\exp(\pm i\,\gamma_j\,z) \\ U_y^{j,\pm} = E_1^{y,j,\pm}\,\exp(\pm i\,\gamma_j\,z) \end{cases} \tag{6.34}$$

and

$$\Phi_j = \begin{bmatrix} U_x^{+,j} \\ U_x^{-,j} \\ U_y^{+,j} \\ U_y^{-,j} \end{bmatrix}. \tag{6.35}$$

Thanks to Eqs. (6.28) and (6.32) and letting $M = B^{-1}$, it comes for the jth layer:

$$\Psi(x, y, z) = \exp(i\,(\alpha_0\,x + \beta_0\,y)) \underbrace{\begin{bmatrix} 1 & 1 & 0 & 0 \\ 0 & 0 & 1 & 1 \\ \gamma_j\,M_{12}^j & -\gamma_j\,M_{12}^j & -\gamma_j\,M_{11}^j & \gamma_j\,M_{11}^j \\ \gamma_j\,M_{22}^j & -\gamma_j\,M_{22}^j & -\gamma_j\,M_{21}^j & \gamma_j\,M_{21}^j \end{bmatrix}}_{\Pi_j} \begin{bmatrix} U_x^{+,j} \\ U_x^{-,j} \\ U_y^{+,j} \\ U_y^{-,j} \end{bmatrix}. \tag{6.36}$$

Finally, the continuity of Ψ at the interface $z = z_j$ leads to:

$$\Phi_{j+1}(z_j) = \Pi_{j+1}^{-1}\,\Pi_j\,\Phi_j(z_j). \tag{6.37}$$

Normal components can be deduced using Eqs. (6.31) and (6.32).

Traveling inside the $j+1^{th}$ layer

Using Eq. (6.28), a simple phase shift allows to travel from $z = z_j$ to $z = z_{j+1} = z_j - e_{j+1}$:

$$\Phi_{j+1}(z_{j+1}) = \underbrace{\begin{bmatrix} \exp(-i\,\gamma_{j+1}\, e_{j+1}) & 0 & 0 & 0 \\ 0 & \exp(+i\,\gamma_{j+1}\, e_{j+1}) & 0 & 0 \\ 0 & 0 & \exp(-i\,\gamma_{j+1}\, e_{j+1}) & 0 \\ 0 & 0 & 0 & \exp(+i\,\gamma_{j+1}\, e_{j+1}) \end{bmatrix}}_{T_{j+1}} \Phi_{j+1}(z_j) \tag{6.38}$$

Thanks to Eqs. (6.38) and (6.37), a recurrence relation can be formulated for the analytical expression of $\boldsymbol{E}_1$ in each layer:

$$\Phi_{j+1}(z_{j+1}) = T_{j+1}\,\Pi_{j+1}^{-1}\,\Pi_j\;\Phi_j(z_j) \tag{6.39}$$

Reflection and transmission coefficients

The last step consists in the determination of the first term Φ_0, which is not entirely known, since the problem definition only specifies $U_x^{0,+}$ and $U_y^{0,+}$, imposed by the incident field $\boldsymbol{E}_0$. Let us use the OWC hypothesis verified by $\boldsymbol{E}_1^d$ (see Eq. (6.10)). This hypothesis directly translates the fact that none of the components of $\boldsymbol{E}_1^d$ can be traveling either down in the superstrate or up in the substrate: $U_y^{N+1,-} = U_x^{N+1,-} = 0$. Therefore, the four unknowns $U_x^{0,-}$, $U_y^{0,-}$, $U_y^{N+1,+}$, and $U_x^{N+1,+}$, i.e., transverse components of the vector fields reflected and transmitted by the stack verify the following equation system:

$$\Phi_{N+1}(z_N) = (\Pi_{N+1})^{-1}\,\Pi_N \prod_{j=0}^{N-1} T_{N-j}\,(\Pi_{N-j})^{-1}\,\Pi_{N-j-1}\,\Phi_0(z_0) \tag{6.40}$$

This allows to extend the definition of transmission and reflection widely used in the scalar case. Finally, Φ_{N+1} is entirely defined.

Using the recurrence relations of Eqs. (6.39) and (6.28) leads to an analytical expression for $\boldsymbol{E}_1^d$ in each layer.

6.2 Multiple Scattering

6.2.1 Introduction

For definiteness, cartesian axes (O, x, y, z) are chosen. We consider a finite set of N scatterers D_n (spheres, ellipsis, cylinders, or any arbitrary shape) numbered $n = 1, 2, \ldots, N$. We denote D the set of scatterers D_n: $D = \bigcup_n D_n$. Each scatterer is characterized by a permittivity function ε_n and a permeability function μ_n. At an arbitrary point P, the permittivity is defined to be $\varepsilon_r = \varepsilon_n(P)$ if P falls inside scatterer n and ε_{ext} otherwise. Similarly, the permeability is $\mu_r(P) = \mu_n(P)$ for $P \in S_n$ and μ_{ext} otherwise. The incident field is supposed to be monochromatic (as everywhere in the book, with a time dependance of $\exp(-i\omega t)$). Let $\mathcal{F}^i = (\mathcal{E}^i, \mathcal{H}^i)$ be that field; it gives rise to a scattered field $\mathcal{F}^s = (\mathcal{E}^s, \mathcal{H}^s)$. The incident field satisfies Maxwell's equations in vacuum, while the total field $\mathcal{F}^t = (\mathcal{E}^t, \mathcal{H}^t) = (\mathcal{E}^i, \mathcal{H}^i) + (\mathcal{E}^s, \mathcal{H}^s)$ satisfies the Maxwell system:

$$\nabla \times \mathcal{E}^t = i\omega\mu_0\mu_r\mathcal{H}^t \tag{6.41}$$

$$\nabla \times \mathcal{H}^t = -i\omega\varepsilon_0\varepsilon_r\mathcal{E}^t \tag{6.42}$$

6.2.2 Multiple Scattering for a Finite Collection of Objects

The multiple scattering approach is based on the following fact: Consider a given scatterer D_n. In the vicinity of this scatterer, the scattered electromagnetic field $\mathcal{F}^s = \mathcal{F}^t - \mathcal{F}^i$ can be decomposed into the sum of two fields: one that satisfies outgoing radiation conditions $\mathcal{F}_n^+$, which is called "field scattered by D_n", and one that is regular $\mathcal{F}_n^-$, called "local incident field on D_n." This is a direct consequences of the time reversal invariance of Maxwell's equations in vacuum, which provides advanced and retarded propagators that translate into outgoing/ingoing Green's functions in the frequency domain. Therefore, there exist projection operators $P_n^\pm$ such that: $\mathcal{F}_n^\pm = P_n^\pm(\mathcal{F}^s)$. The fundamental property of operators P_n^+ is that they add up to the identity matrix of the space of outgoing fields:

$\sum_n P_n^+ = \mathbb{I}$ and they satisfy the relation: $P_n^+ P_m^+ = \delta_{nm}$, where δ_{nm} is the Kronecker symbol.

From the linearity of Maxwell's equations can be deduced the existence of a linear operator $S_n(\omega)$, the scattering matrix, relating $\mathcal{F}_n^-$ to $\mathcal{F}_n^+$: $\mathcal{F}_n^+ = S_n(\omega)\mathcal{F}_n^-$. For each scatterer D_m, there is of course another decomposition $(\mathcal{F}_m^+, \mathcal{F}_m^-)$ and another scattering matrix $S_m(\omega)$. The basic idea of scattering theory is that the field $\mathcal{F}_m^+$ scattered by D_m is seen as an incident field by scatterer D_n: This comes from the relation $\sum_m P_m^+(\mathcal{F}^s) = \mathcal{F}^s$. Indeed, from this relation, it holds: $\mathcal{F} = \mathcal{F}^i + P_n^+(\mathcal{F}^s) + \sum_{m\neq n} P_m^+(\mathcal{F}^s)$ and therefore the local incident field as seen by scatterer D_n is the sum of the incident field F^i and the field scattered by $(D_m)_{m\neq n}$.

$$P_n^+(\mathcal{F}^s) = S_n\mathcal{F}^i + S_n P_n^- \sum_{m\neq n} P_m^+(\mathcal{F}^s), \tag{6.43}$$

which is the fundamental relation of multiple scattering theory.

It is interesting to note that the scatterers D_n were not assumed to be connected, i.e., "in one part." This means that D_n can be a cluster of smaller scatterers. Following this line, it is not difficult to understand that multiple scattering is inherently recursive. This fact can be exploited numerically (Lu et al., 1995).

6.2.3 Multiple Scattering for a Periodic Collection of Objects

When the objects are disposed along a lattice $\mathcal{L}$, the multiple scattering approach can be used to derive the Bloch waves that can propagate in the medium. The projection operators P_n^+ can be indexed with the vectors $\boldsymbol{G}$ of $\mathcal{L}$ instead of an integer n. Moreover, because of the invariance of the medium under a translation $T_{\boldsymbol{G}}$ of vector $\boldsymbol{G} \in \mathcal{L}$, it holds: $T_{\boldsymbol{G}} P_0^+ T_{-\boldsymbol{G}} = P_{\boldsymbol{G}}^+$. Therefore, there is now only one projection operator to consider: that corresponding to the Wigner–Seitz cell. As explained in Section 5.1, the periodicity makes possible the decomposition of the operators with respect to the Bloch vector parameter $\mathbf{k} \in BZ$. Considering an eigenvector $\mathcal{F}_{\mathbf{k}}$ for $T_{\boldsymbol{G}}$, it holds: $T_{\boldsymbol{G}}(\mathcal{F}_{\mathbf{k}}) = e^{i\mathbf{k}\cdot\boldsymbol{G}}\mathcal{F}_{\mathbf{k}}$. Consequently, for $\boldsymbol{G} = 0$ the relation 6.43 becomes:

$$P_0^+(\mathcal{F}_{\mathbf{k}}) = S\mathcal{F}^i + S\Sigma(\mathbf{k})P_0^+(\mathcal{F}_{\mathbf{k}}) \tag{6.44}$$

where the operator $\Sigma(\mathbf{k})$ is: $\Sigma(\mathbf{k}) = P_0^- \sum_{G' \neq 0} e^{-i\mathbf{k}\cdot G'} T_{G'}$. The band structure of the medium is obtained by solving the preceding relation when $\mathcal{F}^i = 0$. The equation $[I - S_n \Sigma(\mathbf{k})]\, P_0^+(\mathcal{F}_{\mathbf{k}}) = 0$ has nontrivial solutions for couples $(\mathbf{k}, \omega)$ such that: $\det [I - S_n \Sigma(\mathbf{k})] = 0$, which is an implicit equation in the variables $(\mathbf{k}, \omega)$.

6.2.4 Modal Representation for Cylinders

A possible realization of these relations is through a partial wave expansion of the fields (Felbacq et al., 1994) or through boundary integral operators (Maystre, 2006). For instance, when the scatterers are cylinders (i.e., invariant along the z axis), then it suffices to know the component $\mathcal{F}_z = (E_z, H_z)$ along the direction of invariance to know all the other components (see the following section for a complete derivation). A partial wave expansion gives (Felbacq et al., 1994):

$$\mathcal{F}_z(\mathbf{r}) = \sum_l \mathbf{b}_l^n H_l(kr^n) e^{il\theta^n} + \sum_l \mathbf{a}_l^n J_l(kr^n) e^{il\theta^n}.$$

The superscript n indicates that the reference system is centered on D_n: The coordinates (r^n, θ^n) of the point $\mathbf{r}$ are given in the polar coordinate system associated with rod n. The function $H_l^{(1)}$ is the Hankel function of first type and order l, and J_l is the Bessel function of order l (Abramowitz and Stegun, 1965). Finally, $\mathbf{b}_l^n = (b_l^{n,E}, b^{n,H_l})$ are the coefficients of the diffracted electric and magnetic fields and $\mathbf{a}_l^n = (a_l^{n,E}, a^{n,H_l})$ are the coefficients of the incident field.

The projection operator P_n^+ is defined as: $P_n^+(\mathcal{F}) = \sum_l \mathbf{b}_l^n H_l(kr^n) e^{il\theta^n}$. The scattering matrix can be represented as the linear operator relating the set of coefficients $\hat{\mathbf{a}} = (\mathbf{a}_l)$ to the set of coefficients $\hat{\mathbf{b}} = (\mathbf{b}_l)$: $\hat{\mathbf{b}} = S(\omega)\hat{\mathbf{a}}$. Explicit expressions are derived in the following section. In order to use the fundamental relation, an explicit form of the operator $P_n^- P_m^+$ is derived. By linearity, it holds: $P_n^- P_m^+(\mathcal{F}^s) = \sum_l b_l^m P_m^-(H_l(kr^m) e^{il\theta^m})$. The expression for $P_n^-(H_l(kr^m) e^{il\theta^m})$ is obtained by using Graf's formula (Abramowitz and Stegun, 1965), which allows to express any Bessel function $\mathcal{C}_l$ in a new coordinate system (see Fig. 6.11):

$$\mathcal{C}_l(kr^m) e^{il\theta^m}) = \sum_p \mathcal{C}_{p-l}(kr^{nm}) e^{-i(p-l)\theta^{nm}} J_p(kr^n) e^{ip\theta^n}.$$

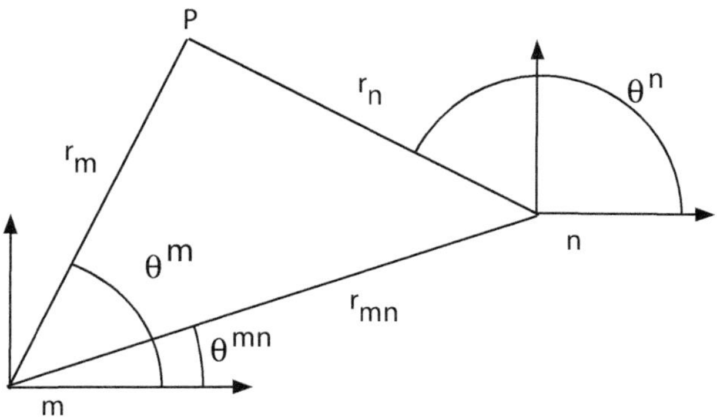

Figure 6.11 Notations for Graf's formula.

The expression of $P_n^- P_m^+(\mathcal{F}^s)$ is then:

$$P_n^- P_m^+(F^s) = \sum_p \left[\sum_l b_l^m H_{p-l}^{(1)}(kr^{nm}) e^{-i(p-l)\theta^{nm}} \right] J_p(kr^n) e^{ip\theta^n}.$$

Denoting by T^{nm} the matrix whose entries are $T_{pl}^{nm} = H_{p-l}^{(1)}(kr^{nm}) e^{-i(p-l)\theta^{nm}}$, the fundamental relation of multiple scattering theory reads as:

$$\hat{\mathbf{b}}_n = S_n \hat{\mathbf{a}}_n + S_n \sum_{m \neq n} T^{nm} \hat{\mathbf{b}}_m. \tag{6.45}$$

6.2.5 Scattering by a Single Object

In this section, we present the theory of scattering by a single object in the 2D case. The main point is the construction of the scattering matrix. Explicit expressions are given for circular cylinders. The scatterer is invariant along the direction Oz. By Fourier transforming the Maxwell system along Oz, the following relations are obtained:

$$i \begin{pmatrix} \gamma & 0 & 0 & Zk \\ 0 & \gamma & -\frac{k\varepsilon}{Z} & 0 \\ 0 & -Zk & \gamma & 0 \\ \frac{k\varepsilon}{Z} & 0 & 0 & \gamma \end{pmatrix} \begin{pmatrix} E_r \\ H_r \\ E_\theta \\ H_\theta \end{pmatrix} = \begin{pmatrix} \partial_r & 0 \\ 0 & \partial_r \\ \frac{1}{r}\partial_\theta & 0 \\ 0 & \frac{1}{r}\partial_\theta \end{pmatrix} \begin{pmatrix} E_z \\ H_z \end{pmatrix}$$

where $Z = \sqrt{\frac{\mu_0}{\varepsilon_0}}$, $\chi^2 = k^2\varepsilon - \gamma^2$. This shows that because of the invariance along Oz, all the field components can be expressed in terms of E_z and H_z. Upon left-inverting the above relation, one gets:

$$\begin{pmatrix} E_\theta \\ H_\theta \end{pmatrix} = \mathbb{T}_{\theta,z} \begin{pmatrix} E_z \\ H_z \end{pmatrix}, \begin{pmatrix} E_r \\ H_r \end{pmatrix} = \mathbb{T}_{r,z} \begin{pmatrix} E_z \\ H_z \end{pmatrix} \tag{6.46}$$

where:

$$\mathbb{T}_{\theta,z} = \begin{pmatrix} \frac{i\gamma}{\chi^2}\frac{1}{r}\partial_\theta & \frac{ikZ}{\chi^2}\partial_r \\ \frac{ik\varepsilon}{Z\chi^2}\partial_r & \frac{i\gamma}{\chi^2}\frac{1}{r}\partial_\theta \end{pmatrix}, \mathbb{T}_{r,z} = \begin{pmatrix} \frac{i\gamma}{\chi^2}\partial_r & -\frac{ikZ}{\chi^2}\frac{1}{r}\partial_\theta \\ -\frac{ik\varepsilon}{Z\chi^2}\frac{1}{r}\partial_\theta & \frac{i\gamma}{\chi^2}\partial_r \end{pmatrix} \tag{6.47}$$

The tangential components of both the electric and magnetic fields are continuous at the boundary of the cylinder:

$$\mathbf{n} \wedge (E^t - E^{\text{int}}) = \mathbf{0} \tag{6.48}$$

$$\mathbf{n} \wedge (H^t - H^{\text{int}}) = \mathbf{0} \tag{6.49}$$

where $\mathbf{n} = (n_x, n_y, 0) = \mathbf{e_r}$ is the outer normal on the boundary of the cylinder. In coordinate, we get: $F_t = F_z e_z + F_\theta e_\theta$ where $F = E, H$. The transmission conditions at the boundary of the cylinder thus lead to the continuity of F_z, F_θ. In the following, we denote with an upperscript $+$ the fields and quantities exterior to the cylinder and an upperscript $-$ the fields inside the cylinder. Because E_z and H_z satisfy the Helmholtz equation outside and inside the cylinder, we have the following expansions:

$$E_z^+ = \sum_m \left(b_n^{E,+} H_n(\chi_+ r) + a_n^{E,+} J_n(\chi_+ r)\right) e^{im\theta} \tag{6.50}$$

$$E_z^- = \sum_m a_n^{E,-} J_n(\chi_- r) e^{im\theta} \tag{6.51}$$

$$H_z^+ = \sum_m \left(b_n^{H,\pm} H_n(\chi_\pm r) + a_n^{H,\pm} J_n(\chi_\pm r)\right) e^{im\theta} \tag{6.52}$$

$$H_z^- = \sum_m a_n^{H,-} J_n(\chi_- r) e^{im\theta} \tag{6.53}$$

Now the boundary conditions read as:

$$\begin{pmatrix} E_z^+ \\ H_z^+ \end{pmatrix} = \begin{pmatrix} E_z^- \\ H_z^- \end{pmatrix}, \begin{pmatrix} E_\theta^+ \\ H_\theta^+ \end{pmatrix} = \begin{pmatrix} E_\theta^- \\ H_\theta^- \end{pmatrix}$$

and the components (E_θ, H_θ) are related to (E_z, H_z) by:

$$\begin{pmatrix} E_\theta^\pm \\ H_\theta^\pm \end{pmatrix} = \mathbb{T}_{\theta,z} \begin{pmatrix} E_z^\pm \\ H_z^\pm \end{pmatrix}$$

Upon expanding the fields in Fourier series:

$$E_{\theta}^{\pm}(r,\theta)=\sum_{n} E_{n,\theta}^{\pm}(r)e^{in\theta},\ H_{\theta}^{\pm}(r,\theta)=\sum_{n} H_{n,\theta}^{\pm}(r)e^{in\theta}\ ,$$

we obtain the block-matrix elements of operator $\mathbb{T}_{\theta,z}$ relating the coefficients $(E_{n,\theta}^{\pm}, H_{n,\theta}^{\pm})$ to the coefficients (a_n, b_n) of the fields (E_z, H_z):

$$\begin{pmatrix} E_{n,\theta}^{+} \\ H_{n,\theta}^{+} \end{pmatrix} = \begin{pmatrix} -\frac{n\gamma}{\chi_+^2 r} J_n(\chi_+ r) & \frac{ikZ}{\chi_+} J_n'(\chi_+ r) \\ \frac{ik\varepsilon}{Z\chi_+} J_n'(\chi_+ r) & -\frac{n\gamma}{\chi_+^2 r} J_n(\chi_+ r) \end{pmatrix} \begin{pmatrix} a_n^{E,+} \\ a_n^{H,+} \end{pmatrix}$$

$$+\begin{pmatrix} -\frac{\gamma n}{\chi_+^2 r} H_n(\chi_+ r) & \frac{ikZ}{\chi_+} H_n'(\chi_+ r) \\ \frac{ik\varepsilon}{Z\chi_+} H_n'(\chi_+ r) & -\frac{\gamma n}{\chi_+^2 r} H_n(\chi_+ r) \end{pmatrix} \begin{pmatrix} b_n^{E,+} \\ b_n^{H,+} \end{pmatrix}$$

and

$$\begin{pmatrix} E_{n,\theta}^{-} \\ H_{n,\theta}^{-} \end{pmatrix} = \begin{pmatrix} -\frac{\gamma n}{\chi_-^2 r} J_n(\chi_- r) & \frac{ikZ}{\chi_-} J_n'(\chi_- r) \\ \frac{ik\varepsilon}{Z\chi_-} J_n'(\chi_- r) & -\frac{\gamma n}{\chi_-^2 r} J_n(\chi_- r) \end{pmatrix} \begin{pmatrix} a_n^{E,-} \\ a_n^{H,-} \end{pmatrix}$$

Let us introduce some notations:

$$\hat{a}_n^{\pm} = \begin{pmatrix} a_n^{E,\pm} \\ a_n^{H,\pm} \end{pmatrix},\ \hat{b}_n^{\pm} = \begin{pmatrix} b_n^{E,\pm} \\ b_n^{H,\pm} \end{pmatrix}$$

The boundary conditions for the azimuthal components read as:

$$\mathbb{Y}_{J,n}^{-}\hat{a}_n^{-} = \mathbb{Y}_{J,n}^{+}\hat{a}_n^{+} + \mathbb{Y}_{H,n}^{+}\hat{b}_n^{+} \tag{6.54}$$

where

$$\mathbb{Y}_{B,n}^{\pm} = \begin{pmatrix} -\frac{\gamma n}{\chi_\pm^2 R} B_n(\chi_\pm R) & -\frac{ikZ}{\chi_\pm} B_n'(\chi_\pm R) \\ \frac{ik\varepsilon_\pm}{Z\chi_\pm} B_n'(\chi_\pm R) & -\frac{\gamma n}{\chi_\pm^2 R} B_n(\chi_\pm R) \end{pmatrix}$$

and B denotes a Bessel function J or H. The boundary conditions for the z components are:

$$\mathbb{J}_n^{-}\widehat{a}_n^{-} = \mathbb{J}_n^{+}\widehat{a}_n^{+} + \mathbb{H}_n^{+}\widehat{b}_n^{+} \tag{6.55}$$

where

$$\mathbb{J}_n^{\pm} = \begin{pmatrix} J_n(\chi_\pm r) & 0 \\ 0 & J_n(\chi_\pm r) \end{pmatrix},\ \mathbb{H}_n^{\pm} = \begin{pmatrix} H_n(\chi_\pm r) & 0 \\ 0 & H_n(\chi_\pm r) \end{pmatrix}$$

From the above relations, it is now straightforward to derive the relations between the coefficients $\hat{a}_n^{\pm}$ and $\hat{b}_n^{+}$:

$$\hat{b}_n^{+} = \mathcal{S}_n^{+}\hat{a}_n^{+} \tag{6.56}$$

$$\hat{a}_n^{-} = \mathcal{S}_n^{-}\hat{a}_n^{+} \tag{6.57}$$

where the outer and inner n^{th} blocks of the scattering matrices are given by:

$$\mathcal{S}_n^+ = \left[\left(\mathbb{Y}_{J,n}^-\right)^{-1}\mathbb{Y}_{H,n}^+ - \left(\mathbb{J}_n^-\right)^{-1}\mathbb{H}_n^+\right]^{-1}\left[\left(\mathbb{J}_n^-\right)^{-1}\mathbb{J}_n^+ - \left(\mathbb{Y}_{J,n}^-\right)^{-1}\mathbb{Y}_{J,n}^+\right] \tag{6.58}$$

$$\mathcal{S}_n^- = \left[\left(\mathbb{Y}_{H,n}^+\right)^{-1}\mathbb{Y}_{J,n}^- - \left(\mathbb{H}_n^+\right)^{-1}\mathbb{J}_n^-\right]^{-1}\left[\left(\mathbb{Y}_{H,n}^+\right)^{-1}\mathbb{Y}_{J,n}^+ - \left(\mathbb{H}_n^+\right)^{-1}\mathbb{J}_n^+\right] \tag{6.59}$$

Simplified expressions can be obtained for z independent fields, i.e., when $\gamma = 0$. In that situation, the electric and magnetic fields are decoupled, which results in the following expression for the impedance matrices:

$$\mathbb{Y}_{B,n}^{\pm} = \begin{pmatrix} 0 & \frac{iZ}{\sqrt{\varepsilon_\pm}} B_n'\left(k\varepsilon_\pm R\right) \\ \frac{i\sqrt{\varepsilon_\pm}}{Z} B_n'\left(k\varepsilon_\pm R\right) & 0 \end{pmatrix}.$$

The scattering matrices blocks become diagonal ones:

$$\mathcal{S}_n^+ = \begin{pmatrix} S_n^{E,+} & 0 \\ 0 & S_n^{H,+} \end{pmatrix}, \mathcal{S}_n^- = \begin{pmatrix} S_n^{E,-} & 0 \\ 0 & S_n^{H,-} \end{pmatrix},$$

where

$$S_n^{E,+} = -\frac{J_n'(k\sqrt{\varepsilon_-}R)J_n(k\sqrt{\varepsilon_+}R) - \sqrt{\frac{\varepsilon_+}{\varepsilon_-}}J_n'(k\sqrt{\varepsilon_+}R)J_n(k\sqrt{\varepsilon_-}R)}{J_n'(k\sqrt{\varepsilon_-}R)H_n(k\sqrt{\varepsilon_+}R) - \sqrt{\frac{\varepsilon_+}{\varepsilon_-}}H_n'(k\sqrt{\varepsilon_+}R)J_n(k\sqrt{\varepsilon_-}R)}, \tag{6.60}$$

$$S_n^{H,+} = -\frac{J_n'(k\sqrt{\varepsilon_-}R)J_n(k\sqrt{\varepsilon_+}R) - \sqrt{\frac{\varepsilon_-}{\varepsilon_+}}J_n'(k\sqrt{\varepsilon_+}R)J_n(k\sqrt{\varepsilon_-}R)}{J_n'(k\sqrt{\varepsilon_-}R)H_n(k\sqrt{\varepsilon_+}R) - \sqrt{\frac{\varepsilon_-}{\varepsilon_+}}H_n'(k\sqrt{\varepsilon_+}R)J_n(k\sqrt{\varepsilon_-}R)}, \tag{6.61}$$

and

$$S_n^{E,-} = \frac{J_n'(k\sqrt{\varepsilon_+}R)H_n(k\sqrt{\varepsilon_+}R) - J_n(k\sqrt{\varepsilon_+}R)H_n'(k\sqrt{\varepsilon_+}R)}{\sqrt{\frac{\varepsilon_-}{\varepsilon_+}}J_n'(k\sqrt{\varepsilon_-}R)H_n(k\sqrt{\varepsilon_+}R) - H_n'(k\sqrt{\varepsilon_+}R)J_n(k\sqrt{\varepsilon_-}R)}, \tag{6.62}$$

$$S_n^{H,+} = \frac{J_n'(k\sqrt{\varepsilon_+}R)H_n(k\sqrt{\varepsilon_+}R) - H_n'(k\sqrt{\varepsilon_+}R)J_n(k\sqrt{\varepsilon_+}R)}{\sqrt{\frac{\varepsilon_+}{\varepsilon_-}}J_n'(k\sqrt{\varepsilon_-}R)H_n(k\sqrt{\varepsilon_+}R) - H_n'(k\sqrt{\varepsilon_+}R)J_n(k\sqrt{\varepsilon_-}R)}. \tag{6.63}$$

6.2.6 Numerical Implementation

For the sake of simplicity, only fields that are invariant along z are considered: This means that the coefficient γ is equal to zero and the Fourier coefficients of the z components of the fields are given by Eqs. (6.60, 6.63). Moreover, the field components E_z and H_z are not coupled and can be determined independently. The cylinders considered have a circular cross section. Numerically, the fields are represented by a finite number of Fourier coefficients with indexes $-N \dots N$. The field scattered by one scatterer is thus described by a $(2N+1)$ vector:

$$\hat{b}^n(N) = \begin{pmatrix} b^n_{-N} \\ \vdots \\ b^n_N \end{pmatrix}, \; n = 1 \dots M \tag{6.64}$$

The scattering matrix $\mathcal{S}_N$ is then a $(2N+1) \times (2N+1)$ diagonal matrix:

$$\mathcal{S}_N = \begin{pmatrix} S^{E,H,\pm}_{-N} & 0 & \dots & 0 \\ 0 & \ddots & & \vdots \\ \vdots & & \ddots & 0 \\ 0 & \dots & \dots 0 & S^{E,+}_N \end{pmatrix},$$

where the entries are given by (6.60 – 6.63). Assuming that there are M scatterers, the entire field is described by an $M \times (2N+1)$ vector with entries:

$$\hat{b}(N) = \begin{pmatrix} \hat{b}^1(N) \\ \vdots \\ b^M(N) \end{pmatrix} \tag{6.65}$$

The incident field is described by its Fourier coefficients obtained in each local coordinate systems attached to each scatterer:

$$\hat{a}(N) = \begin{pmatrix} \hat{a}^1(N) \\ \vdots \\ a^M(N) \end{pmatrix}, \; \hat{a}^n(N) = \begin{pmatrix} a^n_{-N} \\ \vdots \\ a^n_N \end{pmatrix}. \tag{6.66}$$

The fundamental relation (6.45) becomes a relation between vectors of length $2N+1$ and the operator T^{nm}, $n, m \in \{1, 2, \dots, M\}$ becomes a $(2N+1) \times (2N+1)$ matrix with entries:

$$T^{nm}_{pl} = H^{(1)}_{p-l}(kr^{nm})e^{-i(p-l)\theta^{nm}}$$

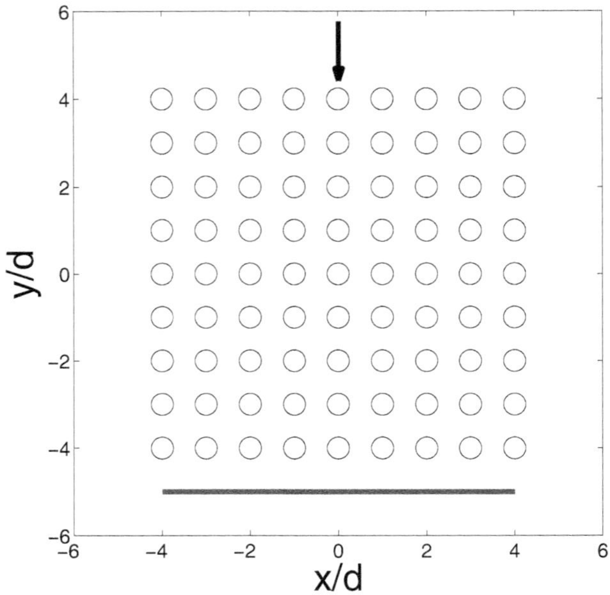

Figure 6.12 A set of dielectric cylinders illuminated by a plane wave in normal incidence. The arrow indicates the direction of the incident field and the segment is the line over which the Poynting vector is integrated.

where $p, l \in \{-N \ldots N\}$. The fundamental relation is then a linear system $\mathbb{S}\,\hat{b}(N) = \mathcal{S}\,\hat{a}(N)$ where $\mathbb{S} = I - \mathcal{S}\mathcal{H}$ and

$$\mathcal{H} = \begin{pmatrix} 0 & T^{1,2} & \cdots & T^{1,M} \\ T^{2,1} & \ddots & \ddots & \vdots \\ \vdots & \ddots & \ddots & T^{M-1,M} \\ T^{M,1} & \cdots & T^{M,M-1} & 0 \end{pmatrix}. \tag{6.67}$$

After inverting $\mathcal{S}$ numerically, the field is known everywhere through the Fourier expansion:

$$F(r,\theta) = \sum_{n=1}^{M} \sum_{p=-N}^{N} \left[a_p^n J_p(kr^n) + b_p^n H_p^{(1)}(kr^n)\right] e^{ip\theta^n} \tag{6.68}$$

For points outside a disc containing the scatterers, it is possible to obtain an expansion of the diffracted field in the main coordinate

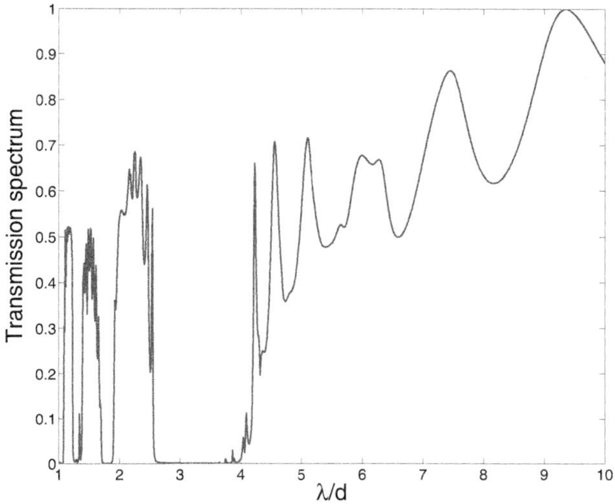

Figure 6.13 Transmission spectrum for the photonic crystal depicted in Fig. 6.12.

system, by means of Graf's formula:

$$F^d(r, \theta) = \sum_{n=1}^{M} \sum_{p=-N}^{N} b_p H_p^{(1)}(kr) e^{ip\theta}. \tag{6.69}$$

When r is very large (as compared to the wavelength), an asymptotic expansion can be obtained from the asymptotic behavior of $H_p^{(1)}(kr) \sim \sqrt{\frac{2}{\pi}} \frac{e^{ikr}}{\sqrt{kr}} e^{-ip\pi/2}$. By inserting this form in the expression of the scattered field, one obtains:

$$F^d(r, \theta) \sim \frac{e^{ikr}}{\sqrt{kr}} g(\theta) \tag{6.70}$$

where

$$g(\theta) = \sqrt{\frac{2}{\pi}} \sum_{p=-N}^{N} b_p (-i)^p e^{ip\theta} \tag{6.71}$$

This function represents the angular behavior of the field far from the scatterers.

As an example, let us consider the set of cylinders depicted in Fig. (6.12). It is a set of dielectric cylinders with relative permittivity $\varepsilon_r = 9$, disposed in vacuum along a square lattice. The period is denoted by d, and the radius of the cylinder is $r/d = 1/4$. The structure is illuminated by a plane wave with the electric field linearly polarized along the axis of the cylinder ($E_{||}$ polarization). The transmission spectrum is calculated as the ratio between the flux of the total Poynting vector through a segment situated below the structure (see Fig. 6.12), normalized by that of the incident field. It is depicted in Fig. 6.13. The presence of forbidden band where the transmission is strongly damped can be clearly seen.

References

Abramowitz, M. and Stegun, I. (1965). *Handbook of Mathematical Functions* (Dover Publication Inc.).

Agha, Y. O., Zolla, F., Nicolet, A., and Guenneau, S. (2008). On the use of PML for the computation of leaky modes: An application to gradient index MOF, *COMPEL* **27**, pp. 95–109.

Arnaud, L. (2008). *Diffraction et diffusion de la lumière: modélisation tridimensionnelle et application à la métrologie de la microélectronique et aux techniques d'imagerie sélective en milieu diffusant*, Ph.D. thesis, Université Paul Cézanne.

Bossavit, A. and Mayergoyz, I. (1989). Edge-elements for scattering problems, *IEEE Trans. Mag.* **25**, pp. 2816–2821.

Bräuer, R. and Bryngdahl, O. (1993). Electromagnetic diffraction analysis of two-dimensional gratings, *Opt. Commun.* **100**.

Bruno, O. P. and Reitich, F. (1993). Numerical solution of diffraction problems: A method of variation of boundaries. III. Doubly periodic gratings, *J. Opt. Soc. Am. A* **10**, pp. 2551–2562.

Bruno, O. P. and Reitich, F. (1998). Boundary-variation solutions for bounded-obstacle scattering problems in three dimensions, *J. Acoust. Soc. Am.* **104**, p. 2579.

Demésy, G., Zolla, F., Nicolet, A., Commandré, M., and Fossati, C. (2007). The finite element method as applied to the diffraction by an anisotropic grating, *Opt. Express* **15**, pp. 18089–18102.

Demésy, G., Zolla, F., Nicolet, A., Commandré, M., Fossati, C., Gagliano, O., Ricq, S., and Dunne, B. (2009). Finite element method as applied to the study of gratings embedded in complementary metal-oxide semiconductor image sensors, *Opt. Eng.* **48**, p. 058002.

Derrick, G. H., McPhedran, R. C., Maystre, D., and Nevière, M. (1979). Crossed gratings: A theory and its applications, *Appl. Phys. B* **18**, 39–52.

Dular, P., Nicolet, A., Genon, A., and Legros, W. (1995). A discrete sequence associated with mixed finite elements and its gauge condition for vector potentials, *IEEE Trans. Mag.* **31**, pp. 1356–1359.

Felbacq, D., Tayeb, G., and Maystre, D. (1994). Scattering by a random set of parallel cylinders, *J. Opt. Soc. Am. A* **11**, p. 2526.

Granet, G. (1998). Analysis of diffraction by surface-relief crossed gratings with use of the chandezon method: Application to multilayer crossed gratings, *J. Opt. Soc. Am. A* **15**, pp. 1121–1131.

Greffet, J. J., Baylard, C., and Versaevel, P. (1992). Diffraction of electromagnetic waves by crossed gratings: A series solution, *Opt. Lett.* **17**, pp. 1740–1742.

Greffet, J. J. and Maassarani, Z. (1990). Scattering of electromagnetic waves by a grating: A numerical evaluation of the iterative-series solution, *J. Opt. Soc. Am. A* **7**, pp. 1483–1493.

Harris, J. B., Preist, T. W., Sambles, J. R., Thorpe, R. N., and Watts, R. A. (1996). Optical response of bigratings, *J. Opt. Soc. Am. A* **13**, pp. 2041–2049.

Ingelstrom, P. (2006). A new set of h (curl)-conforming hierarchical basis functions for tetrahedral meshes, *IEEE Trans. Microw. Theory Tech.* **54**, pp. 106–114.

Li, L. (1997). New formulation of the fourier modal method for crossed surface-relief gratings, *J. Opt. Soc. Am. A* **14**, pp. 2758–2767.

Lu, C. C., Chew, W. C., and Tsang, L. (1995). The application of recursive aggregate T-matrix algorithm in the Monte Carlo simulations of the extinction rate of random distribution of particles, *Radio Sci.* **30**, pp. 25–28.

Maystre, D. (2006). Electromagnetic scattering by a set of objects: An integral method based on scattering properties, *Prog. Electromagnet. Res.* **57**, pp. 55–84.

Maystre, D. and Nevière, M. (1978). Electromagnetic theory of crossed gratings, *J. Optics* , pp. 301–306.

McPhedran, R. C., Derrick, G. H., Nevière, M., and Maystre, D. (1982). Metallic crossed gratings, *J. Optics* **13**, pp. 209–218.

Moharam, M. G., Grann, E. B., Pommet, D. A., and Gaylord, T. K. (1995). Formulation for stable and efficient implementation of the rigorous coupled-wave analysis of binary gratings, *J. Opt. Soc. Am. A* **12**, pp. 1068–1076.

Nicolet, A., Guenneau, S., Geuzaine, C., and Zolla, F. (2004). Modelling of electromagnetic waves in periodic media with finite elements, *J. Comput. Appl. Math.* **168**, pp. 321–329.

Noponen, E. and Turunen, J. (1994). Eigenmode method for electromagnetic synthesis of diffractive elements with three-dimensional profiles, *J. Opt. Soc. Am. A* **11**, 2494–2502.

Popov, E. and Nevière, M. (2001). Maxwell equations in Fourier space: Fast-converging formulation for diffraction by arbitrary shaped, periodic, anisotropic media, *J. Opt. Soc. Am. A* **18**, pp. 2886–2894.

Popov, E., Nevière, M., Gralak, B., and Tayeb, G. (2002). Staircase approximation validity for arbitrary-shaped gratings, *J. Opt. Soc. Am. A* **19**, pp. 33–42.

Schuster, T., Ruoff, J., Kerwien, N., Rafler, S., and Osten, W. (2007). Normal vector method for convergence improvement using the RCWA for crossed gratings, *J. Opt. Soc. Am. A* **24**, pp. 2880–2890.

Vincent, P. (1978). A finite-difference method for dielectric and conducting crossed gratings, *Opt. Commun.* **26**.

Volakis, J. L., Chatterjee, A., and Kempel, L. C. (1994). Review of the finite-element method for three-dimensional electromagnetic scattering, *J. Opt. Soc. Am. A* **11**, pp. 1422–1422.

Wei, X., Wachters, A. J., and Urbach, H. P. (2007). Finite-element model for three-dimensional optical scattering problems, *J. Opt. Soc. Am. A* **24**, pp. 866–881.

Yee, K. (1966). Numerical solution of initial boundary value problems involving Maxwell's equations in isotropic media, *IEEE Trans. Antennas Propagation* **14**, pp. 302–307.

Yee, K. S. and Chen, J. S. (1997). The finite-difference time-domain (FDTD) and the finite-volume time-domain (FVTD) methods in solving Maxwell's equations, *IEEE Trans. Antennas Propagation* **45**, pp. 354–363.

Yioultsis, T. V. and Tsiboukis, T. D. (1996a). Multiparametric vector finite elements: A systematic approach to the construction of three-dimensional, higher order, tangential vector shape functions, *IEEE Trans. Mag.* **32**, pp. 1389–1392.

Yioultsis, T. V. and Tsiboukis, T. D. (1996b). The mystery and magic of whitney elements: An insight in their properties and construction, *Int. Compumag. Soc. Newsletter* **3**, pp. 6–13.

Zolla, F. and Petit, R. (1996). Method of fictitious sources as applied to the electromagnetic diffraction of a plane wave by a grating in conical diffraction mounts, *J. Opt. Soc. Am. A* **13**, pp. 796–802.

SECTION III

APPLICATIONS: EFFECTIVE PROPERTIES OF METAMATERIALS

Chapter 7

Soft Problems: Nonresonant Dielectric Structures

Didier Felbacq,[a] Frédéric Zolla,[b] and Guy Bouchitté[c]

[a]*Laboratory Charles Coulomb UMR CNRS-UM 5221, University of Montpellier, Place Bataillon, 34095 Montpellier Cedex 05, France*
[b]*Institut FRESNEL, University of Aix-Marseille, Avenue Escadrille Normandie Niemen, 13013 Marseille, France*
[c]*Laboratory IMATH, University of Sud-Toulon-Var, BP 20132, 83957 La Garde Cedex, France*
didier.felbacq@umontpellier.fr

7.1 A Brief Foray into the Realm of Two-Scale Homogenization

7.1.1 Two-Scale Homogenization with One Small Parameter

In this section, it is assumed that the characteristic lengths of the obstacles are such that the notions of permittivity, permeability, optical index, and more generally all mesoscopic or macroscopic quantities are relevant (see Chapter 3 for instance). In addition, only periodic structures are addressed; we shall not then tackle random or quasi-periodic structures. Finally, it is assumed that only

Metamaterials Modeling and Design
Edited by Didier Felbacq and Guy Bouchitté

ISBN 978-981-4316-12-5 (Hardcover), 978-1-315-36500-8 (eBook)
www.panstanford.com

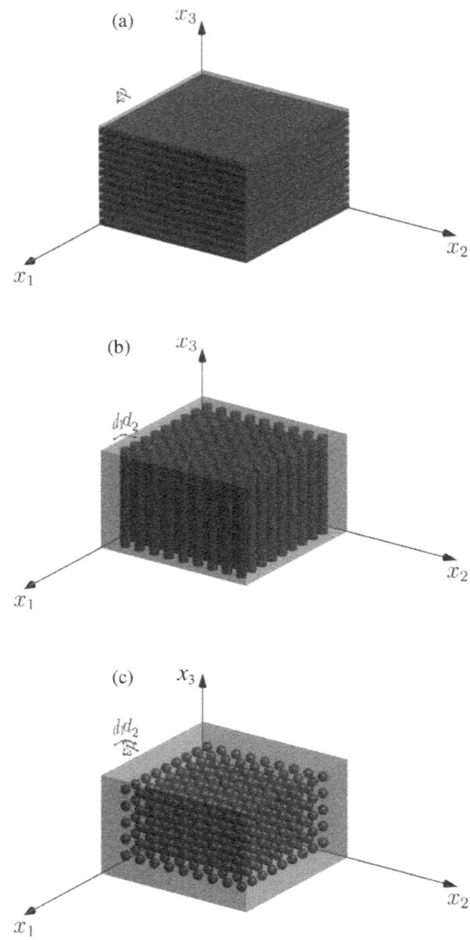

Figure 7.1 In most cases, photonic crystals can be characterized by their permittivity ε, which is represented by a two-valued piecewise function: ε_1 and ε_2. (a) A monoperiodic crystal: Bragg mirror. (b) A two-dimensional crystal: periodic collection of identical circular rods. (c) A three-dimensional crystal: periodic collection of identical spheres.

harmonic regime is considered and, therefore, at least two lengths are at our disposal, namely the wavelength in vacuum, λ_0, of the incident field and one period d_1 (see Fig. 7.1(a)), two periods (d_1, d_2) (see Fig. 7.1(b)), or three periods (d_1, d_2, d_3) (see Fig. 7.1(c)) depending on the very nature of the problem at stake. In this book, the concept of homogenization will be used only if, at least, one period is small compared to λ_0. From a theoretical point of view, it is probably pointless to obtain relevant results when the number of scatterers is finite and this for at least one good reason; strictly speaking, the very notion of periodicity requires that there be an infinite number of scatterers.

Therefore, there are two ways of addressing this problem. The first way, which is often used, consists in assuming that the size of the scatterers is fixed, while the whole obstacle filled up by these scatterers is increased until it covers the overall space $\mathbb{R}^3$. After performing this *Gedanken experiment*, one lets the wavelength tend to infinity (see Fig. 7.2).

It is a powerful and long-past method widely used in solid state physics, where it allows to use Bloch waves decomposition. However, this approach has some weaknesses. First, the boundary of the whole crystal cannot be taken into account. Second, the small parameter in this method is the frequency itself. After obtaining the effective characteristics near the zero frequency, the study of the effects of the time dispersion upon mixing laws is unnatural.

On the contrary, if one considers that the obstacle and the wavelength both remain fixed, while the size of the scatterers goes to zero and their number goes to infinity, the influence of the whole crystal may be studied in a more natural way (see Fig. 7.3). Moreover, the study of the time dispersion becomes more natural and can be integrated into the limit process itself. We are now in a position to describe more precisely the problem we are dealing with. Let us consider a set Ω_f, which represents a simple cubic crystal made of small scatterers periodically arranged within the crystal. This crystal system is solely characterized by the side d_0 (see Fig. 7.4).

The question may then be posed in these terms: If the dimensionless parameter $\eta_0 = \frac{d_0}{\lambda_0}$ is small enough, is it possible to reduce the complexity of the system by replacing the opto-geometric characteristics of a crystal by effective properties? The answer is positive, and in this rather simple case, it will be seen that the effective properties are generally given in the form of an electric anisotropic media. For obtaining the so-called effective properties, one can imagine using the "traditional" method based on the Bloch

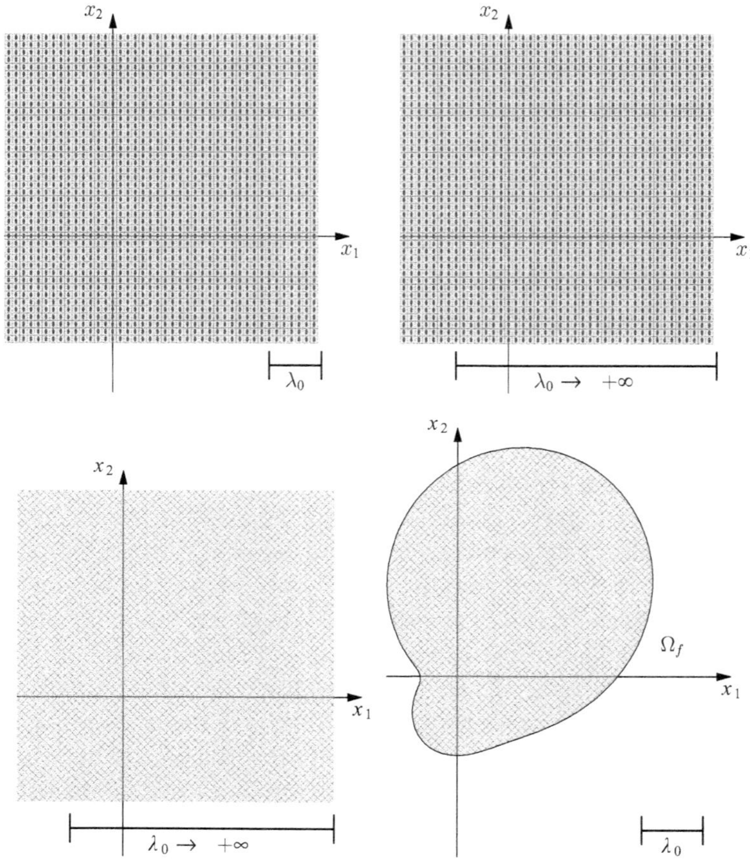

Figure 7.2 The Bloch method for obtaining the effective parameters can be schematically divided into four parts. (a) The whole space (here $\mathbb{R}^2$ for the sake of clarity) is covered by a perfect crystal of infinite extent along all directions. (b) By using Bloch theory, the different modes propagating within the crystal can be obtained and in particular the modes for large wavelengths. (c) For sufficiently large wavelengths, the crystal can be replaced by a homogeneous medium. (d) The last step consists in filling up the set Ω_f by the effective properties found in the previous step. Note that in scattering problem, the wavelength associated with the incoming wave is no longer arbitrarily large.

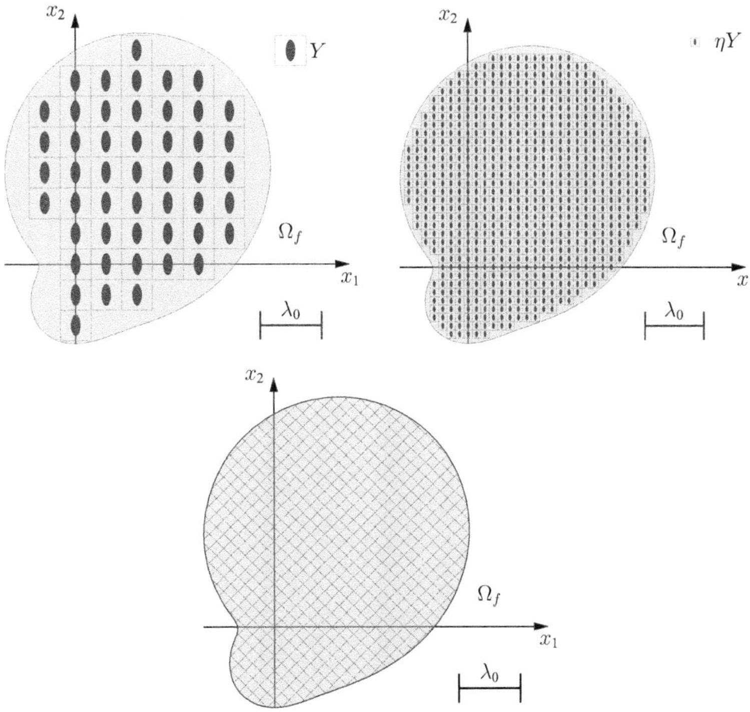

Figure 7.3 In the two-scale homogenization process, the shape of the whole crystal is preserved together with the wavelength λ_0 (Ω_f and λ_0 are unchanged in (a)–(c)). The dynamical variable η is the size of the basic cell ηY, which shrinks as η tends to zero.

decomposition. The main advantage of this approach is that it gives some information for any wavelength irrespective of whether it is large compared to the period of the crystal. It remains, therefore, to study the behavior of the crystal, in particular the dispersion curves, near the zero frequency, that is, for a small parameter η_0 (static procedure). On the other hand, when using the second method, the study of the effective properties is meaningless for a small given parameter η_0; the number of scatterers being finite, the structure cannot be periodic. We, therefore, develop a dynamic procedure (limit analysis) by letting the small parameter η tend toward zero. Let us focus our attention on the permittivity and permeability

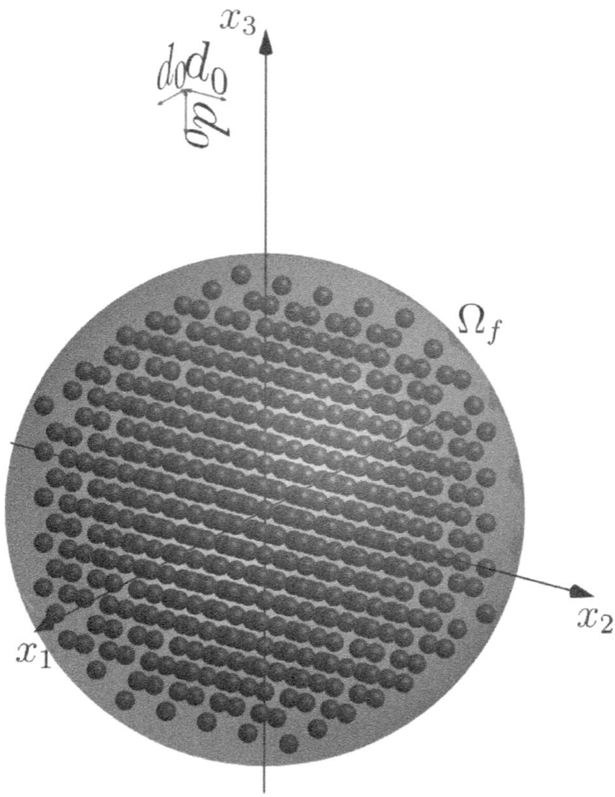

Figure 7.4 In this figure, the obstacle is a spherical simple cubic crystal made itself of a collection of small spherical homogeneous "particles." This crystal is solely determined by the period d_0.

describing the crystal. These optical characteristics depend on η. We denote these characteristics $(\varepsilon_\eta, \mu_\eta)$, and the scattering problem, which depends on η, is denoted $\mathcal{P}_\eta$ and the diffracted field $\boldsymbol{F}_\eta$. Now the issue is to foresee what happens when η tends toward zero and, among other things, to derive the effective optical, possibly tensorial, characteristics denoted by $(\underline{\underline{\varepsilon}}_{\text{hom}}, \underline{\underline{\mu}}_{\text{hom}})$ in the sequel. At first sight, this result may appear as paradoxical: Searching out the effective properties amounts to writing down $(\varepsilon_\eta, \mu_\eta) \underset{\eta \longrightarrow 0}{\overset{E}{\longrightarrow}} (\underline{\underline{\varepsilon}}_{\text{hom}}, \underline{\underline{\mu}}_{\text{hom}})$, where $\underset{\eta \longrightarrow 0}{\overset{E}{\longrightarrow}}$ is a convergence process that remains to be clarified. A type of convergence can be found that transforms a series of infinitely oscillating scalar functions into two constant matrices at the limit. For dispelling this mystery, we must return to physics and recall that both permittivity and permeability are indirectly determined by the diffracted field. It is said that $(\varepsilon_\eta, \mu_\eta) \xrightarrow[\eta \longrightarrow 0]{E} (\underline{\underline{\varepsilon}}_{\text{hom}}, \underline{\underline{\mu}}_{\text{hom}})$ if and only if $\boldsymbol{F}_\eta \longrightarrow \boldsymbol{F}_{\text{hom}}$.

We have seen that the two-scale homogenization theory allows, in principle, to investigate all the usual cases encountered in practice and for periodic materials even if they are lossy. This formulation takes advantage of a limit analysis process, in which a dimensionless parameter η tends to zero. But in practice, the number of scatterers if of course finite, and the ratio of the size of the period of the crystal by the wavelength are not null. In other words, from a practical point of view, the small parameter η is determined by the experiment. What we have to do is to estimate a parameter η_0 in such a way that for every η smaller than η_0, it is legitimate to replace the heterogeneous crystal by the effective properties. To do this, a more accurate theoretical analysis must be considered (corrector theory), but it seems, however, necessary to compare theory with experiment albeit numerical.

7.1.2 Two-Scale Homogenization with Several Small Parameters

In many situations, it may appear that there are several small parameters. For instance, when dealing with highly lossy ohmic materials in the homogenization regime, the ratio of the period to the wavelength $\eta_0 = \frac{d}{\lambda_0}$ and the inverse of the conductivity $\frac{1}{\sigma}$ are both small parameters. This last expression and the parameter η_0 having not the same dimension, a factor has to be found to make the second parameter dimensionless, for instance: $\eta_1 = \frac{1}{Z_0 \sigma \lambda_0}$. Bearing in mind that the two-scale homogenization requires a limit process, the problem amounts to studying the diffracted field $\boldsymbol{F}_{\eta_0, \eta_1}$ as (η_0, η_1) tends to the origin $(0, 0)$. This problem would be clearly posed if the scattered field were continuous at the origin in regard to both the small parameters. It appears, however, that the scattered field strongly depends on the path travelled in the parameters' space. It then remains to find a *critical function* f_c for which the scattered field $\boldsymbol{F}_{\eta_0, f_c(\eta_0)}$ tends toward a non-trivial solution. Choosing this

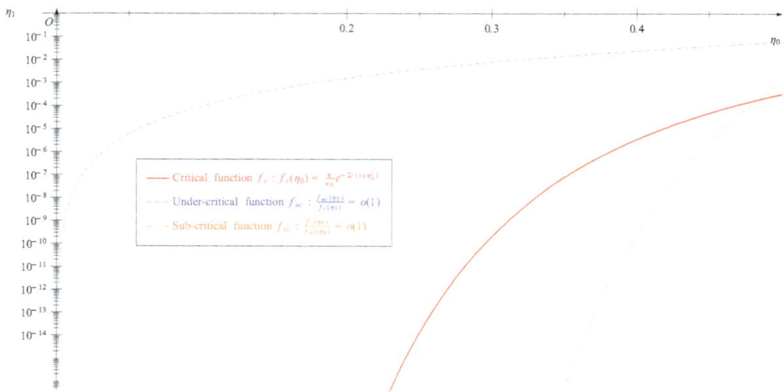

Figure 7.5 In a two-scale homogenization with two parameters, for instance when dealing with small and highly conducting rods, it is necessary to find the so-called critical function f_c, which determines the limit process. This critical function is itself characterized by two parameters, here κ_0 and γ_0 (see chapter 8).

critical function is certainly one of the salient difficulties of the two-scale homogenization with two (or more) small parameters and depends on the problem at stake. If this function is not suitably chosen, the limit process leads to a flimsy result: The homogenized material is either a vacuum (under-critical behavior) or a perfectly conducting metal (over-critical behavior) (see Fig. 7.5).

7.2 Soft Problems: Theory

We consider a photonic nanostructure made of a collection of parallel finite-size rods embedded in a matrix. This covers the case of structures made out of a layer of bulk materials in which holes are made periodically but also the case of structures made out of nanopillars (Benisty et al., 2000; Poborchii et al., 2002). The rods are not supposed to be invariant in the direction of their axes (for instance, they can be cone shaped, see Fig. 7.6). Our point is to derive the effective permittivity and permeability tensors of this structure when the ratio between the period of the structure and the free space wavelength of the incident field is very small.

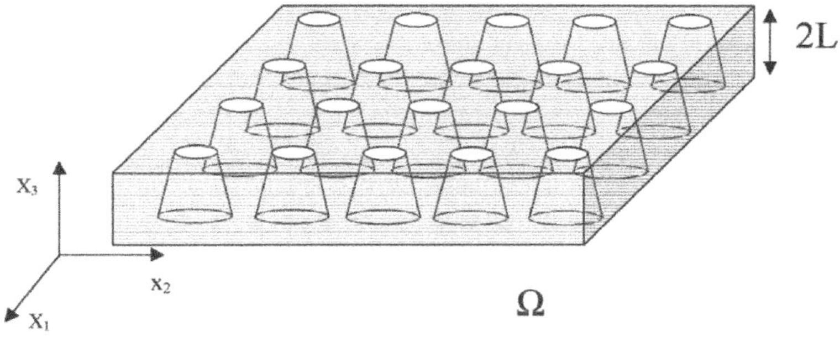

Figure 7.6 Schematic of the structure.

7.3 Two-Scale Approach to Homogenization

7.3.1 Description of the Structure and Methodology

We consider a structure such as that in Fig. 7.6. It is constructed from a basic cell $\widetilde{Y}$ pictured in Fig. 7.7 ($\widetilde{Y} = Y \times (-L, L)$, where $Y = (-1/2, 1/2)^2$). A contraction ratio η is applied to obtain a contracted cell in the horizontal directions ($\widetilde{Y}_\eta = \eta^2 Y \times (-L, L)$). In the units of the free space wavelength, the period of the lattice is, thus, η. The cells are contained in a cylindrical domain $\Omega = \mathcal{O} \times (-L, L)$ (see Fig. 7.6). Thus, the domain Ω is periodically filled with contracted cells. The space coordinates are denoted by $\boldsymbol{x} = (x_1, x_2, x_3)$, and we also denote $\boldsymbol{x}_\perp = (x_1, x_2)$. The coordinates in Y are denoted by $\boldsymbol{y} = (y_1, y_2)$. We consider time harmonic fields; the time dependence is chosen to be $\exp(-i\omega t)$. For a given monochromatic incident field $(\boldsymbol{E}^i, \boldsymbol{H}^i)$, we denote by $(\boldsymbol{E}^\eta, \boldsymbol{H}^\eta)$ the total electromagnetic field.

As explained in the preceding section, our aim is to pass to the limit $\eta \to 0$ and determine the limit of the couple $(\boldsymbol{E}^\eta, \boldsymbol{H}^\eta)$. In our methodology, we get at the limit a true electromagnetic scattering problem, for a given free space wavelength λ and a bounded obstacle Ω characterized by some permittivity and permeability tensors. This situation is quite different from other homogenization techniques, making use of periodization conditions, in which the frequency tends to zero, thus not leading to a diffraction problem but rather to an electrostatic one. Such an approach can sometimes give useful explicit formulas but generally leads

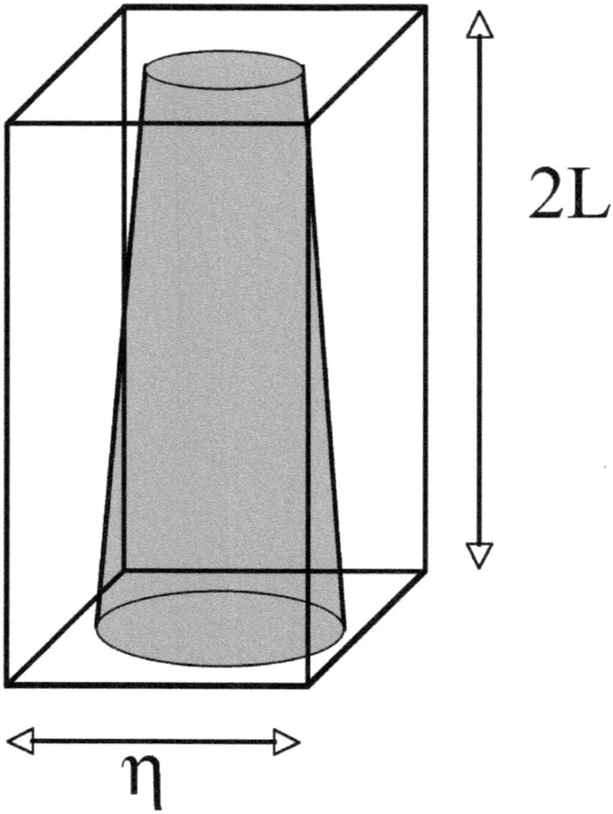

Figure 7.7 Schematic of the basic cell.

to complicated formulations (Erikson et al., 1986; Halevi et al., 1999, 2002; McPhedran et al., 1997, 2000). Moreover, it does not handle the boundary effects, which in some cases may lead to some miscomprehensions (Felbacq, 2000). The relative permittivity tensor $\underline{\underline{\varepsilon}}^{\eta}(\boldsymbol{x})$ and relative permeability tensor $\underline{\underline{u}}^{\eta}(\boldsymbol{x})$ are described by:

$$\begin{cases} \underline{\underline{\varepsilon}}^{\eta}(\boldsymbol{x}) = 1\,,\ \underline{\underline{\mu}}^{\eta}(\boldsymbol{x}) = 1 & \text{for } \boldsymbol{x} \in \mathbb{R}^3\backslash\Omega \\ \underline{\underline{\varepsilon}}^{\eta}(\boldsymbol{x}) = \underline{\underline{\varepsilon}}^{0}\left(\frac{\boldsymbol{x}_{\perp}}{\eta}, x_3\right),\ \underline{\underline{\mu}}^{\eta}(\boldsymbol{x}) = \underline{\underline{\mu}}^{0}\left(\frac{\boldsymbol{x}_{\perp}}{\eta}, x_3\right) & \text{for } \boldsymbol{x}_{\perp} \in \Omega \end{cases} \tag{7.1}$$

where at fixed x_3, the applications $\boldsymbol{y} \to \underline{\underline{\varepsilon}}^{0}(\boldsymbol{y}, x_3) = \left(\varepsilon_{ij}^{0}(\boldsymbol{y}, x_3)\right)$ and $\boldsymbol{y} \to \underline{\underline{\mu}}^{0}(\boldsymbol{y}, x_3) = \left(\mu_{ij}^{0}(\boldsymbol{y}, x_3)\right)$ are Y-periodic 3×3 matrix functions.

The total electromagnetic field $(\boldsymbol{E}^{\eta}, \boldsymbol{H}^{\eta})$ satisfies

$$\begin{cases} \text{curl } \boldsymbol{E}^{\eta} = i\omega\mu_0\underline{\underline{\mu}}^{\eta}\boldsymbol{H}^{\eta} \\ \text{curl } \boldsymbol{H}^{\eta} = -i\omega\varepsilon_0\underline{\underline{\varepsilon}}^{\eta}\boldsymbol{E}^{\eta} \end{cases} \tag{7.2}$$

and $(\boldsymbol{E}^{s,\eta}, \boldsymbol{H}^{s,\eta}) = (\boldsymbol{E}^{\eta} - \boldsymbol{E}^{i}, \boldsymbol{H}^{\eta} - \boldsymbol{H}^{i})$ satisfies Silver–Müller radiation conditions:

$$\lim_{\|\boldsymbol{x}\|\to+\infty} \|\boldsymbol{x}\| \left(Z_0\boldsymbol{H}^{s,\eta} \times \frac{\boldsymbol{x}}{\|\boldsymbol{x}\|} - \boldsymbol{E}^{s,\eta} \right) = 0 \tag{7.3}$$

where Z_0 is the impedance of vacuum.

7.3.2 Derivation of the Microscopic Equations

7.3.2.1 A short account of the two-scale expansion

In order to describe this problem, we will rely on a two-scale expansion of the fields. That way, the physical problem is described by two variables: a macroscopic one $\boldsymbol{x}$ and a microscopic one $\boldsymbol{y}$ representing the rapid variations of the material at the scale of one basic cell, measured by η. By noticing that there are no rapid variations in the vertical direction x_3, the microscopic variable is set to be: $\boldsymbol{y} = \boldsymbol{x}_{\perp}/\eta$. Differential operators with respect to variable $\boldsymbol{y}$ are denoted with a subscript y. The fields are periodic with respect to that microscopic variable (this corresponds to the neighborhood of the center of the first Brillouin zone). The limit problem obtained by letting η tend to 0 will then depend on the macroscopic, physical variable $\boldsymbol{x}$ but also on the microscopic, hidden variable $\boldsymbol{y}$. The total limit fields will read $\boldsymbol{E}^0(\boldsymbol{x}, \boldsymbol{y})$ and $\boldsymbol{H}^0(\boldsymbol{x}, \boldsymbol{y})$ and the observable, physical fields will be given by the mean value over the hidden variable $\boldsymbol{y}$:

$$\boldsymbol{E}(\boldsymbol{x}) = |Y|^{-1} \int_Y \boldsymbol{E}^0(\boldsymbol{x}, \boldsymbol{y}) d\boldsymbol{y} \text{ and } \boldsymbol{H}(\boldsymbol{x}) = |Y|^{-1} \int_Y \boldsymbol{H}^0(\boldsymbol{x}, \boldsymbol{y}) d\boldsymbol{y},$$

where $|Y|$ is the measure area of Y. In order to lighten the notations, we denote by brackets the averaging over Y, hence

$$\boldsymbol{H}(\boldsymbol{x}) = \langle \boldsymbol{H}^0 \rangle \text{ and } \boldsymbol{E}(\boldsymbol{x}) = \langle \boldsymbol{E}^0 \rangle .$$

The main mathematical tool that we use is a mathematically rigorous version of the multiscale expansion, widely used in various areas of physics. More precisely, for a vector field $\boldsymbol{F}^\eta$ in $\left(L^2(\Omega)\right)^3$, we say, by definition, that $\boldsymbol{F}^\eta$ two-scale converges toward $\boldsymbol{F}^0$ if for every sufficiently regular function $\boldsymbol{\phi}(\boldsymbol{x}, \boldsymbol{y})$, Y-periodic with respect to $\boldsymbol{y}$, we have:

$$\int_\Omega \boldsymbol{E}^\eta(\boldsymbol{x}) \cdot \boldsymbol{\phi}(\boldsymbol{x}, \boldsymbol{x}_\perp/\varepsilon)\, d\boldsymbol{x} \to \iint_{\Omega\times Y} \boldsymbol{E}^0(\boldsymbol{x}, \boldsymbol{y}) \cdot \boldsymbol{\phi}(\boldsymbol{x}, \boldsymbol{y}) d\boldsymbol{x} d\boldsymbol{y}, \quad (7.4)$$

as η tends to 0.

The limit field $\boldsymbol{F}^0$ is square integrable with respect to both variables $\boldsymbol{x}$ and $\boldsymbol{y}$ and is Y-periodic in the $\boldsymbol{y}$ variable (this is the definition of the space $L^2\left(\Omega; \left(L^2_{\text{per}}(Y)\right)^3\right)$). A complete analysis of this new mathematical tool can be found in (Allaire, 1992).

We make the physically reasonable assumption that the electromagnetic energy remains bounded when η tends to 0, which is equivalent to assume that $(\boldsymbol{E}^\eta, \boldsymbol{H}^\eta)$ are both locally square integrable. Then it is known (Allaire, 1992) that $(\boldsymbol{E}^\eta, \boldsymbol{H}^\eta)$ two-scale converges toward limit fields denoted $\left(\boldsymbol{E}^0, \boldsymbol{H}^0\right)$. This physical assumption could be justified mathematically; however, it seems quite obvious from the point of view of physics that the limit fields exist. The point is then to give the system of equations that is satisfied by these fields and to derive the effective permittivity and permeability tensors.

7.3.2.2 The equations at the microscopic scale

First of all, we have to determine the set of equations that are microscopically satisfied, that is, satisfied with respect to the hidden variable $\boldsymbol{y}$, because that will give the constitutive relations of the homogenized medium. Multiplying the Maxwell–Faraday equation by a regular test function $\boldsymbol{\phi}\left(\boldsymbol{x}, \frac{\boldsymbol{x}_\perp}{\eta}\right)$, and integrating over Ω, we obtain:

$$\int_\Omega \boldsymbol{E}^\eta(\boldsymbol{x}) \cdot \left[\text{curl}_x(\boldsymbol{\phi}) + \frac{1}{\eta}\text{curl}_y(\boldsymbol{\phi})\right] d\boldsymbol{x}$$

$$= i\omega\mu_0 \int_\Omega \underline{\underline{\mu}}^\eta(\boldsymbol{x})\, \boldsymbol{H}^\eta(\boldsymbol{x})\, \boldsymbol{\phi}(\boldsymbol{x}, \boldsymbol{x}_\perp/\eta)\, d\boldsymbol{x}. \quad (7.5)$$

Multiplying by η and letting η tend to 0, we get using (7.4):

$$\iint_{\Omega\times Y} \boldsymbol{E}^0(\boldsymbol{x}, \boldsymbol{y}) \cdot \text{curl}_y(\boldsymbol{\phi})\, d\boldsymbol{x} d\boldsymbol{y} = 0. \quad (7.6)$$

This is equivalent to:

$$\iint_{\Omega\times Y} \mathbf{curl}_y \, \boldsymbol{E}^0 (\boldsymbol{x}, \boldsymbol{y}) \cdot \boldsymbol{\phi} (\boldsymbol{x}, \boldsymbol{y}) \, d\boldsymbol{x} d\boldsymbol{y} = 0 \tag{7.7}$$

which is nothing else but the variational form for: $\mathbf{curl}_y \, \boldsymbol{E}^0 = 0$. In a very similar way, but now using the Maxwell–Ampere equation, we get $\mathrm{curl}_y \boldsymbol{H}^0 = 0$. On the other hand, since $\underline{\underline{\varepsilon}}^\eta \boldsymbol{E}^\eta$ is divergence free, we have, for every test function

$$\phi(\boldsymbol{x}, \boldsymbol{y}), \int_\Omega \underline{\underline{\varepsilon}}^\eta (\boldsymbol{x}) \, \boldsymbol{E}^\eta (\boldsymbol{x}) \cdot \left[\nabla_x \phi + \frac{1}{\eta} \nabla_y \phi \right] d\boldsymbol{x} = 0.$$

Multiplying by η and letting η tend to 0, we get:

$$\iint_{\Omega\times Y} \underline{\underline{\varepsilon}}^0 (\boldsymbol{y}, x_3) \, \boldsymbol{E}^0 (\boldsymbol{x}, \boldsymbol{y}) \cdot \nabla_y \phi \, d\boldsymbol{x} d\boldsymbol{y} = 0, \tag{7.8}$$

which can be written as (notice that the div_y operator acts only on the transverse components):

$$\mathrm{div}_y \left(\underline{\underline{\varepsilon}}^0 \boldsymbol{E}^0 \right) = 0. \tag{7.9}$$

Similarly, since the magnetic field is divergence free, we derive:

$$\mathrm{div}_y \left(\underline{\underline{\mu}}^0 \boldsymbol{H}^0 \right) = 0. \tag{7.10}$$

Summing up, we have the following microscopic equations holding inside the basic cell Y:

$$\begin{cases} \mathrm{div}_y \left(\underline{\underline{\mu}}^0 \boldsymbol{H}^0 \right) = 0 \\ \mathrm{curl}_y \boldsymbol{H}^0 = 0 \end{cases}, \quad \begin{cases} \mathrm{div}_y \left(\underline{\underline{\varepsilon}}^0 \boldsymbol{E}^0 \right) = 0 \\ \mathrm{curl}_y \boldsymbol{E}^0 = 0 \end{cases} \tag{7.11}$$

7.3.3 Derivation of the Homogenized Parameters

The systems in (Eq. 7.11) are respectively of electrostatic and magnetostatic types. This means that with respect to the hidden variable $\boldsymbol{y}$, the electric field and magnetic field satisfy the static Maxwell system. This property is true only at that scale and not at the macroscopic scale. However, it is these static equations that determine the effective permittivity and permeability. Indeed let us analyze this system starting with the electric field. From the curl relation, we get $\nabla_y E_3^0 = 0$, and so $E_3^0(\boldsymbol{x}, \boldsymbol{y}) \equiv E_3(\boldsymbol{x})$. Besides, the basic cell having the geometry of a flat torus, we get the existence of a regular periodic function $w_E(\boldsymbol{y})$ such that:

$$\boldsymbol{E}_\perp^0 = \boldsymbol{E}_\perp + \nabla_y w_E. \tag{7.12}$$

The function w_E is the electrostatic potential associated with the microscopic electrostatic problem. Inserting (7.9) in Eq. (7.12) and projecting on both horizontal axes, we obtain:

$$\operatorname{div}_y \left[\underline{\underline{\varepsilon}}^0_\perp \left(\boldsymbol{e}_i + \nabla_y w_{E,i} \right) \right] = 0, \; i \in \{1, 2\} \tag{7.13}$$

where $\underline{\underline{\varepsilon}}^0_\perp$ denotes the 2×2 matrix extracted from $\underline{\underline{\varepsilon}}^0$ by removing the last line and last column. By linearity, denoting $\boldsymbol{E}_\perp = (E_1, E_2)$, we derive that the potential w_E is given by $w_E = E_1 w_{E,1} + E_2 w_{E,2}$, where $w_{E,i}$ are the periodic solutions to (7.13). We stress that these potentials depend not only on $\boldsymbol{y}$ but also on x_3. In fact, we get a family of homogenization problems parametrized by the vertical coordinate. By (7.12), we obtain:

$$\boldsymbol{E}^0 (\boldsymbol{x}, \boldsymbol{y}) = \mathcal{E} (\boldsymbol{y}, x_3) \, \boldsymbol{E} (\boldsymbol{x}) \tag{7.14}$$

where:

$$\mathcal{E} (\boldsymbol{y}, x_3) = \begin{pmatrix} 1 + \partial_{y_1} w_{E,1} & \partial_{y_1} w_{E,2} & 0 \\ \partial_{y_2} w_{E,1} & 1 + \partial_{y_2} w_{E,2} & 0 \\ 0 & 0 & 1 \end{pmatrix} \tag{7.15}$$

The magnetic field $\boldsymbol{H}^0$ can be handled in the same way since it satisfies exactly the same kind of equations as $\boldsymbol{H}^0$ (see (7.11)). In particular, we may represent its tranversal component in the form: $\boldsymbol{H}^0_\perp = \boldsymbol{H}_\perp + \nabla_\perp w_H$, where w_H is a periodic magnetic potential (the possibility of this representation is due to the curl-free condition, which means that no microscopic current is present). Analogously as in (7.14, 7.27), we find:

$$\boldsymbol{H}^0 (\boldsymbol{x}, \boldsymbol{y}) = \mathcal{M} (\boldsymbol{y}, x_3) \, \boldsymbol{H} (\boldsymbol{x}) \tag{7.16}$$

where

$$\mathcal{M} (\boldsymbol{y}, x_3) = \begin{pmatrix} 1 + \partial_{y_1} w_{H,1} & \partial_{y_1} w_{H,2} & 0 \\ \partial_{y_2} w_{H,1} & 1 + \partial_{y_2} w_{H,2} & 0 \\ 0 & 0 & 1 \end{pmatrix} \tag{7.17}$$

where:

$$\operatorname{div}_y \left[\underline{\underline{\mu}}^0_\perp \left(\boldsymbol{e}_i + \nabla_y w_{H,i} \right) \right] = 0, \; i \in \{1, 2\} \tag{7.18}$$

and $\underline{\underline{\mu}}^0_\perp$ denotes the 2×2 matrix extracted from $\underline{\underline{\mu}}^0$ by removing the last line and last column. Of course, the same remark as in the case of electric potentials holds: The functions $w_{H,i}$ depend on the vertical coordinate x_3. The above results show that at the microscopic scale, the limit fields $(\boldsymbol{E}^0, \boldsymbol{H}^0)$ are completely determined by the physical fields $(\boldsymbol{E}, \boldsymbol{H})$. Now that the microscopic behavior is precised, we are able to determine the macroscopic system satisfied by $(\boldsymbol{E}, \boldsymbol{H})$. To that aim, let us choose a regular test function $\boldsymbol{\phi}(\boldsymbol{x})$ independent of the variable $\boldsymbol{y}$. From Maxwell's equations, we get:

$$\begin{cases} \int_\Omega \boldsymbol{H}^\eta(\boldsymbol{x}) . \mathrm{curl}(\boldsymbol{\phi})\, d\boldsymbol{x} = -i\omega\varepsilon_0 \int_\Omega \underline{\underline{\varepsilon}}^\eta(\boldsymbol{x})\, \boldsymbol{E}^\eta(\boldsymbol{x})\, \boldsymbol{\phi}(\boldsymbol{x})\, d\boldsymbol{x} \\ \int_\Omega \boldsymbol{E}^\eta(\boldsymbol{x}) \cdot \mathrm{curl}(\boldsymbol{\phi})\, d\boldsymbol{x} = i\omega\mu_0 \int_\Omega \underline{\underline{\mu}}^\eta(\boldsymbol{x})\, \boldsymbol{H}^\eta(\boldsymbol{x})\, \boldsymbol{\phi}(\boldsymbol{x})\, d\boldsymbol{x} \end{cases} \tag{7.19}$$

passing to the limit $\eta \to 0$, we get:

$$\begin{cases} \iint_{\Omega\times Y} \boldsymbol{H}^0(\boldsymbol{x}, \boldsymbol{y}) \cdot \mathrm{curl}(\boldsymbol{\phi})\, d\boldsymbol{x}d\boldsymbol{y} \\ = -i\omega\varepsilon_0 \iint_{\Omega\times Y} \underline{\underline{\varepsilon}}^0(\boldsymbol{y}, x_3)\, \boldsymbol{E}^0(\boldsymbol{x}, \boldsymbol{y})\, \boldsymbol{\phi}(\boldsymbol{x})\, d\boldsymbol{x}d\boldsymbol{y} \\ \iint_{\Omega\times Y} \boldsymbol{E}^0(\boldsymbol{x}, \boldsymbol{y}) \cdot \mathrm{curl}(\boldsymbol{\phi})\, d\boldsymbol{x}d\boldsymbol{y} \\ = i\omega\mu_0 \iint_{\Omega\times Y} \underline{\underline{\mu}}^0(\boldsymbol{y}, x_3)\, \boldsymbol{H}^0(\boldsymbol{x}, \boldsymbol{y})\, \boldsymbol{\phi}(\boldsymbol{x})\, d\boldsymbol{x}d\boldsymbol{y} \end{cases} \tag{7.20}$$

Recalling that $\langle \boldsymbol{E}^0 \rangle = \boldsymbol{E}$ and that $\langle \boldsymbol{H}^0 \rangle = \boldsymbol{H}$, we get:

$$\begin{cases} \mathrm{curl}\, \boldsymbol{E} = i\omega\mu_0 \left\langle \underline{\underline{\mu}}^0 \boldsymbol{H}^0 \right\rangle \\ \mathrm{curl}\, \boldsymbol{H} = -i\omega\varepsilon_0 \left\langle \underline{\underline{\varepsilon}}^0 \boldsymbol{E}^0 \right\rangle \end{cases} \tag{7.21}$$

which, taking into account (7.14, 7.16), brings to the limit system:

$$\begin{cases} \mathrm{curl}\, \boldsymbol{E} = i\omega\mu_0 \left\langle \underline{\underline{\mu}}^0 \mathcal{M} \right\rangle (x_3)\, \boldsymbol{H} \\ \mathrm{curl}\, \boldsymbol{H} = -i\omega\varepsilon_0 \left\langle \underline{\underline{\varepsilon}}^0 \mathcal{E} \right\rangle (x_3)\, \boldsymbol{E} \end{cases} \tag{7.22}$$

7.3.3.1 The special case of a one-dimensional grating

Let us specialize the above results to the case of a one-dimensional grating (Fig. 7.8), that is, the pillars are invariant in the x_2 and x_3 directions (the basic cell Y is depicted in Fig. 7.8). We assume also that the pillars are made on a nonmagnetic material and that the relative permittivity tensor is given by:

$$\underline{\underline{\varepsilon}}^0(y_1) = \begin{pmatrix} \varepsilon_1(y_1) & 0 & 0 \\ 0 & \varepsilon_2(y_1) & 0 \\ 0 & 0 & \varepsilon_3(y_1) \end{pmatrix} \tag{7.23}$$

The invariance of $\underline{\underline{\varepsilon}}^0$ with respect to y_2 and the periodicity condition suggest that we look for solutions that are independent of y_2. Let us first consider $w_{E,2}$, which satisfies: $\partial_{y_1}\left(\varepsilon_1 \partial_{y_1} w_{E,2}\right) = 0$. This implies that $w_{E,2} = \text{cste}$. Next, we turn to $w_{E,1}$. It satisfies

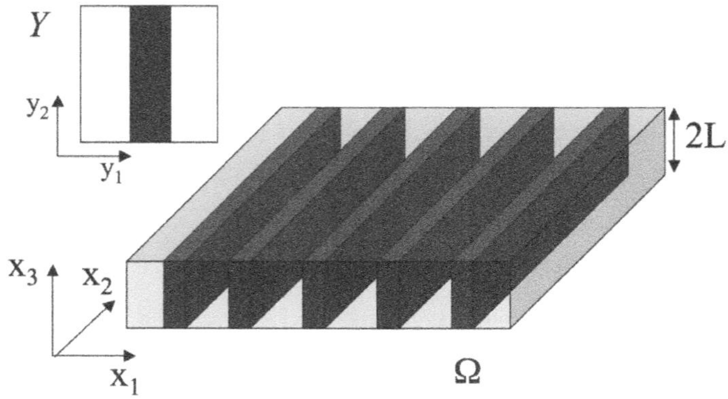

Figure 7.8 Schematic of the one-dimensional photonic crystal. The inset shows the basic cell Y.

$\partial_{y_1}\left[\varepsilon_1\left(1+\partial_{y_1}w_{E,1}\right)\right]=0$. Therefore, $\varepsilon_1\left(1+\partial_{y_1}w_{E,1}\right)=\text{cste}=C$. Let us now average this relation to get: $C\left\langle\frac{1}{\varepsilon_1}\right\rangle=\left\langle 1+\partial_{y_1}w_{E,1}\right\rangle$. Due to the periodicity of $w_{E,1}$, we have: $\left\langle\partial_{y_1}w_{E,1}\right\rangle=0$ and finally: $C=\left\langle\frac{1}{\varepsilon_1}\right\rangle^{-1}$. The homogenized relative permittivity tensor is given by:

$$\left\langle\underline{\underline{\varepsilon}}^0\mathcal{E}\right\rangle=\begin{pmatrix}\left\langle\frac{1}{\varepsilon_1}\right\rangle^{-1} & 0 & 0\\ 0 & \langle\varepsilon_2\rangle & 0\\ 0 & 0 & \langle\varepsilon_3\rangle\end{pmatrix} \tag{7.24}$$

We retrieve a well-established result concerning the homogenization of one-dimensional photonic crystals (Jikov et al., 1994).

In our homogenization result, it is clear that the main numerical problem is the solving of the annex problems (7.13, 7.18) for they give the effective matrices $\mathcal{E}$ and $\mathcal{M}$. In certain simple cases, for instance that of circular isotropic nonmagnetic rods and a permittivity constant in each connected region, it is possible to find an explicit expression for the effective permittivity (it is, in fact, a very old problem).

In this section, we considered a structure periodic in two directions of space. It is not difficult to use the same procedure when the structure is periodic in three directions of space. Let us, thus, consider a periodic set in $\mathbb{R}^3$, contained in a bounded domain Ω. The relative permittivity tensor $\underline{\underline{\varepsilon}}^\eta(\mathbf{x})$ and relative permeability tensor

$\underline{\underline{\mu}}^{\eta}(\mathbf{x})$ are described by:

$$\begin{cases} \underline{\underline{\varepsilon}}^{\eta}(\mathbf{x}) = 1, \ \underline{\underline{\mu}}^{\eta}(\mathbf{x}) = 1 & \text{for } \mathbf{x} \in \mathbb{R}^3 \backslash \Omega \\ \underline{\underline{\varepsilon}}^{\eta}(\mathbf{x}) = \underline{\underline{\varepsilon}}^{0}(\mathbf{x}), \ \underline{\underline{\mu}}^{\eta}(\mathbf{x}) = \underline{\underline{\mu}}^{0}(\mathbf{x}) & \text{for } \mathbf{x} \in \Omega \end{cases} \tag{7.25}$$

Then, as η tends to 0, the following homogenization result holds:

Theorem 7.1. *The homogenized fields $\mathscr{E}$ and $\mathscr{H}$ satisfy the Maxwell system:*

$$\begin{cases} \mathbf{curl}\,\mathscr{E} = i\omega\mu_0 \left\langle \underline{\underline{\mu}}^{0} \mathcal{M} \right\rangle \mathscr{H} \\ \mathbf{curl}\,\mathscr{H} = -i\omega\varepsilon_0 \left\langle \underline{\underline{\varepsilon}}^{0} \mathcal{E} \right\rangle \mathscr{E} \end{cases} \tag{7.26}$$

The matrices $\mathcal{E}$ and $\mathcal{M}$ are defined by:

$$\mathcal{E}(\mathbf{y}) = \begin{pmatrix} 1 + \partial_{y_1} w_{E,1} & \partial_{y_1} w_{E,2} & \partial_{y_1} w_{E,3} \\ \partial_{y_2} w_{E,1} & 1 + \partial_{y_2} w_{E,2} & \partial_{y_2} w_{E,3} \\ \partial_{y_3} w_{E,1} & \partial_{y_3} w_{E,2} & 1 + \partial_{y_3} w_{E,3} \end{pmatrix} \tag{7.27}$$

and

$$\mathcal{M}(y) = \begin{pmatrix} 1 + \partial_{y_1} w_{H,1} & \partial_{y_1} w_{H,2} & \partial_{y_1} w_{H,3} \\ \partial_{y_2} w_{H,1} & 1 + \partial_{y_2} w_{H,2} & \partial_{y_2} w_{H,3} \\ \partial_{y_3} w_{H,1} & \partial_{y_3} w_{H,3} & 1 + \partial_{y_3} w_{H,3} \end{pmatrix} \tag{7.28}$$

In these expressions, the electric potentials w_E satisfy the annex problems:

$$div_y \left[\underline{\underline{\varepsilon}}^{0} \left(\boldsymbol{e}_i + \nabla_y w_{E,i} \right) \right] = 0, \ i \in \{1, 2, 3\} \tag{7.29}$$

and the magnetic potentials w_H satisfy:

$$div_y \left[\underline{\underline{\mu}}^{0} \left(\boldsymbol{e}_i + \nabla_y w_{H,i} \right) \right] = 0, \ i \in \{1, 2, 3\} \tag{7.30}$$

7.4 Soft Problems: Numerical Examples

7.4.1 A Little Vademecum

The theoretical frame is now well established. But it remains various questions on hold. First, contrary to the historical mixing laws (Maxwell Garnett, Clausius–Mossotti, etc.), the theory described above does not give rise to closed formulae. We are, therefore, led to numerically solve the so-called annex problems and then to deduce the effective constants. It should be remarked, however, that the mixing laws can be retrieved from the annex problems through a perturbative approach. For this purpose, two different methods are proposed and compared: the fictitious charges method (FCM) and the finite element method (FEM). In the second step, a special attention will be paid to the case where the permittivity and the permeability are both represented by piecewise constant functions: Such functions describe the inclusions in the scatterers we are dealing with. In this latter case, the results obtained with the theory will be compared in some academic cases with the traditional mixing laws.

7.4.2 Some Prerequisites for Two-Phase Materials

In this section, we restrict our problem to obstacles described by piecewise constant functions (see Fig. 7.9). Moreover, for the sake of simplicity, we assume that the materials are nonmagnetic ($\mu = \mu_0$ everywhere) and only two-phase media will be considered In such conditions, the problem is solely characterized by the two permittivities ε_1 and ε_2 and the boundary shared by the materials denoted by Σ:

$$\varepsilon = \begin{cases} \varepsilon_1 \text{ in } \Omega_1 \\ \varepsilon_2 \text{ in } \Omega_2 \end{cases} \tag{7.31}$$

with $\Sigma = \partial\Omega_1 = \partial\Omega_2$. We now state and prove two lemmas useful for the sequel.

Lemma 7.1.

The solution to the annex problems 7.29 introduced in the fundamental theorem 7.1 amounts to looking for functions V_i solutions of the following system:

$$\begin{cases} \Delta V_i = 0 \,, \text{ on } Y \setminus \Sigma & (7.32a) \\ \left[V_i\right]_\Sigma = 0 & (7.32b) \\ \left[\varepsilon \dfrac{\partial V_i}{\partial n}\right]_\Sigma = -\left[\varepsilon\right]_\Sigma n_i & (7.32c) \end{cases}$$

where $[f]_\Sigma$ denotes the jump of f across the boundary Σ, and n_i, $i \in \{1, 2, 3\}$ denotes the projection on the axis e_i of a normal of Σ.

Proof. Let us denote by e_i, $i \in \{1, 2, 3\}$ the three vectors of the orthonormal base of $\mathbb{R}^3$. As already seen, the problem verified by

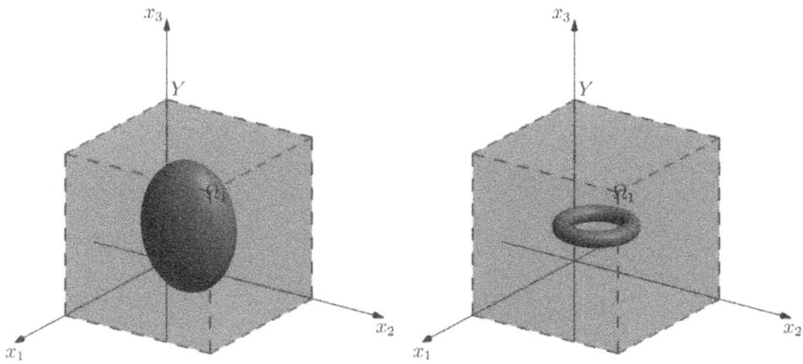

Figure 7.9 In most cases, the metamaterial can be characterized by its permittivity ε which is represented by a two-valued piecewise function: ε_1 and ε_2. (a) A spheroidal scatterer. (b) A toroidal scatterer.

V_i can be written as:

$$\mathrm{div}_y\Big(\varepsilon(\mathbf{y})\big(\mathbf{grad}_y\, V_i + e_i\big)\Big) = 0\ ,\ i \in \{1,\,2,\,3\} \qquad (7.33)$$

Applying the formula of derivatives in the sense of distributions, we get:

$$\begin{aligned}\mathrm{div}(\varepsilon\,\mathbf{grad}\,V_i + \varepsilon e_i) &= \varepsilon\{\Delta V_i\} + \mathrm{div}\big\{[V_i]_\Sigma n\delta_\Sigma\big\}\\ &\quad + [\varepsilon]_\Sigma n\cdot e_i\delta_\Sigma + \underbrace{[\varepsilon\,\mathbf{grad}\,V_i]n}_{\left[\varepsilon\,\frac{\partial V_i}{\partial n}\right]}\delta_\Sigma\end{aligned}$$

Equaling, on the one hand the regular part and on the other hand the singular part of the above distribution, we obtain the expected result. □

These results call for further comments detailed in the following two remarks.

Remark 7.1. (About the physical sense of the annex problem) Coming back for a while to the general case, the annex problem can be rewritten as $\mathrm{div}_y(\varepsilon(y)\,\mathbf{grad}_y\,V_i) = -\,\mathrm{div}_y(\varepsilon\,\mathbf{grad}_y\,y_i)$ which can be seen as an electrostatic problem with a volumic distribution of charges $\rho_i = \mathrm{div}_y(\varepsilon\,\mathbf{grad}_y\,y_i)$.

In a similar way, in the case where ε is a two-valued piecewise constant function (the case encountered in most manufactured optical devices), we can write:

$$\mathrm{div}_y(\varepsilon(\mathbf{y})\,\mathbf{grad}_y\,V_i) = -\sigma_i\delta_\Sigma$$

where σ_i is a surfacic distribution of charges defined by $\sigma_i = [\varepsilon]n_i\delta_\Sigma$.

Solving the annex problem for a two-valued piecewise constant permittivity then reduces to looking for the potential induced by a surfacic density of charges on the edge of the scatterer. This result is not surprising, unlike the following remark, which is far from obvious.

Remark 7.2. (About the numerical calculus of the permittivity) In the fundamental theorem, we have seen that the relative permittivity matrix of the homogenized problem is deduced from six numbers ϕ_{ij} defined as follows:

$$\varphi_{ij} = \int_Y \varepsilon(\mathbf{y})\frac{\partial V_i}{\partial y_j}\,d\mathbf{y}\,, \tag{7.34}$$

The permittivity ε being a two-valued piecewise function, we then deduce that:

$$\varphi_{ij} = \varepsilon_1\int_{\Omega_1}\frac{\partial V_i}{\partial y_j}\,d\mathbf{y} + \varepsilon_2\int_{\Omega_2}\frac{\partial V_i}{\partial y_j}\,d\mathbf{y} \tag{7.35}$$

It then follows that:

$$\varphi_{ij} = \varepsilon_1\int_{\Omega_1}\mathrm{div}(V_i e_j)\,d\mathbf{y} + \varepsilon_2\int_{\Omega_2}\mathrm{div}(V_i e_j)\,d\mathbf{y} \tag{7.36}$$

Applying Green's formula, we obtain:

$$\varphi_{ij} = \varepsilon_1\int_\Sigma V_i\,n_j^+\,ds + \varepsilon_2\int_\Sigma V_i\,n_j^-\,ds \tag{7.37}$$

Choosing the convention $n_j = n_j^+$, we have $n_j^- = -n_j$, which leads to the equality:

$$\varphi_{ij} = -\,[\varepsilon]_\Sigma\int_\Sigma V_i\,n_j\,ds\,. \tag{7.38}$$

Let us remark that V_i is well defined on Σ because it does not suffer a jump across the boundary of the scatterer. This last formula is of prime importance for numerical implementations: It is not necessary to compute the gradient of V (which gives rise to numerical inaccuracy) to perform the calculus of the homogenized permittivity.

Although the potential V is defined up to an additive constant, the permittivity is well defined thanks to the nullity of $\int_\Sigma n_i\,ds$.

7.4.3 Fictitious Charges Method as Applied to the Annex Problem

This method requires the same restrictions as those mentioned in Section 7.4.2. The method of fictitious charges consists in representing the potential by approximate potentials created by two families of fictitious charges. The first ones are located in the scatterer Ω_1, and they radiate in its complement Ω_2. Conversely, the second ones are located in Ω_2, and they radiate in Ω_1. Each of the approximate potentials satisfies a Laplace equation in Y with periodic conditions. The amplitudes of the sources are chosen in such a way that the approximate potentials are continuous across the boundary of Σ (see Eq. (7.32b)) and they satisfy a condition involving the normal derivative of the true potential (see Eq. (7.32c)).

7.4.3.1 Introduction to the column space $\mathcal{V}$

To begin with, it is convenient to recall an important property of functions that are solution to the Laplace equation. Let us consider a regular harmonic Y-periodic function u satisfying **one of the two** following problems:

$$\begin{cases} \Delta u = 0 & \text{in } \Omega_1 \\ u_{|\Sigma} = \phi \end{cases} \tag{7.39}$$

$$\begin{cases} \Delta u &= 0 \quad \text{in } \Omega_1 \\ D_n u_{|\Sigma} &= \psi \end{cases} \tag{7.40}$$

where ϕ and ψ are given functions belonging to $L^2(\Sigma)$ and where $u_{|\Sigma}$ denote the trace of u on Σ and $D_n u_{|\Sigma}$ the normal trace of **grad** u on Σ. Then, the Lax–Milgram lemma insures the existence and uniqueness of the solution in the Sobolev space $H^1_\sharp$ of periodic functions (up to an additive constant for the second problem).

Remark 7.3. Thanks to the Kirchhoff–Helmholtz relation, the function u is fully defined in whole Ω_1 and can also be easily computed if we know the values ϕ and ψ on Σ.

In fact, it can be shown that there are two operators K and L such that:

$$\phi = K(\psi) \tag{7.41}$$

and conversely

$$\psi = L(\phi) \tag{7.42}$$

It is now convenient to introduce the following notations:

$$W_1 = \begin{pmatrix} -V_{1|\Sigma} \\ -\varepsilon_1 \, D_n V_{1|\Sigma} \end{pmatrix}, \; W_2 = \begin{pmatrix} V_{2|\Sigma} \\ \varepsilon_2 \, D_n V_{2|\Sigma} \end{pmatrix} \text{ and } W_0 = \begin{pmatrix} 0 \\ [\varepsilon] \; n_i \end{pmatrix}. \tag{7.43}$$

With these notations, our problem is reduced to finding the columns W_1 or W_2—the column W_0 being known—such that:

$$W_0 = W_1 + W_2 \,. \tag{7.44}$$

7.4.3.2 The spaces $\mathcal{V}$, $\mathcal{V}_1$, and $\mathcal{V}_2$

We consider the space $\mathcal{V} = L^2(\Sigma) \times L^2(\Sigma)$ of pairs of functions $\Phi = \begin{pmatrix} \phi \\ \psi \end{pmatrix}$ defined on Σ. The space $\mathcal{V}$ is a Hilbert space for the following inner product:

$$(\Phi_1, \Phi_2)_{\mathcal{V}} = \int_\Sigma \phi_1 \overline{\phi_2} \, dl + \int_\Sigma \psi_1 \overline{\psi_2} \, dl \tag{7.45}$$

The associated norm is:

$$||\Phi||_{\mathcal{V}} = (\Phi, \Phi)_{\mathcal{V}}^{1/2} \tag{7.46}$$

We will now define two subspaces of $\mathcal{V}$: the spaces $\mathcal{V}_1$ and $\mathcal{V}_2$. The space $\mathcal{V}_j$ is defined as the subspace of $\mathcal{V}$ of columns $\Phi = \begin{pmatrix} \phi \\ \psi \end{pmatrix}$ satisfying the following property:

> The functions ϕ and ψ are such that there is a regular Y-periodic harmonic function f defined on Ω_j (i.e., $\triangle f = 0$) such that $\phi = f_{|\Sigma}$ and $\psi = \varepsilon_j D_n f_{|\Sigma}$.

Consequently, $\mathcal{V}_1$ can be said to be associated with the field in Ω_1, whereas $\mathcal{V}_2$ is associated with the field in Ω_2. Returning now to the columns W_0, W_1, and W_2, it is clear that W_0 is in $\mathcal{V}$, W_1 is in $\mathcal{V}_1$, and W_2 is in $\mathcal{V}_2$. Besides W_0 being given in $H^1_\sharp$, the uniqueness of W_1 and W_2 is guaranteed (the solution to system 7.44 is unique) and, therefore, the decomposition

$$W_0 = W_1 + W_2 \tag{7.47}$$

is unique. In other words, $\mathcal{V}$ is the direct sum of $\mathcal{V}_1$ and $\mathcal{V}_2$:

$$\mathcal{V} = \mathcal{V}_1 \oplus \mathcal{V}_2 \,. \tag{7.48}$$

7.4.3.3 Solution to the annex problem

The problem is now to find the column W_1 (or the column W_2) such as

$$W_0 = W_1 + W_2 \tag{7.49}$$

with a known column W_0. To find this column W_1, there is a general procedure based on looking for total families in $\mathcal{V}_1$ and $\mathcal{V}_2$ and on the use of least squares method (Petit and Zolla, 1994; Zolla and Petit, 1996; Zolla et al., 1994). Let us assume that we know in each vector space $\mathcal{V}_j$ ($j \in \{1, 2\}$) a total family $\{W_{j,n}\}$. This means that each vector W_j in $\mathcal{V}_j$ is the limit of a linear combination of the $W_{j,n}$. Then, if the complex numbers $\{c_{j,n}\}$ are such that

$$\lim_{N \longrightarrow +\infty} ||W_0 - (W_1^N + W_2^N)||_{\mathcal{V}} = 0\,, \quad \text{with} \quad W_j^N = \sum_{n=1}^{N} c_{j,n}(N) W_{j,n} \tag{7.50}$$

then

$$\lim_{N \longrightarrow +\infty} ||W_j^N - W_j)||_{\mathcal{V}} = 0\,. \tag{7.51}$$

For numerical calculations, we consider spaces $\mathcal{V}_j^N$ of finite dimension N instead of the spaces $\mathcal{V}_j$. The spaces $\mathcal{V}_j^N$ are generated by N columns W_j^N. By and large, there are no column W_j^N, which belongs to $\mathcal{V}_j^N$ such that the norm $||W_0 - (W_1^N + W_2^N)||_{\mathcal{V}} = 0$ is null. The problem is therefore, for a fixed N, to find the coefficients $c_{j,n}(N)$, which minimize the positive real Δ_N defined as follows:

$$\Delta_N = ||W_0 - (W_1^N + W_2^N)||_{\mathcal{V}}\,. \tag{7.52}$$

Having found the coefficients $\tilde{c}_{j,n}(N)$, which minimize Δ_N, we obtain the approximation $\tilde{W}_j^N$ of W_j with:

$$\tilde{W}_j^N = \sum_{n=1}^{N} \tilde{c}_{j,n}(N) W_{j,n}\,. \tag{7.53}$$

The method just described converges in the following sense:

- $\lim_{N \longrightarrow +\infty} \tilde{\Delta}_N = 0$, with $\tilde{\Delta}_N = ||W_0 - (\tilde{W}_1^N + \tilde{W}_2^N)||_{\mathcal{V}}$
- $\lim_{N \longrightarrow +\infty} \tilde{\Delta}_N^j = 0$, with $\tilde{\Delta}_N^j = ||W_j - \tilde{W}_j^N||_{\mathcal{V}}$.

In summary, we have shown that $\tilde{W}_j^N$ is a good approximation of W_j. This approximation will be even better as N increases and can be checked using the value of $\tilde{\Delta}_N$. In fact, we should be point out that $\tilde{\Delta}_N$ is a sequence which decreases with N. We now write:

$$\check{\Delta}_N = \frac{\tilde{\Delta}_N}{||W_0||_{\mathcal{V}}} \quad \text{, with} \quad ||W_0||_{\mathcal{V}} = |\,[\varepsilon]_\Sigma\,| \left(\int_\Sigma |n_i|^2 \, ds \right)^{1/2} . \quad (7.54)$$

This sequence is a decreasing sequence and is such that $1 \leq \check{\Delta}_N \leq 0$. This number $\check{\Delta}_N$ represents a relative error on Σ and will make it possible during calculation to check the quality of the result. It will be sufficient, for instance, to take the integer N_0 such that $\check{\Delta}_{N_0} \leq 10^{-3}$.

7.4.3.4 An example of total family in $\mathcal{V}_1$ and $\mathcal{V}_2$

We consider two smooth curves γ_1 and γ_2. The curve γ_1 (γ_2) must be in Ω_2 (Ω_1) so that they lie on either side and "all along" the boundary Σ.

Theorem 7.2. *Let us consider a dense and countable set of points $P_{j,k}$ of coordinates $\mathbf{x}_{j,k} = (x_{j,k}, y_{j,k})$ on γ_j. Denoting by $V_{j,k}^\sharp$ the unique solution in $H_\sharp^1$ to the equation:*

$$\triangle V_{j,k}^\sharp(\mathbf{x}) = \sum_{\mathbf{m}\in\mathbb{Z}^2} \delta(\mathbf{x}-\mathbf{x}_{j,k}-\mathbf{m})-1, \quad \textit{with } \mathbf{x} = (x, y) \textit{ and } \mathbf{m} = (m_1, m_2) \quad (7.55)$$

and denoting by $W_{j,k}$ the column defined on Σ (Class C^2), as follows:

$$W_{j,k} = \begin{pmatrix} V_{j,k|\Sigma}^\sharp \\ \varepsilon_j \, D_n V_{j,k|\Sigma}^\sharp \end{pmatrix} \quad (7.56)$$

then the family $\{W_{j,k}\}_{k\in\mathbb{N}}$ is a total family in $\mathcal{V}_j$.

Remark 7.4. If we write $G_\sharp$ the unique function in $H_\sharp^1$ defined by:

$$\triangle G_\sharp(\mathbf{x}) = \sum_{\mathbf{m}\in\mathbb{Z}^2} \delta(\mathbf{x} - \mathbf{m}) - 1 \quad (7.57)$$

$V_{j,k}^\sharp$ is nothing but $G_\sharp(\mathbf{x}-\mathbf{x}_{j,k})$ and $D_n V_{j,k}^\sharp$ nothing but $\nabla G_\sharp(\mathbf{x}-\mathbf{x}_{j,k})\cdot\mathbf{n}$.

7.4.3.5 Fine estimation of the uniform bound of the error

Let us recall the expression of the coefficients $\phi_{i,j}$:

$$\varphi_{i,j} = [\varepsilon]_\Sigma \int_\Sigma V_i n_j \, ds \, . \tag{7.58}$$

In practice, we only have access to the estimation $\phi_{i,j}^N$ of $\varphi_{i,j}$ with:

$$\tilde{\phi}_{i,j}^N = [\varepsilon]_\Sigma \int_\Sigma \tilde{V}_i^N n_j \, ds \, . \tag{7.59}$$

which leads to

$$\begin{aligned} |\phi_{i,j} - \tilde{\varphi}_{i,j}^N| &= |\, [\varepsilon]_\Sigma \,| \left| \int_\Sigma (V_i - \tilde{V}_i^N) n_j \, ds \right| \\ &\leq |\, [\varepsilon]_\Sigma \,| \int_\Sigma |(V_i - \tilde{V}_i^N)| \, |n_j| \, ds \end{aligned} \tag{7.60}$$

By using the Cauchy-Schwartz inequality, we deduce the following inequality:

$$|\varphi_{i,j} - \tilde{\varphi}_{i,j}^N| \leq |\, [\varepsilon]_\Sigma \,| \left(\int_\Sigma |(V_i - \tilde{V}_i^N)|^2 \, ds \right)^{1/2} \left(\int_\Sigma |n_j|^2 \, ds \right)^{1/2} \tag{7.61}$$

Besides, it is straightforward that we have $\left(\int_\Sigma |(V_i - \tilde{V}_i^N)|^2 \, ds \right)^{1/2} < \tilde{\Delta}_N^j$ and remarking that $|\, [\varepsilon]_\Sigma \,| \left(\int_\Sigma |n_j|^2 \, ds \right)^{1/2}$ is nothing but $||W_0||_\mathcal{V}$, we finally deduce the following estimation:

$$|\varphi_{i,j} - \tilde{\varphi}_{i,j}^N| \leq ||W_0||_\mathcal{V} \, \tilde{\Delta}_N^j \, . \tag{7.62}$$

This last inequality proves that the estimate $\tilde{\varphi}_{i,j}^N$ does converge to the exact value $\phi_{i,j}$ and gives an excellent criterion to evaluate the accuracy of the estimate.

7.4.4 Closed Formulae for Small Spherical and Cylindrical Scatterers

For small scatterers, we assume that the potentials V_i are small enough in the vicinity of ∂Y to replace them by approximate potentials V_i^0, which are computed not in the unit cell Y but in the whole space: More precisely, V_i^0 are solutions to the following modified annex problem:

$$\begin{cases} \Delta V_i^0 = 0\,, \text{in } \mathbb{R}^3 & (7.63a) \\ \left[\varepsilon \dfrac{\partial V_i^0}{\partial n}\right]_{\Sigma_R} = -\left[\varepsilon\right]_{\Sigma_R} n_i & (7.63b) \\ \left[V_i^0\right]_{\Sigma_R} = 0 & (7.63c) \\ \lim\limits_{r \longrightarrow +\infty} V_i^0 = 0 & (7.63d) \end{cases}$$

where Σ_R denotes the sphere of radius R.

First, let us remark that the function V_3^0 defined by:

$$V_3^0 = \begin{cases} \xi x_3 & , \text{for } r \leq R \\ \xi R^2 \frac{x_3}{r^2} & , \text{for } r \geq R \end{cases} \tag{7.64}$$

satisfies the criteria (7.63a–7.63c) for any complex number ξ. Now we have to determine this complex number in order to fulfill the second jump condition (7.63d). Expressed in spherical coordinates, the normal derivative $D_n V_3^0$ yields:

$$\frac{\partial V_3^0}{\partial r} = \begin{cases} \xi \cos\theta & , \text{for } r < R \\ -\xi R^2 \frac{\cos\theta}{r^2} & , \text{for } r > R \end{cases} \tag{7.65}$$

On the other hand, $n_3 = \mathbf{n} \cdot \mathbf{x_3} = \mathbf{r} \cdot \mathbf{x_3} = \cos\theta$. The jump condition (7.63d) leads, therefore, to:

$$\varepsilon_2 \frac{dV_3^0}{dr}_{|r=R^+} - \varepsilon_1 \frac{dV_3^0}{dr}_{|r=R^-} = (\varepsilon_1 - \varepsilon_2)\cos\theta \tag{7.66}$$

that is

$$\xi = \frac{\varepsilon_1 - \varepsilon_2}{\varepsilon_1 + \varepsilon_2}\,. \tag{7.67}$$

Remark 7.5. Note that the expression of V_3 does work for every component of the multiscalar $\mathbf{V}$ and in $\mathbb{R}^3$ ($\mathbb{R}^2$) provided that the scatterers are spherical (cylindrical):

$$V_i^0 = \begin{cases} \xi x_i & , \text{for } r \leq R \\ \xi R^2 \frac{x_i}{r^2} & , \text{for } r \geq R \end{cases} \tag{7.68}$$

with ξ given above and with $r = \sqrt{x_1^2 + x_2^2 + x_3^2}$ (resp. $r = \sqrt{x_1^2 + x_2^2}$).

Remark 7.6. The reader should probably remember the expression found in 7.64: It is the potential radiated by a dipole pointing in the x_3 direction.

7.4.4.1 Computation of $\varphi_{1,1}$ (cylindrical case)

What we have to do is use Eq. (7.38):

$$\varphi_{1,1} = -[\varepsilon]\int_{\Sigma_R} V_1^0\, n_1\, dl \tag{7.69}$$

$$= -[\varepsilon]\int_0^{2\pi} \xi R\cos\theta\, \cos\theta\, R\, d\theta \tag{7.70}$$

$$= -[\varepsilon]\,\xi f \tag{7.71}$$

where f denotes the filling ratio. We are now in a position to obtain the expression of ε_{hom}:

$$\varepsilon_{\text{hom}} = \langle\varepsilon\rangle_Y - \varphi_{1,1} \tag{7.72}$$

$$= \varepsilon_1 f + \varepsilon_2(1-f) + \frac{(\varepsilon_1+\varepsilon_2)^2}{\varepsilon_1+\varepsilon_2} f \tag{7.73}$$

$$= \varepsilon_2 + \frac{2\varepsilon_2(\varepsilon_1-\varepsilon_2)}{\varepsilon_1+\varepsilon_2} f\,. \tag{7.74}$$

In order to illustrate our results and estimate the domain of validity, four figures are represented, showing the rigorous effective permittivity ε_{hom}, the approximate effective permittivity $\varepsilon^0_{\text{hom}}$ together with the lower and the upper bounds, namely the harmonic ($\langle\varepsilon^{-1}\rangle_Y^{-1}$) and the arithmetic mean values ($\langle\varepsilon\rangle_Y$) versus the filling ratio f.

7.4.4.2 Computation of $\varphi_{3,3}$ (spherical case)

$$\varphi_{3,3} = -[\varepsilon]\int_{\Sigma_R} V_3^0\, n_3\, ds \tag{7.75}$$

$$= -[\varepsilon]\int_{\theta=0}^{\pi}\int_{\phi=0}^{2\pi} \xi R\cos\theta\, \cos\theta\, \sin\theta\, R^2\, d\theta d\phi \tag{7.76}$$

$$= -[\varepsilon]\,\xi\, 2\pi R^3 \int_{\theta=0}^{\pi} \cos^2\theta\, \sin\theta\, d\theta \tag{7.77}$$

$$= -[\varepsilon]\,\xi f\,. \tag{7.78}$$

By doing so, the expression (7.71) found in a two-dimensional context still holds for the spherical scatterers!

7.4.5 Closed Formulae for Foliated and Checkerboard-Like Media

In this paragraph, we seek to determine the coefficients of the tensor effective permittivity $\underline{\underline{\varepsilon}}_{\text{hom}}$ in the case where the permittivity ε does

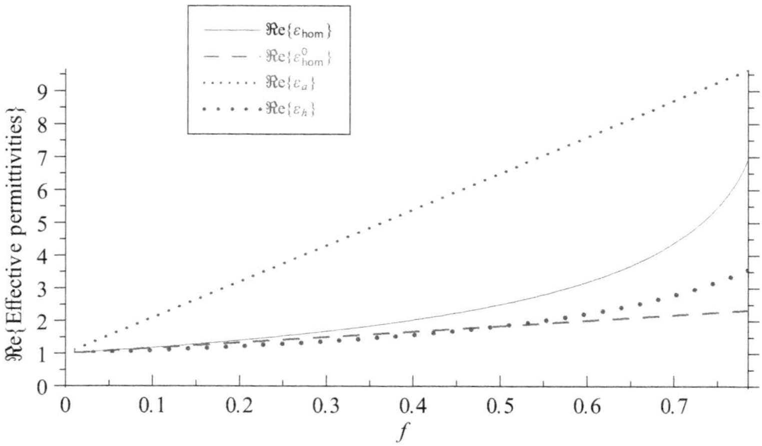

Figure 7.10 Effective permittivities versus the filling ratio f. (a) The circular empty holes $\varepsilon_1 = 1$ are drilled within a bulk of high permittivity $\varepsilon_2 = 12$. (b) The circular scatterers are of high permittivity and $\varepsilon_1 = 12$ are supposed to be surrounded by a vacuum $\varepsilon_2 = 1$.

not depend on a variable; we assume, therefore, that $\frac{\partial \varepsilon}{\partial x_3} = 0$. The problem we are dealing with is then substantially simplified, as the coefficients $\varphi_{i,3}$ and $\varphi_{3,i}$ vanishes. By doing so, the determination of $\underline{\underline{\varepsilon}}_{\text{hom}}$ amounts to looking for the 2×2 matrix $\underline{\underline{e}}_{\text{hom}}$ such that:

$$\underline{\underline{\varepsilon}}_{\text{hom}} = \langle \varepsilon \rangle_Y Id - \begin{pmatrix} \underline{\underline{e}}_{\text{hom}} & 0 \\ & 0 \\ 0 & 0\ 0 \end{pmatrix}, \tag{7.79}$$

where the elements $\underline{\underline{e}}_{\text{hom,i,j}}$ $(i, j \in \{1, 2\}^2)$ are deduced from potentials $V_i(y_1, y_2)$, unique solutions in $H^1_{0,\sharp}(Y^2)$ to

$$\text{div}_y\Big(\varepsilon(y_1, y_2)\big(\mathbf{grad}_y\, V_i + e_i\big)\Big) = 0\,,\ i \in \{1, 2\} \tag{7.80}$$

Despite this simplification, in general, closed formulae cannot be found for such structures. Nevertheless, there are two remarkable exceptions: foliated structures (see Fig. 7.16(a)) and checkerboard (see Fig. 7.16(b)). The first of the two configurations is obviously the simplest. Indeed ε depends only on one variable, y_1, for instance. Equation (7.80) becomes a simple differential equation, namely:

$$\frac{d}{dy_1}\left(\varepsilon(y_1)\frac{dV_1}{dy_1}\right) = -\frac{d\varepsilon}{dy_1} \tag{7.81}$$

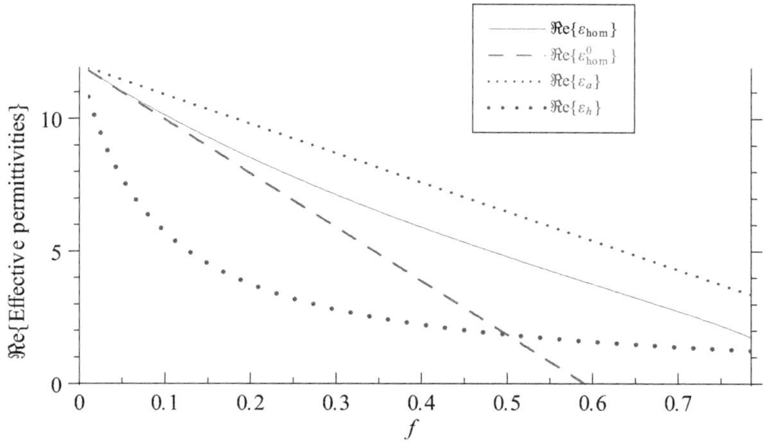

Figure 7.11 Effective permittivities versus the filling ratio f. (a) The circular empty holes $\varepsilon_1 = 1$ are drilled within a bulk of high permittivity $\varepsilon_2 = 12$. (b) The circular scatterers are of high permittivity $\varepsilon_1 = 12$ and are surrounded by a vacuum $\varepsilon_2 = 1$.

which leads to:

$$\varepsilon(y_1)\frac{dV_1}{dy_1} = -\varepsilon + C \tag{7.82}$$

where C is a constant possibly complex that remains to be determined with the aid of the periodicity of V_1. On one hand, we have:

$$\int_0^1 \frac{dV_1}{dy_1}\,dy_1 = V_1(1) - V_1(0) = 0 \tag{7.83}$$

and on the other:

$$\int_0^1 \frac{dV_1}{dy_1}\,dy_1 = -1 + C\int_0^1 \varepsilon^{-1}\,dy_1 \tag{7.84}$$

As a result, C is nothing but the harmonic mean value of ε:

$$C = \left(\int_0^1 \varepsilon^{-1}\,dy_1\right)^{-1} =: \varepsilon_h \tag{7.85}$$

Coming back to Eq. (7.82) and integrating it, we are led to:

$$\varphi_{1,1} = \int_0^1 \varepsilon\frac{dV_1}{dy_1}\,dy_1 \tag{7.86}$$

$$= -\varepsilon_a + \varepsilon_h \tag{7.87}$$

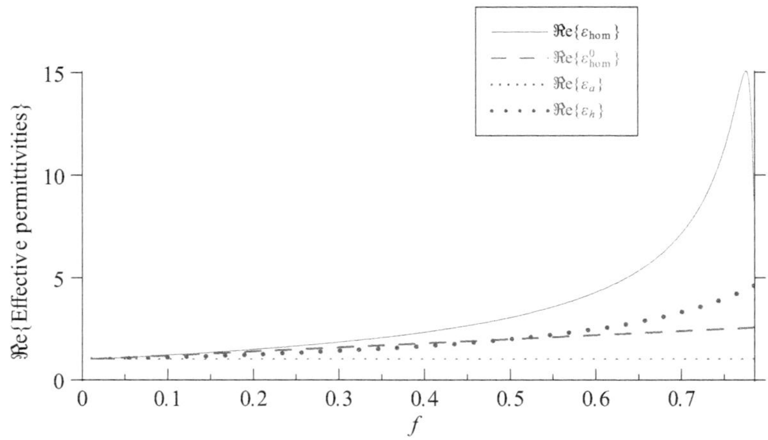

Figure 7.12 Effective permittivities versus the filling ratio f. The circular scatterers are of high complex permittivity $\varepsilon_1 = 1+50\,i$ and are surrounded by a vacuum $\varepsilon_2 = 1$. Note that the real part of ε_{hom} behaves anomalously when the scatterers are in quasi-contact and possesses a maximum around $f = 0.775$, whereas the imaginary part drops abruptly for filling ratio greater than 0.7. (a) Real part of effective permittivities. (b) Imaginary part of effective permittivities.

We then find a well-known result:

$$\underline{\underline{\varepsilon}}_{\text{hom}} = \begin{pmatrix} \varepsilon_h & 0 & 0 \\ 0 & \varepsilon_a & 0 \\ 0 & 0 & \varepsilon_a \end{pmatrix} . \tag{7.88}$$

The results concerning the homogenization of the checkerboard-like structures depicted in Fig. 7.16(b) are based essentially on an intriguing theorem whose statement follows:

Theorem 7.3 (Theorem of duality). *Let* $\underline{\underline{e}}_{hom}(\varepsilon)$ *be the* 2×2 *matrix introduced above. For a given permittivity* ε*, there exists a relation between* $\underline{\underline{e}}_{hom}(\varepsilon)$ *and* $\underline{\underline{e}}_{hom}(\varepsilon^{-1})$*:*

$$\underline{\underline{e}}_{hom}(\varepsilon^{-1}) = \frac{\underline{\underline{e}}_{hom}(\varepsilon)}{\det \underline{\underline{e}}_{hom}(\varepsilon)} . \tag{7.89}$$

The second property that is used for establishing the mixing law is the homogeneity (of first order) of the homogenization process. To

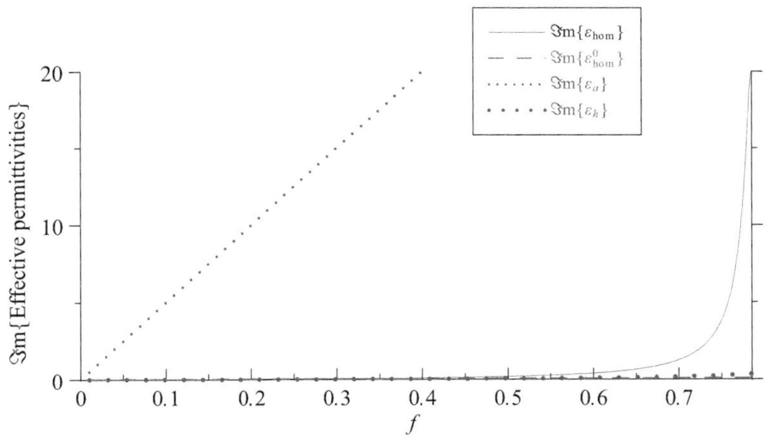

Figure 7.13 Effective permittivities versus the filling ratio f. The circular scatterers are of high complex permittivity $\varepsilon_1 = 1 + 50\,i$ are supposed to be surrounded by a vacuum $\varepsilon_2 = 1$. Note that the real part of ε_{hom} behaves anomalously when the scatterers are in quasi-contact and possesses a maximum around $f = 0.775$, whereas the imaginary part drops abruptly for filling ratio greater than 0.7. (a) Real part of effective permittivities. (b) Imaginary part of effective permittivities.

be clear, for any complex number λ, we have:

$$\underline{\underline{e}}_{hom}(\lambda\varepsilon) = \lambda \underline{\underline{e}}_{hom}(\varepsilon)\,. \tag{7.90}$$

Now if we denote the matrix $\underline{\underline{e}}_{hom}$ for two-phase media by $\underline{\underline{e}}_{hom}([\varepsilon_1, \varepsilon_2])$ and by using the two aforementioned properties, we must have:

$$\underline{\underline{e}}_{hom}([\lambda\varepsilon_1^{-1}, \lambda\varepsilon_2^{-1}]) = \lambda \frac{\underline{\underline{e}}_{hom}([\varepsilon_1, \varepsilon_2])}{\det \underline{\underline{e}}_{hom}([\varepsilon_1, \varepsilon_2])} \tag{7.91}$$

If we chose $\lambda = \varepsilon_1 \varepsilon_2$, we find a relation between the effective properties of a problem and its dual, that is to say, in inverting the permeability of the host with that of the guest.

$$\underline{\underline{e}}_{hom}([\varepsilon_2, \varepsilon_1]) = \varepsilon_1 \varepsilon_2 \frac{\underline{\underline{e}}_{hom}([\varepsilon_1, \varepsilon_2])}{\det \underline{\underline{e}}_{hom}([\varepsilon_1, \varepsilon_2])} \tag{7.92}$$

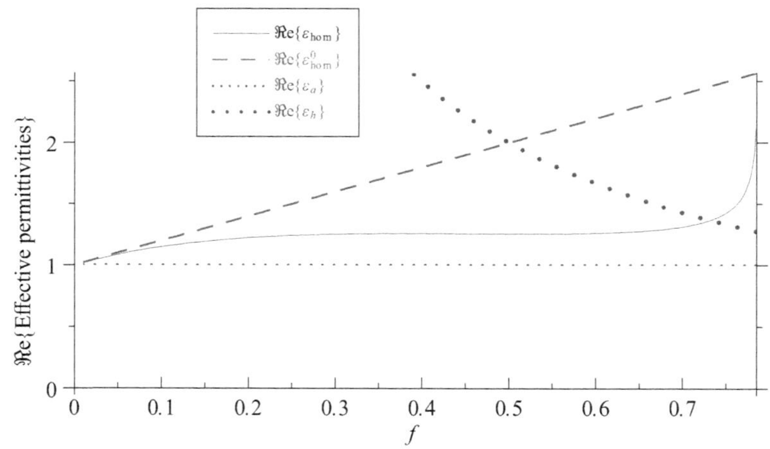

Figure 7.14 Effective permittivities versus the filling ratio f. The circular empty holes $\varepsilon_1 = 1$ are drilled within a bulk of high complex permittivity $\varepsilon_2 = 1 + 50\,i$. (a) Real part of effective permittivities. (b) Imaginary part of effective permittivities.

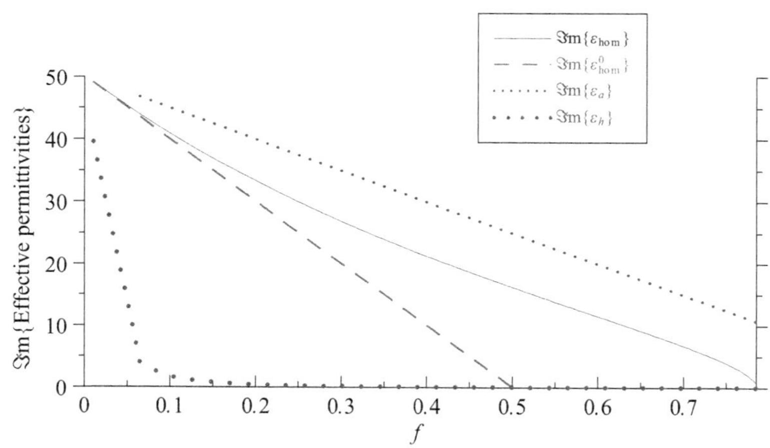

Figure 7.15 Effective permittivities versus the filling ratio f. The circular empty holes $\varepsilon_1 = 1$ are drilled within a bulk of high complex permittivity $\varepsilon_2 = 1 + 50\,i$. (a) Real part of effective permittivities. (b) Imaginary part of effective permittivities.

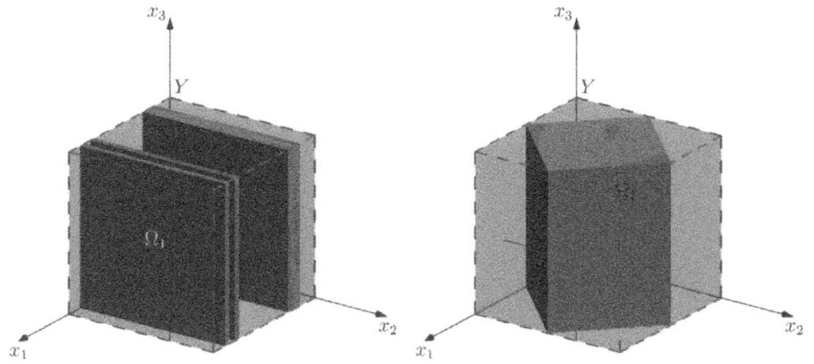

(a) (b)

Figure 7.16 The effective permittivities of these two two-dimensional structures have simple expressions.

Moreover, if the crystal and its dual are indistinguishable, the effective parameters for these two configurations are, therefore, the same.

$$\underline{\underline{e}}_{\text{hom}}([\varepsilon_1, \varepsilon_2]) = \varepsilon_1\,\varepsilon_2 \frac{\underline{\underline{e}}_{\text{hom}}([\varepsilon_1, \varepsilon_2])}{\det \underline{\underline{e}}_{\text{hom}}([\varepsilon_1, \varepsilon_2])} \tag{7.93}$$

For such crystals, if we denote by $e_{1,1}$ and $e_{2,2}$ the diagonal terms and by e_a the off-diagonal term of $\underline{\underline{\varepsilon}}_{\text{hom}}$, these terms are linked by:

$$e_{1,1}\,e_{2,2} - e_a^2 = \varepsilon_1\,\varepsilon_2 \tag{7.94}$$

For instance, when dealing with two-phase foliated materials of same thickness, they satisfy the above relation with:

$$e_{1,1} = \varepsilon_a = (\varepsilon_1 + \varepsilon_2)/2 \;\; e_{2,2} = \varepsilon_h = \left(\left(\frac{1}{\varepsilon_1} + \frac{1}{\varepsilon_2}\right)/2\right)^{-1} \text{ and } e_a = 0. \tag{7.95}$$

At least, if a crystal and the same crystal tilted by a rotation of $\pi/2$ are indistinguishable, the crystal in question is isotropic and

$$\underline{\underline{e}}_{\text{hom}} = e_0\, Id \quad \text{with} \quad e_0 = \sqrt{\varepsilon_1\,\varepsilon_2}\,. \tag{7.96}$$

Table 7.1 Table of the main mixing laws for spherical scatterers: In all cases, the effective permittivity ε_{hom} is given versus ε_1, ε_2, f, where ε_1 is the permittivity of the bulk, ε_2 is the permittivity the sphere, and f is the filling ratio

Name	Effective permittivity
Rayleigh (1892)	$\varepsilon_{\text{hom}}^{\text{R}} = \varepsilon_2 + \frac{3f\varepsilon_2}{\frac{\varepsilon_1+2\varepsilon_2}{\varepsilon_1-\varepsilon_2} - f - 1.305\frac{\varepsilon_1-\varepsilon_2}{\varepsilon_1-4/3\varepsilon_2} f^{10/3}}$
Maxwell Garnett (1904)	$\varepsilon_{\text{hom}}^{\text{MG}} = \varepsilon_2 + 3\varepsilon_2 f \frac{(\varepsilon_1/\varepsilon_2-1)}{\varepsilon_1/\varepsilon_2+2-f(\varepsilon_1/\varepsilon_2-1)}$
Bruggeman (1935)	$(1-f)\frac{\varepsilon_2-\varepsilon_{\text{hom}}^{\text{B}}}{\varepsilon_2+2\varepsilon_{\text{hom}}^{\text{B}}} + f\frac{\varepsilon_1-\varepsilon_{\text{hom}}^{\text{B}}}{\varepsilon_1+2\varepsilon_{\text{hom}}^{\text{B}}} = 0$
Two-scale homogenization	ε_{hom}

Table 7.2 Comparison with the main mixing laws for different filing ratios f. The structure is a simple cubic crystal made of homogeneous spherical dielectric inclusions embedded in a vacuum $\varepsilon_2 = 1$. The permittivity associated with the spheres is $\varepsilon_1 = 3$

f	ε_{hom}	$\varepsilon_{\text{hom}}^{\text{R}}$	$\varepsilon_{\text{hom}}^{\text{MG}}$	$\varepsilon_{\text{hom}}^{\text{B}}$
0.1	1.1241	1.1250	1.1250	1.1288
0.2	1.2600	1.2612	1.2609	1.2758
0.3	1.4097	1.4111	1.4091	1.4410
0.4	1.5778	1.5793	1.5714	1.6238
0.5	1.7738	1.7731	1.7500	1.8229

7.4.6 Numerical Examples and Comparisons

7.4.6.1 Spherical inclusions: comparison with the main mixing laws

We cannot, of course, redo the history of mixing laws and to peruse all of them; a whole monography would not cover it. In this paragraph, our purpose is more modest: The most famous formulae are simply recalled together with the context. It is worth noting that one could loosely distinguish two categories of laws: The first ones come under the theory of random media, whereas the second ones are a subject for the theory of periodic media.

Table 7.3 Comparison with the main mixing laws for different filing ratios f. The structure is a simple cubic crystal made of homogeneous spherical bubbles $\varepsilon_2 = 1$ embedded in a dielectric material $\varepsilon_2 = 3$

f	ε_{hom}	$\varepsilon_{\text{hom}}^{\text{R}}$	$\varepsilon_{\text{hom}}^{\text{MG}}$	$\varepsilon_{\text{hom}}^{\text{B}}$
0.1	2.7483	2.7500	2.7500	2.7462
0.2	2.5118	2.5113	2.5135	2.5000
0.3	2.2866	2.2877	2.2895	2.2629
0.4	2.0701	2.0711	2.0769	2.0365
0.5	1.8590	1.8602	1.8750	1.8229

Table 7.4 Comparison with the main mixing laws for different filing ratios f. The structure is a simple cubic crystal made of homogeneous spherical lossy inclusions embedded in a vacuum $\varepsilon_2 = 1$. The permittivity associated with the spheres is $\varepsilon_1 = 1 + 50 * i$

f	ε_{hom}	$\varepsilon_{\text{hom}}^{\text{R}}$	$\varepsilon_{\text{hom}}^{\text{MG}}$	$\varepsilon_{\text{hom}}^{\text{B}}$
0.1	1.331+i*0.022	1.332+i*0.022	1.332+i*0.022	1.4224+i*0.046
0.2	1.749+i*0.057	1.751+i*0.057	1.746+i*0.056	2.3453+i*0.369
0.3	2.316+i*0.119	2.320+i*0.119	2.276+i*0.109	3.6702+i*2.313
0.4	3.216+i*0.264	3.198+i*0.257	2.980+i*0.198	3.5036+i*6.534
0.5	5.530+i*1.103	4.921+i*0.698	3.957+i*0.355	2.3923+i*12.81

7.4.6.2 Non-spherical inclusions giving rise to isotropic metamaterials

In this short paragraph, we give some examples of mixing laws of non-spherical inclusions (Figs. 7.17–7.19) giving rise nonetheless to isotropic metamaterials due to the cubic symmetry of the crystals at stake.

7.5 Soft Problems: Toward Resonance (Metal–Dielectric Mixing)

Up to now, we have disregarded the dispersive behavior of the media in the computation of the effective characteristics. However,

Table 7.5 Comparison with the main mixing laws for different filing ratios f. The structure is a simple cubic crystal made of homogeneous spherical bubbles $\varepsilon_2 = 1$ embedded in a lossy material $\varepsilon_2 = 1 + 50 * i$

f	ε_{hom}	$\varepsilon_{\text{hom}}^{\text{R}}$	$\varepsilon_{\text{hom}}^{\text{MG}}$	$\varepsilon_{\text{hom}}^{\text{B}}$
0.1	1.061+i*42.84	1.061+i*42.86	1.061+i*42.86	1.0881+i*42.50
0.2	1.100+i*36.32	1.101+i*36.34	1.099+i*36.36	1.2137+i*35.02
0.3	1.129+i*30.28	1.129+i*30.29	1.119+i*30.44	1.4067+i*27.53
0.4	1.157+i*24.50	1.157+i*24.51	1.125+i*25.00	1.7378+i*20.08
0.5	1.214+i*18.69	1.199+i*18.79	1.120+i*20.00	2.3923+i*12.81

as mentioned before, the two-scale homogenization is well suited to the homogenization of frequency-dependent materials: The reader has to bear in mind that the small parameter in our limit process is not the frequency but the period of the crystal. Hence, the frequency becomes a simple parameter in the annex problems and the recipe is simple: it suffices to resume our theory done with the relative permittivity $\varepsilon(\mathbf{r})$ in the annex problems by the permittivity $\varepsilon(\mathbf{r}, \omega)$. The awareness of the time dispersion is, therefore, very natural contrary to the quasi-static approach. In this theory, the small parameter is the frequency itself and the homogenization process consists in studying the Maxwell equation when the frequency tends to zero. After obtaining the effective characteristics by this method, it is awkward to study the effects of the dispersion on mixing laws. Besides, all the examples encountered in the previous sections were performed either with lossless or lossy dielectrics or with ohmic metals. For such materials, the real part of the permittivity remained non-negative as required by the Lax–Millgram theorem, which is crucial in the two-step homogenization techniques. Now it is well known that the metals in optical range are no more ohmic and the real part of the permittivity becomes negative with a possibly weak leakage (see Fig. 7.20).

Nowadays, the achievements in nanotechnology are such that we can consider mixing metal–dielectric at sub-wavelength dimension at least in the infrared. With this mixing seemingly harmless, we are leaving the quiet realm of the mixing addressed in previous paragraphs because as we shall see, this mixing has the ability to not only shift the resonant frequencies but also multiply them.

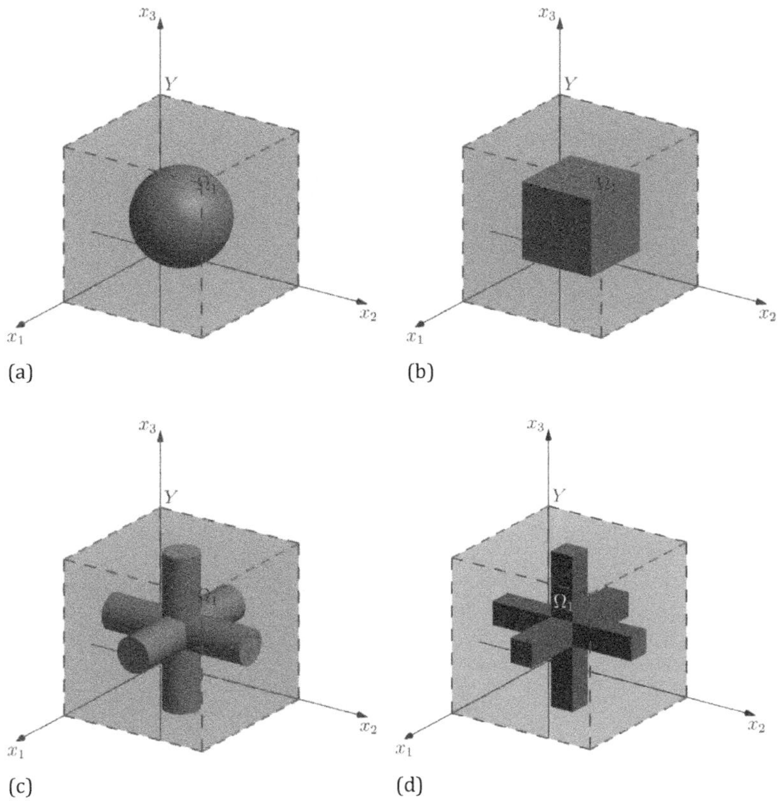

Figure 7.17 Three cases giving rise to isotropic metamaterials. The three scatterers are of the same filling ratio $f = \frac{1}{8}$. (a) Spherical scatterer, (b) cubic scatterer, (c) a three-dimensional circular grid shape scatterer, and (d) a three-dimensional squared grid.

In order to illustrate this remarkable property, let us consider the following two-phase composite problem, which is illustrated diagrammatically in Fig. 7.21. Circular rods made out of dispersive media characterized by the permittivity $\varepsilon_i(\omega)$ are embedded in a matrix, which is considered being dispersless and defined by its relative permittivity ε_e. It is then clear that the tensor effective permittivity does depend on ω:

$$\underline{\underline{\varepsilon}}_{\text{hom}}(\omega) = \text{diag}\left(\varepsilon_H(\omega), \varepsilon_H(\omega), \varepsilon_E(\omega)\right) , \tag{7.97}$$

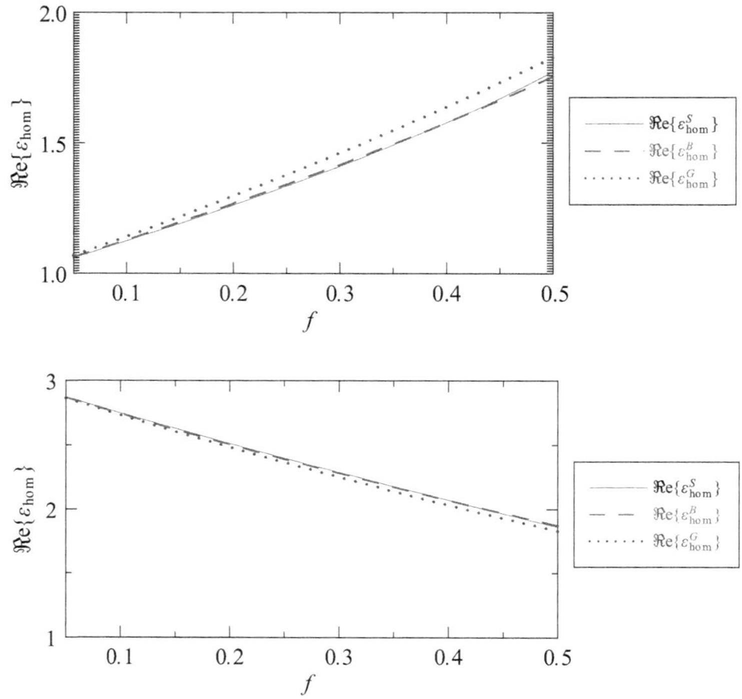

Figure 7.18 The effective permittivities of these two two-dimensional structures have simple expression.

where $\varepsilon_E(\omega)$ is simply given by the mean value of ε in the cell Y:

$$\varepsilon_E(\omega) = \langle \varepsilon(\omega) \rangle_Y = f(\varepsilon_i(\omega) - \varepsilon_e) + \varepsilon_e \,. \tag{7.98}$$

Suppose now that $\varepsilon_i(\omega)$ is given by a Lorentz-like model. Under such a hypothesis, ε_i is solely characterized by two real numbers $\varepsilon_{i,\infty}, \omega_{i,p}$ and a complex number (complex pole) ω_0:

$$\varepsilon_i(\omega) = \varepsilon_{i,\infty} - \frac{\omega_{i,p}^2}{2\Re e\{\omega_0\}} \left(\frac{1}{\omega - \omega_0} - \frac{1}{\omega + \omega_0^*} \right) \tag{7.99}$$

The point now is to find $\underline{\underline{\varepsilon}}_{\text{hom}}$ against ω. Let us start by ε_E:

$$\varepsilon_E(\omega) = \varepsilon_{E,\infty} - \frac{\omega_{E,p}^2}{2\Re e\{\omega_0\}} \left(\frac{1}{\omega - \omega_0} - \frac{1}{\omega + \omega_0^*} \right) \tag{7.100}$$

with

$$\varepsilon_{E,\infty} = \varepsilon_e + (\varepsilon_\infty - \varepsilon_e) f \quad \text{and} \quad \omega_{E,p} = \sqrt{f}\omega_p \,. \tag{7.101}$$

It then turns out that ε_E and ε_i are very similar and in particular the position of the pole in the complex plane ω. Now what of ε_H? In that

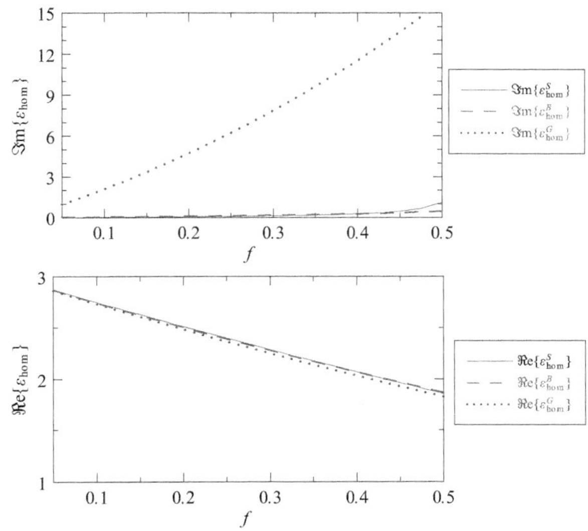

Figure 7.19 The effective permittivities of these two two-dimensional structures have simple expression.

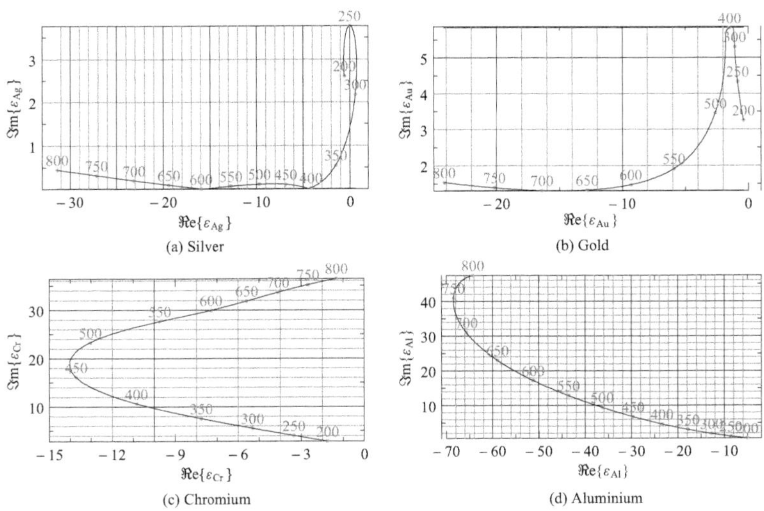

Figure 7.20 The Cole–Cole plot of the complex permittivity of metal in optical range. The data are given in nanometer. Note that the leakage becomes negligible for silver at wavelengths $\lambda_1^{\text{Ag}} = 400$ nm and $\lambda_2^{\text{Ag}} = 600$ nm and for gold at wavelength $\lambda_1^{\text{Au}} = 680$ nm. As for chromium and aluminum, the corresponding wavelengths are found in UV range near 200 nm. Moreover, the real parts of the permittivities are negative in optical range (visible and ultraviolet).

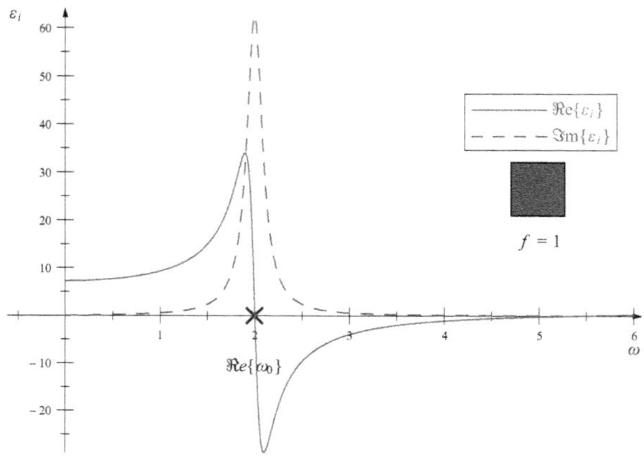

Figure 7.21 The resonant structure is characterized by its permittivity $\varepsilon_i(\omega)$ given in Eq. 7.99 and solely represented by three parameters $\varepsilon_{i,\infty} = 1$, $\omega_{i,p} = 5$ and the complex pole $\omega_0 = 1.9975 - 0.1\,i$.

case, the answer cannot be given by a simple closed formula except for small scatterers (see 7.74):

$$\varepsilon_H(\omega) \approx \varepsilon_H^0(\omega) = \varepsilon_e + \frac{2\varepsilon_e(\varepsilon_i(\omega) - \varepsilon_e)}{\varepsilon_i(\omega) + \varepsilon_e} f \tag{7.102}$$

This equation can be recast in the following manner:

$$\varepsilon_H^0(\omega) = \varepsilon_{H,\infty} - \frac{\omega_{H,p}^2}{2\Re\mathrm{e}\{\omega_1\}}\left(\frac{1}{\omega - \omega_1} - \frac{1}{\omega + \omega_1^*}\right) \tag{7.103}$$

where

$$\varepsilon_{H,\infty} = \varepsilon_e + 2\varepsilon_e f - \frac{4\varepsilon_e^2}{\varepsilon_\infty + \varepsilon_e} f \quad , \quad \omega_{H,p} = \frac{2\,\varepsilon_e}{\varepsilon_e + \varepsilon_\infty}\sqrt{f}\omega_p\,, \tag{7.104}$$

and

$$\omega_1 = \sqrt{\Re\mathrm{e}\{\omega_0\}^2 + \frac{\omega_p^2}{\varepsilon_e + \varepsilon_\infty}} + i\,\Im\mathrm{m}\{\omega_0\}\,. \tag{7.105}$$

Once again, the latter formula looks similar to that found previously except that the pole is shifted from ω_0 to ω_1, which seems insensitive to the filling ratio at least for small scatterers as shown in Figs. 7.22 and 7.23.

Now as f grows, the discrepancy between $\varepsilon_{\mathrm{hom}}$ and $\varepsilon_{\mathrm{hom}}^0$ becomes manifest and the resonance frequency is then shifted toward

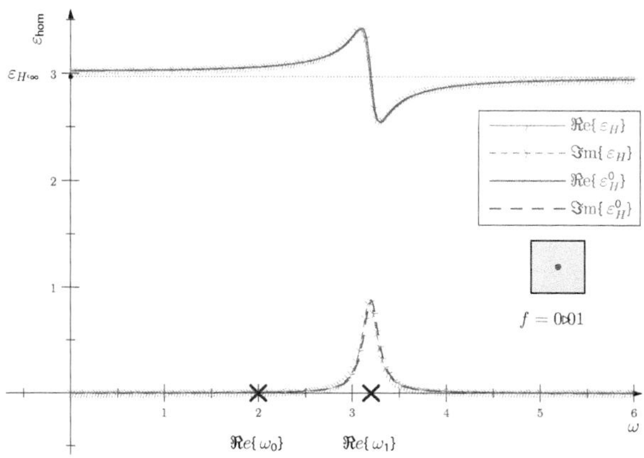

Figure 7.22 Comparison between ε_H and the approximation ε_H^0. A periodic collection of resonant rods made of a material characterized by a permittivity $\varepsilon_i(\omega)$ described by the law (7.99) and diagrammatically illustrated in Fig. 7.21 is embedded in an ideal dispersiveless bulk characterized by its permittivity $\varepsilon_e = 3$. For this very sparse mixture (filling ratio 1%), the permittivity ε_H derived from the two-step homogenization and the approximated permittivity ε_H^0 given in Eq. 7.102 for small scatterers are in perfect agreement.

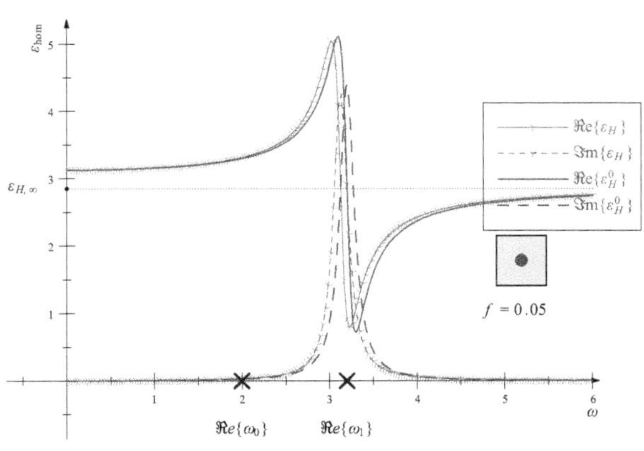

Figure 7.23 Comparison between ε_H and the approximation ε_H^0. A periodic collection of resonant rods made of a material characterized by a permittivity $\varepsilon_i(\omega)$ described by the law (7.99) and diagrammatically illustrated in Fig. 7.21 is embedded in an ideal dispersiveless bulk characterized by its permittivity $\varepsilon_e = 3$. For this sparse mixture (filling ratio 5%), the permittivities ε_H and ε_H^0 are in fair agreement. However, the resonance of ε_H is slightly shifted toward the low frequencies.

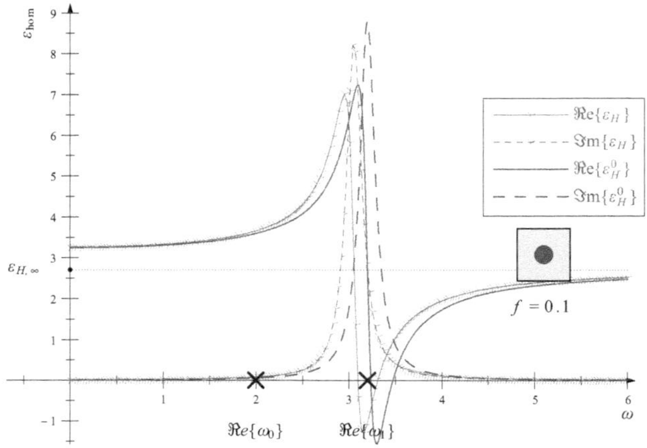

Figure 7.24 Comparison between ε_H and the approximation ε_H^0. For this structure (filling ratio 10%), the two resonances are clearly different.

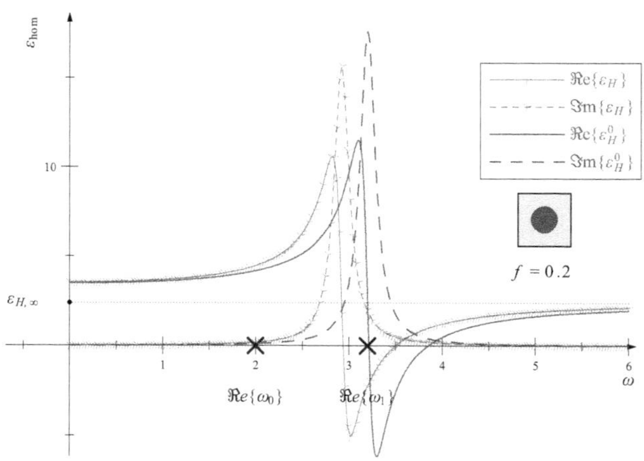

Figure 7.25 Comparison between ε_H and the approximation ε_H^0. The discrepancy is all the more evident as the filling ratio (here 20%) is high.

the resonance frequency of ε_i, namely ω_0: The resonance frequency Ω_1 of the mixture is then filling ratio-dependent.

Surprisingly enough, it turns out that new poles come out between ω_0 and ω_1 as seen in Figs. 7.26 and 7.27.

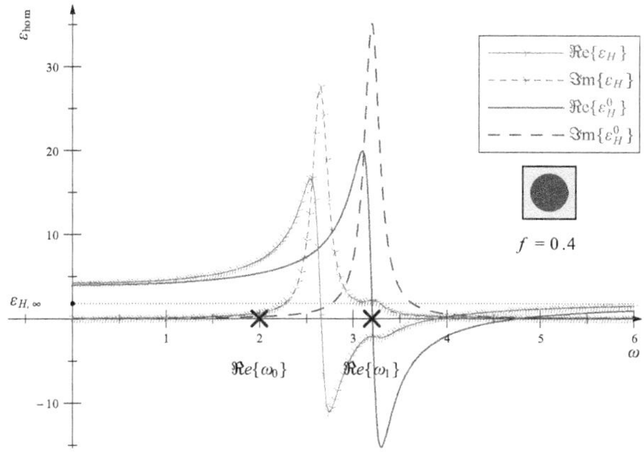

Figure 7.26 Comparison between ε_H and the approximation ε_H^0. For rods of such a size (filling ratio 40%), one supernumerary pole is discernible near ω_1.

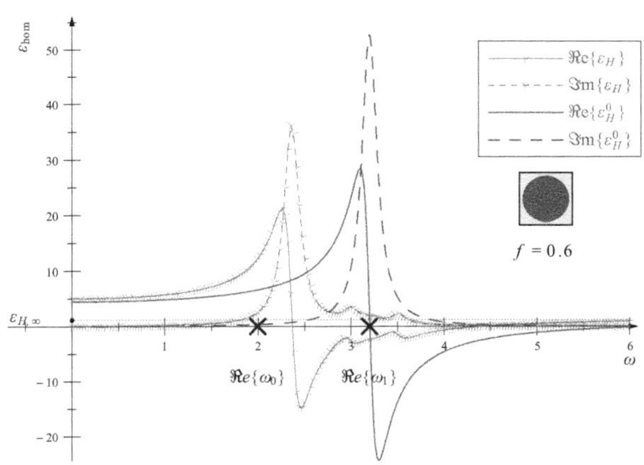

Figure 7.27 Comparison between ε_H and the approximation ε_H^0. For rods of such a size (filling ratio 60%), two supernumerary poles are now visible from both sides of ω_1.

7.6 Tiny Enough to Be Homogeneous?

7.6.1 Introduction

It is commonly admitted that homogenization processes apply when the characteristic lengths of periodically arranged obstacles are "much smaller" than the incident wavelength. The goal of this section is resolutely pragmatical: How small compared to the incident wavelength periodically arranged obstacles should be to form a homogeneous metamaterial? We are now in position to answer this practical question: On the one hand, a general homogenization methodology was derived in this chapter. On the other hand, several accurate numerical methods were introduced in Section 7.4.

Let us consider the following problem: An incident plane wave is illuminating a slab made of a slice of infinite crystal with cubic symmetry, with lattice constants denoted by d_x, d_y, and d_z. This slab has a fixed thickness e equal to an integer (N) multiple of d_z. The FEM, through the rigorous computation of the diffracted field, allows to deduce the transmitted, reflected, and absorbed energy by such an object, as shown in Chapter 6. But as N increases for a fixed thickness e of the slab, it is tempting to assume the complex crystal homogeneous and calculate its homogeneous mesoscopic characteristics, to finally compare the result of the real problem obtained by the FEM to the one of the classical one-dimensional homogeneous slab problem obtained analytically.

In this section, we propose to confront the two approaches for various permittivities ε_1 (inside the obstacle) and ε_2 (background) of the real problem: lossless dielectric, metals, non-symmetric shapes. The study is restricted to ellipsoidal obstacle arranged in a three-dimensional crystal with cubic symmetry ($d_x = d_y = d_z = a$), but could be easily extended to other types of geometries (grids, cubes) and crystal symmetries (diamond, fcc).

7.6.2 Lossless Dielectric

7.6.2.1 Convergence

First, the obstacles are considered to be spherical ($r_x = r_y = r_z = r_0$), filled with a lossless dielectric of permittivity ε_1. Each obstacle occupies a filling fraction 0.35 of the total volume of the unit cell, that is, its radius is set to $0.437\,a$. A bulk slab of thickness $3\,\lambda_0$ and permittivity ε_2 is progressively filled with an integer number N of previously described spheres arranged in a cubic symmetry of lattice constant a. Under these hypotheses, the mass per unit surface of each material remains constant as N increases.

Figure 7.28 shows reflectivity (the specular diffraction efficiency $R_{0,0}$, defined reff gratings, since the periodicity is sub-wavelength) of the structure as N increases from 7 (consider $\lambda_0 = 660$ nm, the diameter of the spheres is $2r_0 = 247.3$ nm) to 171 (diameter 10.2 nm), for different values of ε_1:4, (red stars on Fig. 7.28), 9 (green stars) and 16 (blue stars), keeping ε_2 set to 1. An energy balance criterion was used for each FEM calculation to ensure numerical validity (red circles, green and black crosses). Respectively, the horizontal lines represent the reflectivity of a slab of homogeneous permittivity 1.643 (red line), 2.054 (green line), and 2.286 (blue line). As N increases, the reflectivity of the structured device converges toward the value of the homogeneous slab. Indeed, for values of N above 50 (resp. $N > 40$, resp. $N > 20$), the reflectivity slightly fluctuates above and below the value given by the homogeneous problem for $\varepsilon_1 = 16$ (resp. $\varepsilon_1 = 9$, resp. $\varepsilon_1 = 4$). These different convergences are not surprising and shed light on an important parameter when tackling homogenization: the wavelength inside a material ($\lambda_0/\sqrt{\varepsilon}$). For dielectric materials, even with a relatively high index of refraction ($n = 4$), one can safely consider a material homogeneous (for $\varepsilon_2 = 16$ and $N = 50$, the diameter of the sphere is 34 nm, and the periodicity $a = 6\%\,\lambda_0$); the fluctuations around the expected value mentioned above can be attributed to the discretization of the computation volume into tetrahedrons.

7.6.2.2 Angular response

The angular responses (p-polarization case) confirm the previous result. The angular behavior of the homogeneous slab and the one of the realistic models with 171 spheres are in quite good agreement for the three studied lossless cases $\varepsilon_1 = 4$ (see Fig. 7.29), $\varepsilon_1 = 9$ (see Fig. 7.30), and $\varepsilon_1 = 16$ (see Fig. 7.31).

In lossless cases, even the homogenization scheme presented is a convenient tool to predict the optical response of lossless materials. It should be noted that other homogenization techniques allow to retrieve the homogeneous permittivity with a comparable accuracy

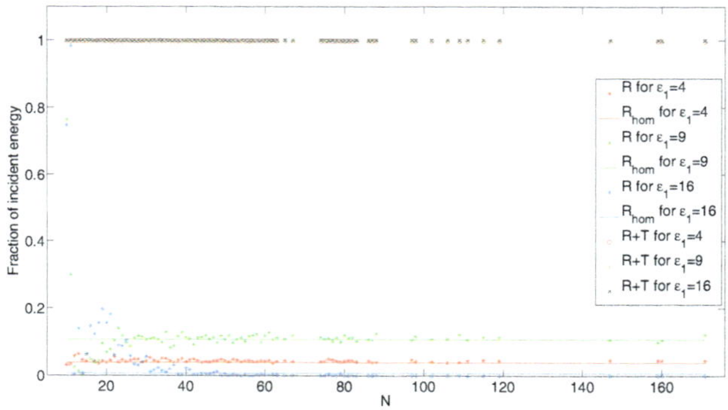

Figure 7.28 Reflection of the 3λ thick slab, as a function of the number of spheres depending on the permittivity ε_1 inside the spheres. The structure is illuminated by a plane wave at normal incidence: $\lambda_0 = 660$ nm and $\theta_0 = \phi_0 = \psi_0 = 0°$. The filling fraction is 0.35 ($r_0 =$). $\varepsilon_1 \in \{4, 9, 16\}$ while $\varepsilon_2 = 1$. This leads to an isotropic $\varepsilon_{\text{hom}} \approx \{1.643, 2.054, 2.286\}$.

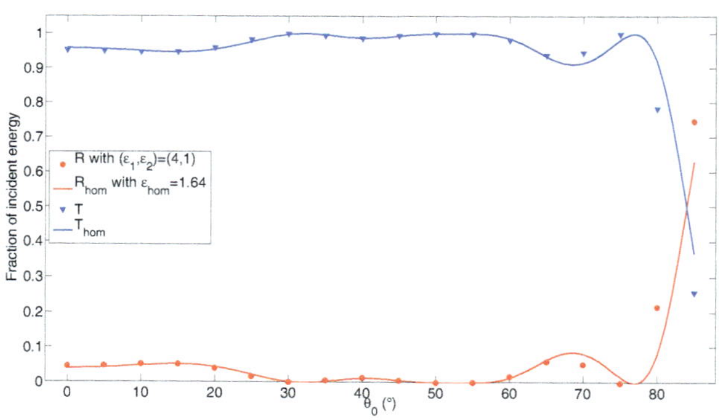

Figure 7.29 Angular response of the 3λ thick slab, consisting of 171 spheres (filling fraction 0.35, corresponding to diameters of 10.2 nm). The incident plane wave is p-polarized, at oblique incidence: $\lambda_0 = 660$ nm, $\theta_0 \in \{0°, 5°, \ldots, 85°\}$ and $\phi_0 = \psi_0 = 0°$. $\varepsilon_1 = 4$ while $\varepsilon_2 = 1$. Leads to an isotropic $\varepsilon_{\text{hom}} \approx 1.643$.

(see column "Rayleigh" in Table 7.6). As previously established, the Rayleigh formula leads to homogeneous characteristics close to those obtained with the proposed homogenization scheme. In the next section, we will see that this is not always the case, even in the case of simple spheres.

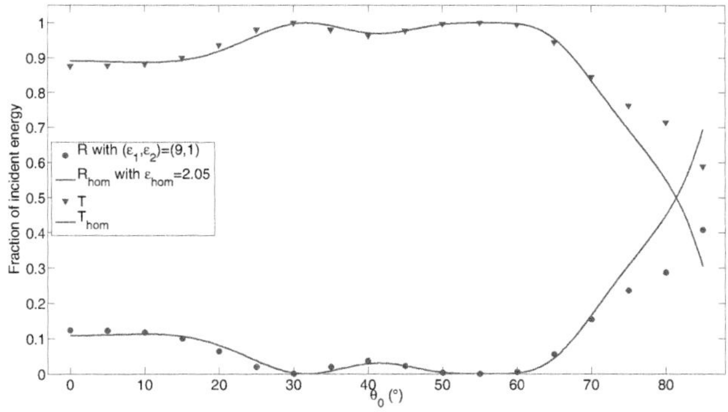

Figure 7.30 Angular response of the 3λ thick slab, consisting of 171 spheres (filling fraction 0.35, corresponding to diameters of 10.2 nm). The incident plane wave is p-polarized, at oblique incidence: $\lambda_0 = 660$ nm, $\theta_0 \in \{0°, 5°, \ldots, 85°\}$ and $\phi_0 = \psi_0 = 0°$. $\varepsilon_1 = 16$ while $\varepsilon_2 = 1$. Leads to an isotropic $\varepsilon_{\text{hom}} \approx 2.054$.

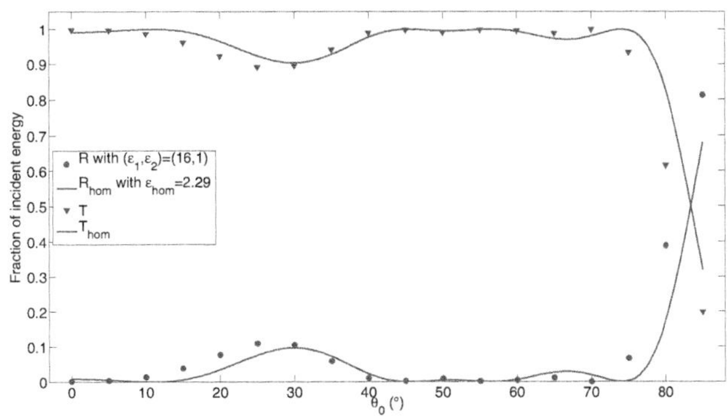

Figure 7.31 Angular response of the 3λ thick slab, consisting of 171 spheres (filling fraction 0.35 corresponding to diameters of 10.2 nm). The incident plane wave is p-polarized, for various off-normal incidence angles: $\lambda_0 = 660$ nm, $\theta_0 \in \{0°, 5°, \ldots, 85°\}$ and $\phi_0 = \psi_0 = 0°$. $\varepsilon_1 = 16$ while $\varepsilon_2 = 1$. Leads to an isotropic $\varepsilon_{\text{hom}} \approx 2.286$.

7.6.3 Metals

7.6.3.1 Convergence

In this section, we are now dealing with lossy materials. The challenging case commonly encountered when tackling plasmonics

Table 7.6 Table of the main mixing laws for spherical scatterers with $f = 0.35$, $\varepsilon_2 = 1$, and $\varepsilon_1 \in \{4, 9, 16\}$

moy ari	moy har	eps0	Clauss–Moss	Rayleigh	HOM
2.050	1.356	1.420	1.288	1.645	1.643
3.800	1.452	1.560	1.121	2.056	2.054
6.250	1.488	1.618	1.067	2.287	2.286
−4.139+	1.601+	1.810+	0.926−	3.622+	3.636+
0.362 i	0.005 i	0.009 i	0.005 i	0.098 i	0.100 i
−6.396+	1.581+	1.773+	0.949−	3.264+	3.2713+
0.157 i	0.001 i	0.002 i	0.001 i	0.015 i	0.016 i

is deliberately selected: a high negative real part of the permittivity with relatively low imaginary part. For instance, in the visible range, let us focus on silver and gold at 660 nm ($\varepsilon_{\text{gold}}^{660\,\text{nm}} =\approx -13.683 + 1.036\,i$ and $\varepsilon_{\text{silver}}^{660\,\text{nm}} = -20.132 + 0.448\,i$). The same convergence study as in the previous section is conducted. The results in reflection, transmission, and absorption obtained with the FEM for gold (and silver) spheres are presented (∗) in Figs. 7.32 and 7.33, respectively. The realistic problem is, again, to be compared to the bulk $3\,\lambda$-thick slab one-dimensional problem with permittivity $\varepsilon_{\text{hom}} = 3.636 + 0.100\,i$ for gold (resp. $3.271 + 0.015\,i$. for silver). For a large number of spheres $N = 171$ (i.e., spheres of diameter 10.2 nm), convergence is not reached in both metallic cases.

7.6.3.2 Angular response

Even if convergence has not been reached, the angular responses are in reasonable agreement with homogeneous material.

In the next section, we reduce the size of the slab for taking into account smaller lattice constants compared to the wavelength.

7.6.3.3 Comparison between the different homogenization approaches

In this case, the spheres are reaching a diameter of 3.6 nm, which is less than 20 times the diameter of gold and silver atoms, probably reaching the classical limit. Table 7.6 shows clearly that the

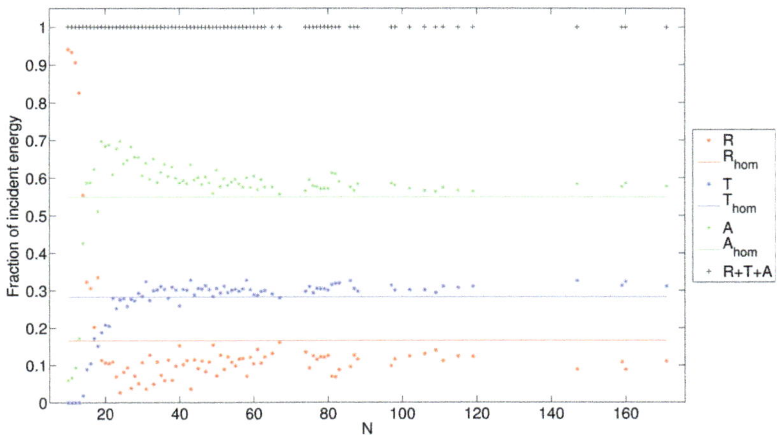

Figure 7.32 Reflection of the 3λ thick slab, as a function of the number of spheres. Incident plane wave: $\lambda_0 = 660\,\text{nm}$ and $\theta_0 = \phi_0 = \psi_0 = 0°$. Filling fraction 0.35. $\varepsilon_1 = \varepsilon_{\text{gold}}(660\,\text{nm}) \approx -13.683 + 1.036\,i$ while $\varepsilon_2 = 1$. Leads to an isotropic lossy (diagonal complex-valued permittivity tensor) $\varepsilon_{\text{hom}} \approx 3.636 + 0.100\,i$.

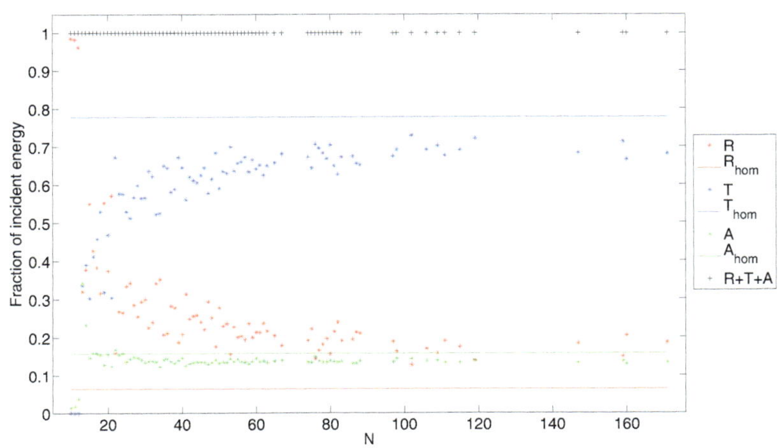

Figure 7.33 Reflection of the 3λ thick slab, as a function of the number of spheres. Incident plane wave: $\lambda_0 = 660\,\text{nm}$ and $\theta_0 = \phi_0 = \psi_0 = 0°$. Filling fraction 0.35. $\varepsilon_1 = \varepsilon_{\text{silver}}(660\,\text{nm}) \approx -20.132 + 0.448\,i$ while $\varepsilon_2 = 1$. Leads to an isotropic lossy $\varepsilon_{\text{hom}} \approx 3.271 + 0.015\,i$.

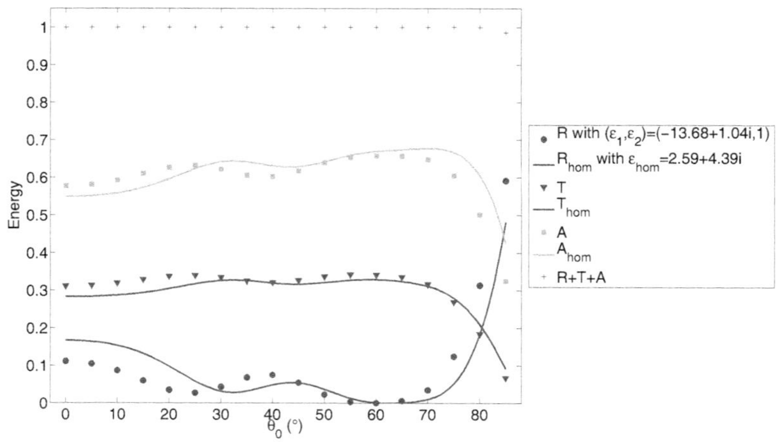

Figure 7.34 Reflection (in red), transmission (in blue), and absorption (in green) of the 3λ thick slab, consisting of 171 spheres (filling fraction 0.35, corresponding to a radius of 5.1 nm). Incident plane wave: $\lambda_0 = 660$ nm, $\theta_0 \in \{0°, 5°, \ldots, 85°\}$, and $\phi_0 = \psi_0 = 0°$. $\varepsilon_1 = \varepsilon_{\text{gold}}(660\text{ nm}) \approx -13.68 + 1.036\,i$ while $\varepsilon_2 = 1$. Leads to an isotropic $\varepsilon_{\text{hom}} \approx 3.636 + 0.100\,i$.

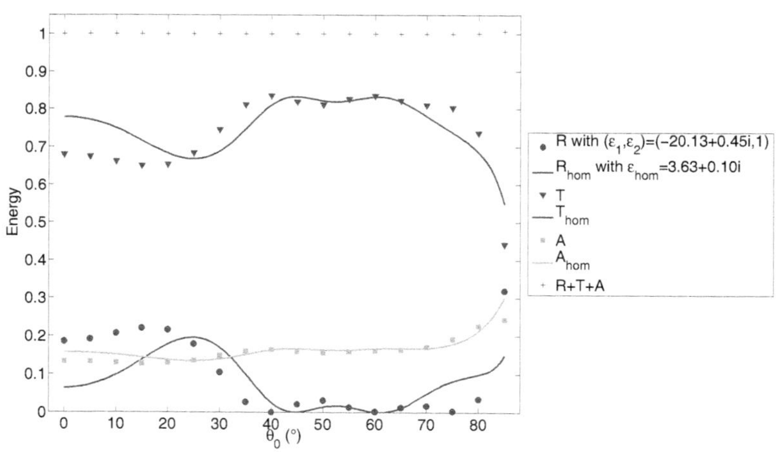

Figure 7.35 Reflection (in red), transmission (in blue), and absorption (in green) of the 3λ thick slab, consisting of 171 spheres (filling fraction 0.35, corresponding to a radius of 5.1 nm). Incident plane wave: $\lambda_0 = 660$ nm, $\theta_0 \in \{0°, 5°, \ldots, 85°\}$, and $\phi_0 = \psi_0 = 0°$. $\varepsilon_1 = \varepsilon_{\text{silver}}(660\text{ nm}) \approx -20.132 + 0.448\,i$ while $\varepsilon_2 = 1$. Leads to an isotropic $\varepsilon_{\text{hom}} \approx 3.271 + 0.015\,i$.

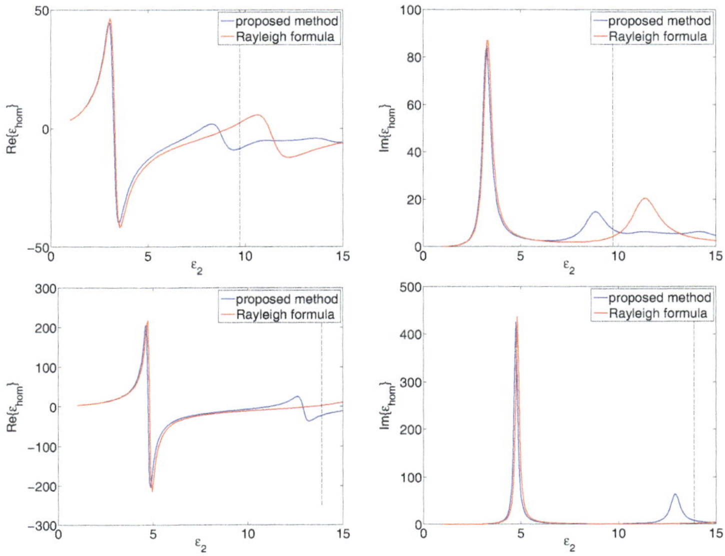

Figure 7.36 Real and imaginary parts of homogeneous permittivities ε_{hom} and $\varepsilon^R_{\text{hom}}$ as functions of the background permittivity ε_2, for spherical inclusions, with cubic symmetry, of filling fraction 0.35 and made of gold or silver ($\varepsilon^{660\,\text{nm}}_{\text{silver}} \approx -20.132 + 0.448\,i$ and $\varepsilon^{660\,\text{nm}}_{\text{gold}} \approx -13.68 + 1.036\,i$). (a) $\text{Re}\{\varepsilon^R_{\text{hom}}\}$ (red solid line) and $\text{Re}\{\varepsilon_{\text{hom}}\}$ (blue solid line) as functions of ε_2 for gold spherical inclusions. (b) $\text{Im}\{\varepsilon^R_{\text{hom}}\}$ (red solid line) and $\text{Im}\{\varepsilon_{\text{hom}}\}$ (blue solid line) as functions of ε_2 for gold spherical inclusions. (c) $\text{Re}\{\varepsilon^R_{\text{hom}}\}$ (red solid line) and $\text{Re}\{\varepsilon_{\text{hom}}\}$ (blue solid line) as functions of ε_2 for silver spherical inclusions. (d) $\text{Im}\{\varepsilon^R_{\text{hom}}\}$ (red solid line) and $\text{Im}\{\varepsilon_{\text{hom}}\}$ (blue solid line) as functions of ε_2 for silver spherical inclusions.

Rayleigh analytical formula for homogenization leads to results close to those obtained using our proposed approach.

Figure 7.36 shows the real and imaginary parts of homogeneous permittivities ε_{hom} and $\varepsilon^R_{\text{hom}}$ as functions of the background permittivity ε_2, for spherical inclusions, with cubic symmetry, of filling fraction 0.35 and made of gold or silver ($\varepsilon^{660\,\text{nm}}_{\text{silver}} \approx -20.132 + 0.448\,i$ and $\varepsilon^{660\,\text{nm}}_{\text{gold}} \approx -13.68 + 1.036\,i$). The two homogenization techniques clearly come to a disagreement at higher values of the background permittivity—for gold (resp. silver) inclusions, $\varepsilon_2 > 7$ (resp. $\varepsilon_2 > 11$). Again let us consider the following numerical experiment: a one wavelength-thick slab, made of gold (resp. silver)

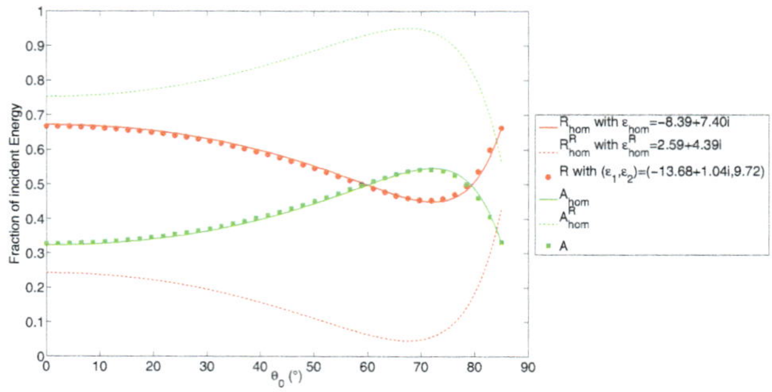

Figure 7.37 Reflection and absorption of the λ thick slab, consisting of 157 gold spheres (filling fraction 0.35, corresponding to a diameter of spheres of 3.7 nm). Incident plane wave: $\lambda_0 = 660\,\text{nm}$, $\theta_0 \in \{0°, 5°, \ldots, 85°\}$ and $\phi_0 = \psi_0 = 0°$. $\varepsilon_1 = \varepsilon_{\text{gold}}^{660\,\text{nm}} \approx -13.683 + 1.036\,i$ while $\varepsilon_2 = 9.72$. Leads to an isotropic $\varepsilon_{\text{hom}}^{\text{R}} \approx 2.585 + 4.389\,i$, while $\varepsilon_{\text{hom}} \approx -8.389 + 7.398\,i$.

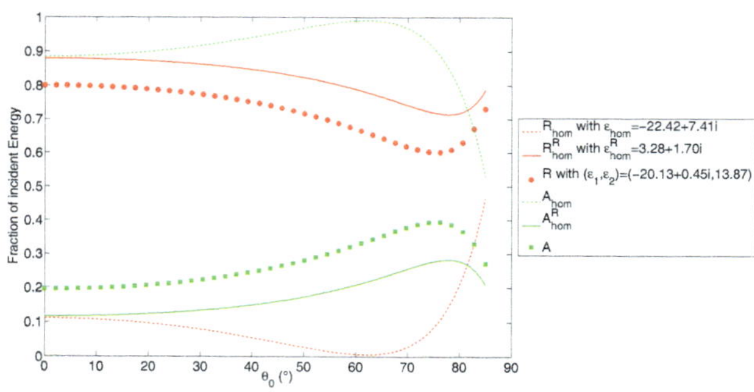

Figure 7.38 Reflection and absorption of the λ thick slab, consisting of 157 silver spheres (filling fraction 0.35, corresponding to a diameter of spheres of 3.7 nm). Incident plane wave: $\lambda_0 = 660\,\text{nm}$, $\theta_0 \in \{0°, 5°, \ldots, 85°\}$ and $\phi_0 = \psi_0 = 0°$. $\varepsilon_1 = \varepsilon_{\text{silver}}^{660\,\text{nm}} \approx -20.132 + 0.448\,i$ while $\varepsilon_2 = 13.87$. Leads to an isotropic $\varepsilon_{\text{hom}}^{\text{R}} \approx 3.281 + 1.703\,i$, while $\varepsilon_{\text{hom}} \approx -22.420 + 7.413\,i$.

spherical inclusions (FF $= 0.35$) arranged in a cubic symmetry and merged in a lattice of permittivity ε_2. Let us set ε_2 to a challenging value, for instance 9.72 (resp. 13.87), as represented by the vertical black dashed lines in Fig. 7.36. Both real structures seem to follow what predicts the proposed homogenization rather than the Rayleigh formula.

References

Allaire, G. (1992). Homogenization and two-scale convergence, *SIAM J. Math. Anal.* **23**, pp. 1482–1518.

Benisty, H., Weisbuch, C., Labilloy, D., and Rattier, M. (2000). Photonic crystals in two-dimensions based on semiconductors: Fabrication, physics and technology, *Appl. Surf. Sci.* **164**, pp. 205–218.

Bruggeman, D. A. G. (1935). Berechnung verschiedener physikalischer konstanten von heterogenen substanzen i. dielektrizitätskonstanten und leitfähigkeiten der mischkörper aus isotropen substanzen, *Annalen der Physik* **24**, pp. 636–664.

Erikson, J. L., Kinderlehrer, D., Kohn, R., and Lions, J.-L. (1986). *Homogenization and Effective Moduli of Materials and Media* (Springer-Verlag, New York).

Felbacq, D. (2000). Anomalous homogeneous behaviour of metallic photonic crystals, *J. Phys. A: Math. Gen.* **33**, pp. 815–821.

Halevi, P., Krokhin, A. A., and Arriaga, J. (1999). Photonic crystal optics and homogenization of 2d periodic composites, *Phys. Rev. Lett.* **82**, p. 719722.

Halevi, P., Krokhin, A. A., and Arriaga, J. (2002). Long-wavelength limit (homogenization) for two-dimensional photonic crystals, *Phys. Rev. Lett.* **65**, p. 115208.

Jikov, V. V., Kozlov, S. M., and Oleinik, O. A. (1994). *Homogenization of Differential Operators and Integral Functionals* (Springer, New York).

Lord Rayleigh Sec. R. S. (1892). LVI. On the influence of obstacles arranged in rectangular order upon the properties of a medium, *Philos. Mag. Series 5*, **34**, pp. 481–502. doi 10.1080/14786449208620364.

Maxwell Garnett, J. (1904). Colours in metal glasses and in metallic films, *Phil. Trans. R. Soc. Lond. A* **203**, p. 385.

McPhedran, R. C., Botten, L. C., and Nicorovici, N. A. (1997). Effective dielectric constant of arrays of elliptical cylinders, *Phys. A* **241**, pp. 173–178.

McPhedran, R. C., Botten, L. C., and Nicorovici, N. A. (2000). Homogenization of composites: Dynamic and static theories, *Phys. B* **279**, pp. 1–3.

Petit, R. and Zolla, F. (1994). The method of fictitious sources as applied to the conical diffraction by a homogeneous rod, *J. Electr. Waves Appl.* **8**, pp. 1–18.

Poborchii, V. V., Tada, T., and Kanayama, T. (2002). Si pillar photonic crystal slab with linear defects: Transmittance and waveguide properties, *Opt. Comm.* **210**, pp. 285–290.

Zolla, F. and Petit, R. (1996). Method of fictitious sources as applied to the electromagnetic diffraction of a plane wave by a grating in conical diffraction mounts, *J. Opt. Soc. Am. A* **13**, pp. 796–802.

Zolla, F., Petit, R., and Cadilhac, M. (1994). Electromagnetic theory of diffraction by a system of parallel rods, *J. Opt. Soc. Am. A* **11**, pp. 1087–1096.

Chapter 8

Stiff Problems: High Contrast Objects

Didier Felbacq,[a] Frédéric Zolla,[b] André Nicolet,[b] and Guy Bouchitté[c]

[a] *Laboratory Charles Coulomb UMR CNRS-UM 5221, University of Montpellier, Place Bataillon, 34095 Montpellier Cedex 05, France*
[b] *Institut FRESNEL, University of Aix-Marseille, Avenue Escadrille Normandie Niemen, 13013 Marseille, France*
[c] *Laboratory IMATH, University of Sud-Toulon-Var, BP 20132, 83957 La Garde Cedex, France*
didier.felbacq@umontpellier.fr

8.1 Introduction: Metallic Metamaterials and Metasurfaces

Metamaterials or their two-dimensional analogues, metasurfaces, are generally made of basic elements containing metallic parts. The latter are generally very thin and conductive, which makes a theoretical analysis rather subtle: One cannot just let the thickness of the elements tend to zero, because the metamaterial would simply disappear in the end. In this chapter, the problem of the effective properties of a wire medium is investigated. This problem was largely discussed in the physical literature (Belov et al., 2003, 2002; Simovski and Belov, 2004). Two different approaches are

Metamaterials Modeling and Design
Edited by Didier Felbacq and Guy Bouchitté

ISBN 978-981-4316-12-5 (Hardcover), 978-1-315-36500-8 (eBook)
www.panstanford.com

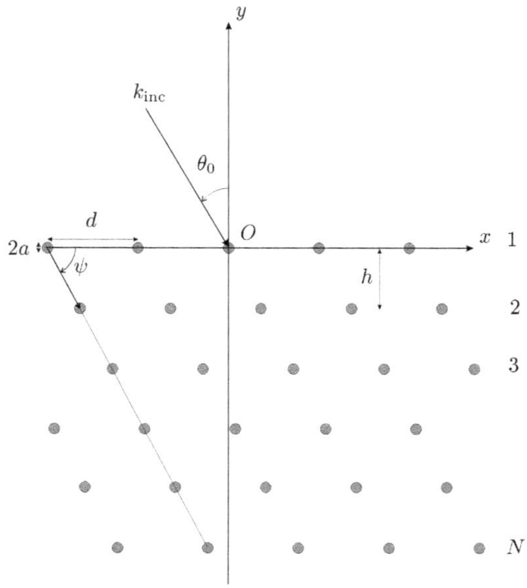

Figure 8.1 Grid composed of thin infinitely metallic circular rods.

proposed here. The first one deals with infinitely long wires and uses explicit calculations: It relies on the use of Green's functions, and it can be used to deal with metasurfaces. The other approach considers the case of wires of finite length and is based on a variational approach. The main difficulty here is to identify correctly the boundary conditions at the end of the wired medium. The variational approach is very interesting because it does not require additional boundary conditions (ABC) as sometimes suggested in the electric engineering community (Maslovski et al., 2010).

8.2 Infinitely Long Wires

A bidimensional metamaterial made of a biperiodic arrangement of infinitely long and very conducting wires is considered. It can be seen as a stack of metasurfaces made of thin rods (radius a) with period d (see Fig. 8.2). It is illuminated by an incident monochromatic wave U^i invariant along z. The incident field

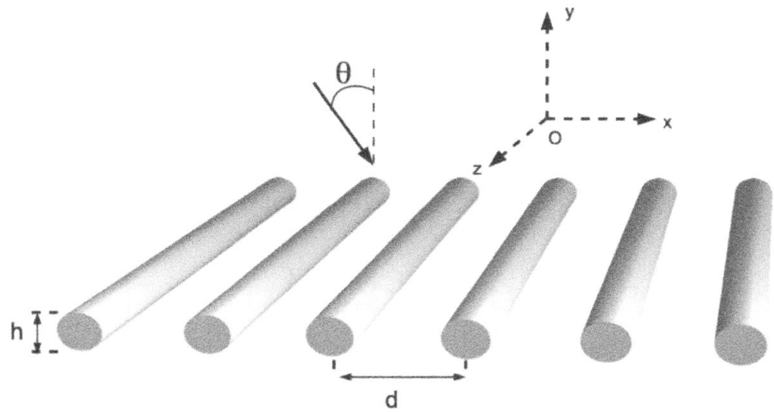

Figure 8.2 Sketch of the structure under study.

is linearly polarized along z. Both polarizations $E_{||}$ and $H_{||}$ are treated simultaneously. The wavelength in vacuum is λ, and the wavenumber is $k_0 = 2\pi/\lambda$. The total field is denoted by U^t, and the diffracted field is $U^d = U^t - U^i$. In order to provide a modal analysis, the incident field is further Fourier transformed along x, so that the incident field is a plane wave $U^i(x, y) = e^{i(\alpha_0 x \pm \beta_0 y)} = \sum_n a_n J_n(k_0 r) e^{in\theta}$, where $\alpha_0 = k_0 \sin\phi$, ϕ is an angle of incidence and $k_0 = \frac{\omega}{c}$. We denote $\beta_0(k_0, \alpha_0) = \sqrt{k_0^2 - \alpha_0^2}$, $\beta_n = \beta_0(k_0, \alpha_n)$ where $\alpha_n = \alpha_0 + nK$ and $K = \frac{2\pi}{d}$. The following result will be used:

Lemma 8.1.

$$\sum_n e^{in\alpha_0 d} H_0^{(1)}(k_0 |\mathbf{r} - nd\mathbf{e}_x|) = \frac{2}{d} \sum_n \frac{1}{\beta_n} e^{i(\alpha_n x + \beta_n |y|)} \tag{8.1}$$

8.2.1 Expression of the Scattered Field

The metamaterial being decomposed as a stack of metasurfaces (or, more simply, gratings), an asymptotic analysis is performed on a single grating. The structure under study (see Fig. 8.2) is made of an infinite number of wires periodically disposed at points $x = pd$, $y = 0$, d is the period and $p \in \mathbb{Z}$. Each scatterer at position M_p is characterized in the frequency domain by a scattering matrix S.

For one wire alone, the incident field gives rise to a diffracted field $U_p^d(M) = \sum_n b_n^p \varphi_n(M - M_p)$, where $\varphi_n(r, \theta) = H_n^{(1)}(k_0|r|)e^{in\theta}$ and $H_n^{(1)}$ is the nth Hankel function of order 1. For the infinite set of scatterers, this gives a diffracted field that reads as:

$$U^d(M) = \sum_{p,n} b_n^p \varphi_n(M - M_p). \tag{8.2}$$

The diffracted field can then be written as a Rayleigh series:

$$U^s(x, y) = \sum_n U_n^s e^{i(k_n x + \beta_n(k_0, \alpha_0)|y|)}, \tag{8.3}$$

where $U_n^s = \frac{2}{d} \sum_p b_n^p \hat{\phi}_n(k + \frac{2\pi p}{d})$, $\hat{}$ denotes the Fourier transform along x and Poisson formula was used. Multiple scattering theory (see Section 6.2) allows to write that for $p = 0$:

$$\widehat{b^0} = (1 - S\Sigma)^{-1} S\widehat{a} \tag{8.4}$$

where $\widehat{b^0} = (\dots, b_{-n}^0, \dots, b_n^0, \dots)^T$ and $\widehat{a} = (\dots, a_{-n}, \dots, a_n, \dots)^T$. The matrix Σ is given by: $\Sigma(k_0, \alpha_0) = \sum_{m \neq 0} e^{i\alpha_0 m} T_{0m}$, thanks to the pseudo-periodicity of the incident field. In this expression, $(T_{0m})_{pq} = e^{i(p-q)\theta_0^m} H_{p-q}^{(1)}(k_0|m|d)$, i.e.,

$$T_{0m} = \begin{pmatrix} \ddots & \vdots & \vdots & \vdots & \dots \\ \dots & H_0(k_0|m|d) & -\epsilon_m H_1(k_0|m|d) & H_2(k_0|m|d) & \dots \\ \dots & \epsilon_m H_1(k_0|m|d) & H_0(k_0|m|d) & -\epsilon_m H_1(k_0|m|d) & \dots \\ \dots & H_2(k_0|m|d) & \epsilon_m H_1(k_0|m|d) & H_0(k_0|m|d) & \dots \\ \dots & \vdots & \vdots & \vdots & \ddots \end{pmatrix}$$

where $\epsilon_m = sign(m)$ (note that $e^{i\theta_0^m} = -sign(m)$. The following series indexed by p appear:

$$\Sigma_p = \sum_{m \neq 0} e^{im\alpha_0} \epsilon_m^p H_p(k_0|m|d). \tag{8.5}$$

and the entries of the matrix $\Sigma(k_0, \alpha_0)$ are: $(\Sigma(k_0, \alpha_0))_{pq} = \Sigma_{p-q}$

8.2.2 Asymptotic Analysis of the Scattered Field

For infinitely circular, infinitely conducting rods, the scattering matrix is diagonal, with entries: $S(k_0) = \mathrm{diag}(S_{-1}(k_0), S_0(k_0), S_1(k_0))$. For circular wires of radius a, the following asymptotics for the scattering matrix hold, where the superscript E or H denotes the corresponding polarization:

$$S_0^E(k_0a) = -\frac{H_0^{(1)}(k_0a)}{J_0(k_0a)} = \frac{1}{-1 - \frac{2i}{\pi}(\gamma + \ln(k_0a/2))} + \mathcal{O}((k_0a)^2) \tag{8.6}$$

$$S_{\pm 1}^E(k_0a) = -\frac{H_1^{(1)}(k_0a)}{J_1(k_0a)} = -\frac{i\pi}{4}(k_0a)^2 + \mathcal{O}((k_0a)^4) \tag{8.7}$$

and

$$S_0^H(k_0a) = -\frac{H_1^{(1)}(k_0a)}{J_1(k_0a)} = -\frac{i\pi}{4}(k_0a)^2 + \mathcal{O}((k_0a)^4) \tag{8.8}$$

$$S_{\pm 1}^H(k_0a) = -\frac{H_1^{'(1)}(k_0a)}{J_1'(k_0a)} = \frac{i\pi}{4}(k_0a)^2 + \mathcal{O}((k_0a)^4) \tag{8.9}$$

This shows that in the regime where $k_0a \ll 1$, the cylinder can be described by a 3×3 scattering matrix (this corresponds to an electric dipole and a magnetic dipole) and the field by three coefficients $b_{-1},\ b_0,\ b_1$. Therefore, only three series are involved: $\Sigma_0,\ \Sigma_1,\ \Sigma_2$. It holds:

$$\Sigma(k_0, \alpha_0) = \begin{pmatrix} \Sigma_0 & -\Sigma_1 & \Sigma_2 \\ \Sigma_1 & \Sigma_0 & -\Sigma_1 \\ \Sigma_2 & \Sigma_1 & \Sigma_0 \end{pmatrix}$$

In the extreme limit ($a \ll d$) where the scatterers are very small as compared to the wavelength and the period, the scattering matrix $S(\omega)$ reduces to a scalar matrix $S_0(\omega)$: The scatterers are thus dipoles with a dipole moment along e_z and the only involved series is Σ_0. The multiple scattering relation (8.4) then becomes:

$$b_0^0(k_0, \alpha_0) = (1 - S_0\Sigma_0)^{-1} S_0\,. \tag{8.10}$$

In the limit $k_0d \ll 1$, the following asymptotics hold:

Proposition 8.1.

$$\begin{aligned} \Sigma_0(k_0, \alpha_0) &\sim -1 - \tfrac{2i}{\pi}\gamma + \tfrac{2i}{\pi}\ln\left(\tfrac{2K}{k_0}\right) + \tfrac{K}{\pi\beta_0} \\ \Sigma_1(k_0, \alpha_0) &\sim \tfrac{\alpha_0}{\pi k_0}\left(-2 + i\tfrac{K}{\beta_0}\right) \\ \Sigma_2(k_0, \alpha_0) &\sim \tfrac{K}{\pi k_0^2}\tfrac{\beta_0^2-\alpha_0^2}{\beta_0} - \tfrac{i}{\pi k_0^2}\left(\tfrac{K^2}{3} - \beta_0^2 + \alpha_0^2\right) \end{aligned} \tag{8.11}$$

Proof. We study in detail the series Σ_0; the other series can be dealt with in the same way. The point is to let $\mathbf{r} = (x, y)$ tend to 0 in Eq. (8.1). First, setting $y = 0$, one obtains:

$$\sum_{n\neq 0} e^{in\alpha_0 d} H_0^{(1)}\left(k_0 \left|x - nd\right|\right) = -H_0^{(1)}(k_0|x|) + \frac{2}{d}\sum_n \frac{1}{\beta_n} e^{i\alpha_n x} \quad (8.12)$$

The term $H_0^{(1)}(k_0|x|)$ and the series are both singular at $x = 0$. However, both singularities compensate. This can be seen by analyzing the asymptotic behavior of the terms of the series: as n tends to infinity, it holds $\beta_n \sim 2i\pi|n|/d$; the series is, therefore, logarithmic:

$$\frac{2}{d}\sum_n \frac{1}{\beta_n} e^{i\alpha_n x} = \frac{2}{d\beta_0} e^{i\alpha_0 x} + \frac{2}{d}\sum_{n\neq 0}\left(\frac{1}{\beta_n} - \frac{d}{2i\pi|n|}\right) e^{i\alpha_n x}$$
$$+ \sum_{n\neq 0} \frac{1}{i\pi|n|} e^{i\alpha_n x}$$

and the last series is equal to: $\frac{2}{i\pi} e^{i\alpha_0 x} \ln[2\sin(\pi x/d)]$. By using the expansion of $H_0^{(1)}(k_0|x|)$ near $x = 0$, we get (γ is the Euler constant):

$$-H_0^{(1)}(k_0|x|) + e^{i\alpha_0 x}\frac{2}{i\pi}\ln[2\sin(\pi x/d)] \sim -1 - \frac{2i}{\pi}\gamma + \frac{2i}{\pi}\ln\left(\frac{2\lambda}{d}\right) \quad (8.13)$$

which shows that:

$$\sum_{n\neq 0} e^{in\alpha_0 d} H_0^{(1)}\left(k_0 \left|n\right| d\right) = -1 - \frac{2i}{\pi}\gamma + \frac{2i}{\pi}\ln\left(\frac{2\lambda}{d}\right)$$
$$+ \frac{2}{d\beta_0} + \frac{2}{d}\sum_{n>0}\left(\frac{1}{\beta_n} + \frac{1}{\beta_{-n}} - \frac{d}{i\pi|n|}\right). \quad (8.14)$$

Finally, when $k_0 d$ is small, the last series is equivalent to $\frac{i[-3+2\cos^2(\theta)]}{\pi^3}\zeta(3)(k_0 d)^2$ and consequently

$$\sum_{n\neq 0} e^{in\alpha_0 d} H_0^{(1)}\left(k_0 \left|n\right| d\right) = -1 - \frac{2i}{\pi}\gamma + \frac{2i}{\pi}\ln\left(\frac{2\pi}{k_0 d}\right) + \frac{2}{d\beta_0} + \mathcal{O}((k_0 d)^2) \quad (8.15)$$

The other expansion can be obtained in a similar fashion (Cabuz, 2007). □

The diffracted field (8.3) simplifies to the following form:

$$\begin{aligned} y > 0: \; U(x, y) &= e^{i(\alpha_0 x - \beta_0 y)} + r\, e^{i(\alpha_0 x + \beta_0 y)} \\ y < 0: \; U(x, y) &= t\, e^{i(\alpha_0 x - \beta_0 y)} \end{aligned} \quad (8.16)$$

where:

$$r = \frac{2}{d}\left(\frac{b_0^0}{\beta_0} + i\frac{\alpha_0}{\beta_0}(b_1^0 + b_{-1}^0) + (b_{-1}^0 - b_1^0)\right),$$

$$t = 1 + \frac{2}{d}\left(\frac{b_0^0}{\beta_0} + i\frac{\alpha_0}{\beta_0}(b_1^0 + b_{-1}^0) - (b_{-1}^0 - b_1^0)\right). \quad (8.17)$$

8.2.3 Asymptotic Form of the Transfer Operator

The scattering matrix of the metasurface is:

$$S_m = \begin{pmatrix} r & t \\ t & r \end{pmatrix}$$

The transfer matrix reads as:

$$T_m = \begin{pmatrix} \frac{t^2 - r^2 + 1}{2t} & \frac{(r+1)^2 - t^2}{2ik} \\ -ik\frac{t^2 - (r-1)^2}{2t} & \frac{t^2 - r^2 + 1}{2t} \end{pmatrix} \quad (8.18)$$

In the homogenization limit of large wavelength $k_0 d \ll 1$, the reflection and transmission coefficients become:

$$r^E = \frac{2b_0^0}{d\beta_0}, \; t^E = 1 + \frac{2b_0^0}{d\beta_0} \quad (8.19)$$

and

$$r^H = \mathcal{O}((k_0 a)^2), \; t^H = \mathcal{O}((k_0 a)^2). \quad (8.20)$$

That is, the scattered magnetic field is null in the first order of $k_0 a$. This means that the medium is basically transparent for $H_{||}$ fields.

From these expressions and (8.15), where the terms that tend to 0 with $k_0 d$ are removed, it is obtained from (8.10):

$$b_0^0 \sim -\frac{\beta_0 d}{2}\frac{1}{1 - i\chi}E^i(0) \quad (8.21)$$

where

$$L = \frac{d}{\pi}\ln\left(\frac{d}{2\pi a}\right), \; \chi = \beta_0 L \quad (8.22)$$

The reflection and transmission coefficients now read:

$$r^E = \frac{-1}{1 - i\chi}, \; t^E = 1 + r = \frac{\chi}{\chi + i} \tag{8.23}$$

and the transfer operators reduce to:

$$\mathcal{T}^E = \begin{pmatrix} 1 & 0 \\ \frac{2}{L} & 1 \end{pmatrix}, \; \mathcal{T}^H = \begin{pmatrix} 1 & 0 \\ 0 & 1 \end{pmatrix} \tag{8.24}$$

The expression (8.23) shows that the reflection coefficient tends to -1 as $k_0 d$ tends to 0. This is quite a striking result if one thinks of the extremely low concentration of material in this scattering experiment. The behavior of such a grid is equivalent to a perfect mirror for $\lambda \gg d$. For instance, for $a/d = 1/1000$ and for $k_0 d = 1/100$, in normal incidence we find a theoretical reflection coefficient worthy of the best mirrors: $R = |r|^2 = 0.999739$. The expression of the reflection coefficient r gives us the critical dimension of the radii of the wires. If $a(k_0)$ is related to k_0 in such a way that

$$\frac{k_0 d \cos\theta_0}{\pi} \log\left(\frac{d}{2\pi a(k_0)}\right) = \Gamma \tag{8.25}$$

where Γ is some constant, i.e.,

$$a(k_0) = \frac{d}{2\pi} e^{-\frac{\pi\Gamma}{k_0 d \cos\theta_0}} \tag{8.26}$$

then the grid, at the limit, behaves neither as vacuum nor as a perfect mirror because the reflection coefficient is equal to $r = \frac{-1}{1+i\Gamma}$.

Let us remark that the evanescent part of the field can be expressed as (Zolla et al., 2006):

$$U^{d,E}_{\text{evan}} \sim \frac{-4i}{\log\left(\frac{d}{2\pi a}\right)} \log\left(1 - e^{iK(x+i|y|)}\right), \tag{8.27}$$

8.2.4 Derivation of the Transfer Matrix and Effective Parameters

In the preceding section, we have derived an explicit expression for the field diffracted by the metasurface and its transfer matrix. In order to model the wire mesh metamaterial, we derive the total dressed T matrix by adding a slab of air below and above the metasurface. The T_h matrix for a homogeneous slab of width $h/2$ of dielectric material with permittivity 1 is given by:

$$T_h = \begin{pmatrix} \cos(\beta_0 h/2) & \beta_0^{-1}\sin(\beta_0 h/2) \\ -\beta_0 \sin(\beta_0 h/2) & \cos(\beta_0 h/2) \end{pmatrix} \tag{8.28}$$

so that the total T-matrix for the grating sandwiched between two slabs of height $h/2$ is:

$$T = \begin{pmatrix} \cos(\beta_0 h) + \frac{1}{\beta_0 L}\sin(\beta_0 h) & \beta_0^{-1}\sin(\beta_0 h) + \frac{2}{\beta_0^2 L}\sin^2(\beta_0 h/2) \\ -\beta_0 \sin(\beta_0 h) + \frac{2}{L}\cos^2(\beta_0 h/2) & \cos(\beta_0 h) + \frac{1}{\beta_0 L}\sin(\beta_0 h) \end{pmatrix} \tag{8.29}$$

A basic layer of the metamaterial has now been characterized. A general device is made of a stack of N such layers. Due to the very weak evanescent fields, the transfer matrix of such a layered device is very well approximated by T^N. From this transfer matrix formulation, explicit formulas for the effective permittivity can be derived (Cabuz, 2007).

The main feature of the low-frequency behavior is the existence of a photonic bandgap down to the null frequency. The rest of this section is devoted to the derivation of an implicit equation for the cut-wavelength λ_c characterizing the edge of the band gap. A band gap is characterized by the fact that $|tr(T)| \geq 2$. When the wavelength is very large with respect to d, the transfer matrix is very near the identity matrix; consequently, the equation for the edge λ_c of the last gap reads as: $tr(T(\lambda, \theta)) = 2$. This amounts to looking for $\beta_0^c = \frac{2\pi}{\lambda_c}\cos\theta$, which is solution to:

$$\cos(\beta_0^c h) + \frac{1}{\beta_0^c L}\sin(\beta_0^c h) = 1. \tag{8.30}$$

Denoting $X = \tan(\beta_0^c h/2)$, the following relation holds:

$$\frac{1 - X^2}{1 + X^2} + \frac{1}{\chi}\frac{2X}{1 + X^2} = 1 \tag{8.31}$$

whose solution is $\beta_0^c L = X$. Finally, the plasmon frequency is given by the following implicit dispersion relation:

$$\beta_0^c L \tan\left(\beta_0^c h/2\right) = 1\ . \tag{8.32}$$

Now letting $x^c = \frac{h\beta_0^c}{2\pi}$, this dimensionless number is the solution to:

$$2\pi x^c \frac{L}{h}\tan\left(\pi x^c\right) = 1\ , \tag{8.33}$$

depending on the parameter $\frac{L}{h}$.

The following asymptotic results hold:

- $\frac{L}{h} \ll 1$ (i.e., $h \gg d$, for realistic size of rods, say $a = 10^{-3}d$). In that case, $x^c \sim 1/2$ and, as a result, the cut-wavelength λ_0^c is given by:

$$\lambda_0^c = \frac{2\pi \cos\theta}{\beta_0^c} \sim 2h\cos\theta \,. \tag{8.34}$$

  which is nothing but the Bragg condition. As a conclusion, $\lambda_0^c \gg d$ (except for grazing incidence), which is compatible with the homogenization process.
- $\frac{L}{h} \gg 1$. In that case, we can make Eq. (8.33) explicit by using the expansion $\tan(\pi x^c) \sim \pi x^c$. We find:

$$x^c = \frac{1}{\sqrt{2}\pi}\sqrt{\frac{h}{L}} \tag{8.35}$$

  and we deduce an approximation $\lambda_0^{c,1}$ of λ_0^c:

$$\lambda_0^{c,1} = \pi\sqrt{2Lh}\cos\theta \tag{8.36}$$

8.3 Finitely Long Wires: The Bed of Nails

8.3.1 Setup of the Problem

In this section, the case of wires of finite length is considered. The structure is called "bed of nails" (Fig. 8.3). An effective medium model based on a two-scale renormalization approach is derived. A detailed discussion of the model's domain of applicability is included.

The structure is a square bi-periodic array of thin wires of length L, radius r, and conductivity σ. The period is d, and the wavelength is λ. The renormalization (depicted in Fig. 8.3) involves a limiting process whereby the three quantities r, d, and $1/\sigma$ tend simultaneously to zero. The period, which is the parameter governing the limiting process, is noted $\eta = d$.

The asymptotics of the other two parameters, σ and r, with respect to η are described by *fixed* parameters κ and γ according

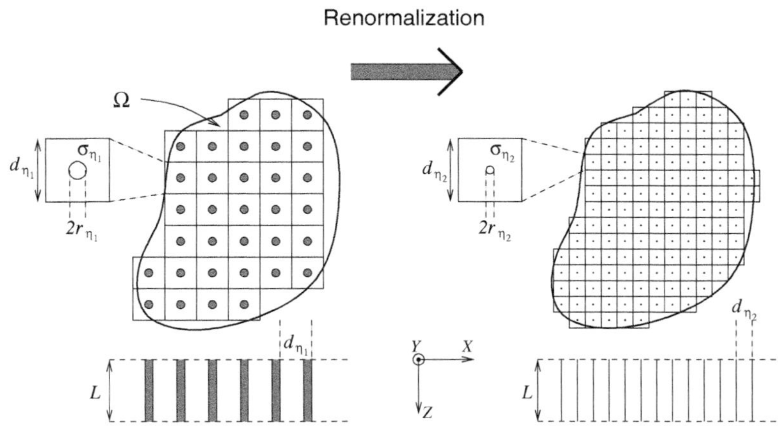

Figure 8.3 The bed-of-nails structure and the renormalization process. The conducting fibers occupy a region $\Omega \subseteq \mathbb{R}^2$, are oriented in the z direction, and the structure is periodic in the xy plane. Two renormalized structures are shown, corresponding to η_1 and η_2, respectively, with $\eta_1 > \eta_2$, $d_{\eta 1} > d_{\eta 2}$, $\sigma_{\eta 1} < \sigma_{\eta 2}$ and $r_{\eta 1}/d_{\eta 1} > r_{\eta 2}/d_{\eta 2}$. The physical structure corresponds by definition to $\eta = 1$: $d_{\eta=1} = d$. The length L and the wavelength λ remain fixed.

to the following relations:

$$\kappa = \frac{\pi r_\eta^2 \sigma_\eta}{\varepsilon_0 \omega \eta^2} \tag{8.37}$$

$$\frac{1}{\gamma} = \eta^2 \log\left(\frac{r_\eta}{\eta}\right) \tag{8.38}$$

where ω is the angular frequency of the electromagnetic field. In other words, the conductivity is renormalized inversely to the fill factor $\theta_\eta = \frac{\pi r_\eta^2}{\eta^2}$, while the radius is renormalized such that the expression $\eta^2 \log(\frac{r_\eta}{\eta})$ remains constant.

These expressions have simple intuitive interpretations. The first requires the current density to remain constant during the renormalization. The parameter κ is the volume average of the imaginary part of the permittivity. Also recall that the *static* admittance per unit length of a circular wire is given by

$$Y_{\text{wire}} = \pi r^2 \sigma \tag{8.39}$$

and that the number of wires per unit area is given by $1/\eta^2$. The second expression requires the average internal capacitance of the wires to remain constant during renormalization. This feature is known to be essential for their asymptotic behavior (see, for instance Refs. (Pendry et al., 1996, 1998)). The essential quantities in the rescaling process are the geometric quantities r_η, η, the material quantity σ_η, and the field quantities $\mathscr{E}_\eta$ and $\mathscr{H}_\eta$. To these, we add a quantity characterizing the all-important electric field in the wires. This is noted as F_η; it is nonzero only *inside* the wires and is given by

$$\mathbf{F}_\eta = \frac{\kappa}{\theta_\eta}\mathscr{E}_\eta = \frac{\sigma_\eta}{\varepsilon_0 \omega}\mathscr{E}_\eta.$$

$\mathbf{F}_\eta$ has the units of electric field, and in the microscopic, inhomogeneous picture, it is proportional to the current density. In the macroscopic, homogeneous picture, however, it will correspond to the polarization density $\mathbf{P}$. More precisely, it holds $\lim_{\eta\to 0}\mathbf{F}_\eta = \mathbf{P}/i\varepsilon_0$.

In the limit $\eta \to 0$, the fields converge (in a precise sense described in Ref. (Bouchitté and Felbacq, 2006)) to the unique solution to the following system:

$$\begin{cases} \mathbf{curl}\,\mathscr{E} & = i\omega\mu_0\mathscr{H} \\ \mathbf{curl}\,\mathscr{H} & = -i\omega\varepsilon_0(\mathscr{E} + \frac{\mathbf{P}}{\varepsilon_0}\hat{z}) \\ \frac{\partial^2 \mathbf{P}_z}{\partial z^2} + \left(k_0^2 + \frac{2i\pi\gamma}{\kappa}\right)\mathbf{P}_z & = -2\pi\gamma\varepsilon_0\mathscr{E}_z,\ z \in [-L/2, L/2] \\ \frac{\partial \mathbf{P}_z}{\partial z} & = 0,\ z \in \{-L/2, L/2\} \end{cases} \tag{8.40}$$

All field quantities above are effective, homogeneous quantities, which have meaning when the wires have been replaced with a homogeneous effective medium with an electric polarization density equal to $\mathbf{P}$. The equation that gives $\mathbf{P}$ is an inhomogeneous Helmholtz equation where the source term is given by the z component of the electric field $\mathscr{E}_z$. The polarization satisfies Neumann conditions at the upper and lower interfaces of the slab. It is not, in general, continuous there because Maxwell's equations impose the continuity of the normal component of the displacement field $\mathscr{D} \equiv \varepsilon_0\mathscr{E} + \mathbf{P}$; consequently, any jump in $\mathscr{E}$ must be canceled by an equivalent jump in $\mathbf{P}/\varepsilon_0$. The dependence of $\mathbf{P}$ on $\mathscr{E}$, i.e., the constitutive relation, takes the form of an integral. The polarization field has the form

$$\mathbf{P}(x, z_0) = -2\pi\gamma\varepsilon_0 \int_{-L/2}^{L/2} g(z, z_0)\mathscr{E}_z(x, z)dz \tag{8.41}$$

where $g(z, z_0)$ is the Green's function of the Helmholtz operator on the bounded domain $z \in \left(-\frac{L}{2}, \frac{L}{2}\right)$. It takes the form (see Appendix A)

$$g(z, z_0) = \frac{1}{K \sin(KL)} \cos\left[K\left(z_< + \frac{L}{2}\right)\right] \cos\left[K\left(z_> - \frac{L}{2}\right)\right]$$

where $K^2 = k_0^2 + \frac{2i\pi\gamma}{\kappa}$, $z_< = \min(z, z_0)$, and $z_> = \max(z, z_0)$. Relation 8.41 is a non-local constitutive relation because the value of the polarization field at a position z_0 depends on values of the electric field at positions different from z_0.

When the imaginary part of K is large, the integral above drops off quickly. In the limit of small conductivity (and hence small κ), the polarization becomes local for sufficiently large wavelengths. In the opposite limit, for infinite conductivity and infinitely long wires, the integral covers all space (in the z direction) and the material is non-local, even in the long-wavelength regime. In fact, this can be seen immediately by performing a Fourier transform on the third equation of system (8.40) (with $\kappa \to \infty$):

$$\widehat{\mathbf{P}}_z = \frac{-2\pi\gamma\varepsilon_0}{k_z^2 - k_0^2} \widehat{\mathbf{E}}_z$$

which gives

$$\varepsilon = 1 + \frac{2\pi\gamma}{k_0^2 - k_z^2}$$

This is consistent with the findings of Belov et al. (Belov et al., 2003, 2002; Simovski and Belov, 2004).

Until now, the discussion has been independent of the actual shape of the domain Ω (Fig. 8.3). From this point on, however, for purposes of illustration, we specialize to the case $\Omega = \mathbb{R}^2$, which is an infinite two-dimensional bed of nails, of thickness L, period d, wire radius r, and conductivity σ. The effective medium is, therefore, a homogeneous slab parallel to the xy plane and of thickness L.

8.3.2 Numerical Results

The homogeneous model is tested by comparing it with three-dimensional full vector simulations of the structure. The reflection, transmission, and absorption coefficients and the current distribution of the homogeneous problem are compared with those

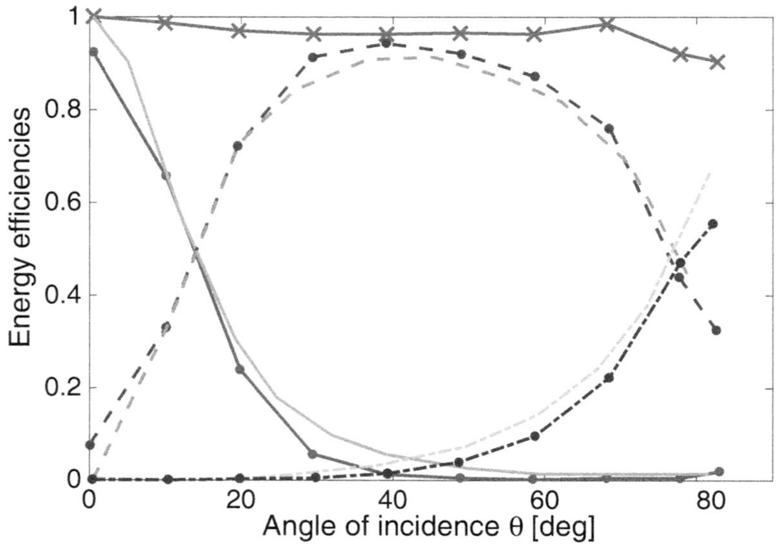

Figure 8.4 Transmission (solid), reflection (dot-dashed), and absorption (dashed) efficiency curves comparing the finite element solution (dot markers) and the effective medium solution (no markers) as a function of angle of incidence. The structure has a conductivity $\sigma = 8\ (\Omega\text{m})^{-1}$, period $d = 0.01$ m, and dimensionless parameters $L/d = 120$, $\lambda/d = 20$, $r/d = 0.1$, and $\delta/d = 4.6$. Computational constraints forced us to use a very coarse mesh, which explains the approximate nature of the energy conservation ($\times$ markers) of the finite element model.

of the original bed-of-nails metamaterial. The solution to the homogeneous problem is obtained by integrating system 8.40 as described in Appendix B.

The three-dimensional full vector simulations of the bed-of-nails metamaterial were done using the COMSOL Multiphysics® finite element method (Dular et al., 1995) software package. The periodicity was implemented using Floquet–Bloch conditions (Nicolet et al., 2004) in the two periodic directions (x and y), and absorbing perfectly matched layers (Agha et al., 2008) in the positive and negative z directions. The linearity of the materials in the structure was used to treat the incident field as a localized

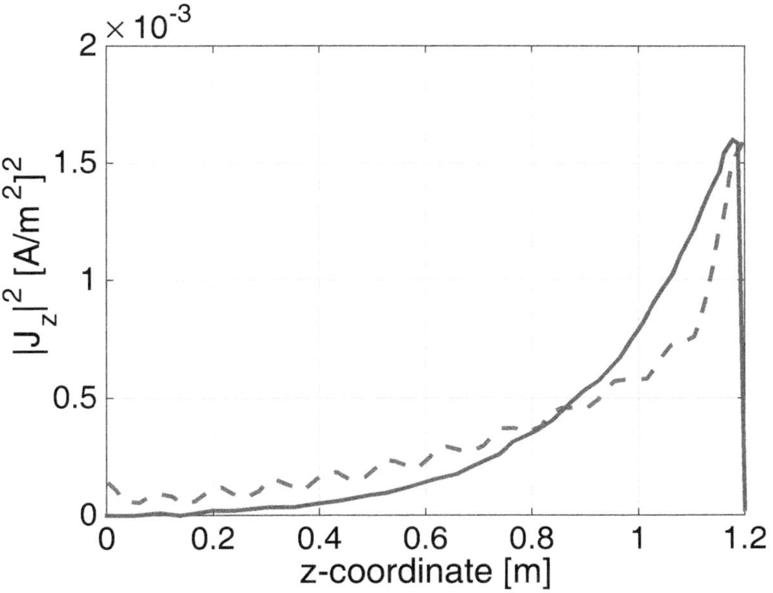

Figure 8.5 Square of the current density for the effective medium solution to Eqs. B.1 (dashed) and the finite element solution (solid) as a function of position within the bed-of-nails structure (which is positioned in $z \in (0, L)$). The structure is the same as in Fig. 8.4, illuminated at an angle of incidence $\theta = 40°$ from the top.

source within the obstacle, as detailed in Ref. (Demésy et al., 2007; 2009).

Figures 8.4 and 8.5 show good agreement between the effective medium model and the finite element simulation. Note that the current density behavior near the boundaries differs between the effective medium model and the finite element model. This is due to the fact that in the macroscopic, homogeneous scenario, one speaks of a polarization field obeying Neumann boundary conditions, as discussed earlier. In the microscopic scenario, however, there is a free conductor carrying current induced by an external electric field. Since in the geometry at the given wavelength, the capacitance of the wire endpoints is very small, the accumulation of charge will be correspondingly small, leading to an almost continuous normal component of the electric field (and, therefore, also current). Numerically, it seems as if the current goes to zero at the wire endpoints, even though this is not strictly exact. Nevertheless, since in the homogeneous limit the boundary condition of the current is of Neumann type, the convergence of the renormalization process is clearly non-uniform near the boundaries.

In the numerical simulations, advantage was taken of the fact that the structures under consideration have $r \ll d$ and $\delta \gg r$. Such thin conducting structures can be simulated very efficiently as lines of zero thickness (i.e., *edges*, in the finite element formulation) carying current and exhibiting an equivalent *linear impedance*. This approach gives excellent results with a fraction of the computing power required for a finite element meshing of thin long circular wires.

For instance, Figs. 8.6 and 8.7 show the results of calculations for a structure of Toray T300® carbon fibers with a conductivity of $\sigma = 5.89 \cdot 10^4 \ (\Omega\text{m})^{-1}$ and a radius of 3.5 microns. The wires have an aspect ratio $L/r = 2.28 \times 10^5$, which is far beyond what would have been accessible by meshing the interior of the wires. The finite element model of Fig. 8.4 (curves with markers), in which the interior of the wires is meshed, is a problem with approximately 2.8 million degrees of freedom, which requires at least 42 GB of available RAM to solve. By comparison, the model of Fig. 8.6 (curves with markers), in which the wires are modeled as current-carrying *edges*, is a problem of approximately 62,000 degrees of freedom, which requires less than 1 GB of available RAM.

Figures 8.4–8.7 illustrate the behavior typical of the model. The agreement remains good up to high incidence angles, and over a large wavelength domain (Fig. 8.9). The structure is transparent in normal incidence. For increasingly oblique angles of incidence, the absorption increases more or less gradually, depending on the thickness L. The reflection is generally low, though it increases when approaching grazing incidence. The low reflection may be explained by the small radii of the wires: Their extremities have low capacitance, hence they exhibit very little charge accumulation, leading to an almost continuous normal component of the electric field. Certain configurations exhibit very low reflection for almost all angles of incidence (see Figs. 8.8 and 8.9 around $\lambda = 1.2$ m).

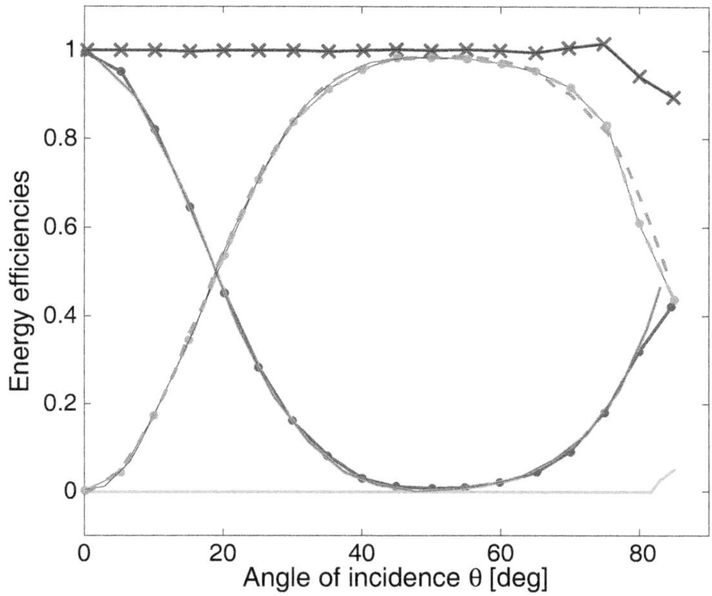

Figure 8.6 Transmission (solid), reflection (dot-dashed), and absorption (dashed) efficiency curves comparing the finite element solution (dot markers) and the effective medium (no markers) as a function of angle of incidence. The wire conductivity is that of Toray T300® carbon fibers $\sigma = 5.89 \cdot 10^4\ (\Omega\mathrm{m})^{-1}$. The structure has period $d = 0.01$ m, and dimensionless parameters $L/d = 80$, $\lambda/d = 20$, $r/d = 3.5 \cdot 10^{-4}$, and $\delta/r = 15$. Energy conservation of the finite element model ($\times$ markers) is respected to within better than 1% for most angles of incidence. The departure around 80° is explained by the poor performance of the PML absorbing layers when close to grazing incidence.

The current density decreases roughly exponentially within the structure due to absorption.

8.3.3 Domain of Validity

The boundaries of the domain of validity of the model are given by four dimensionless parameters: the ratio δ/r of the skin depth to the radius in the wires, the ratio L/d of the wire length to the period, the ratio λ/d of the wavelength to the period, and the ratio r/d of the wire radius to the period.

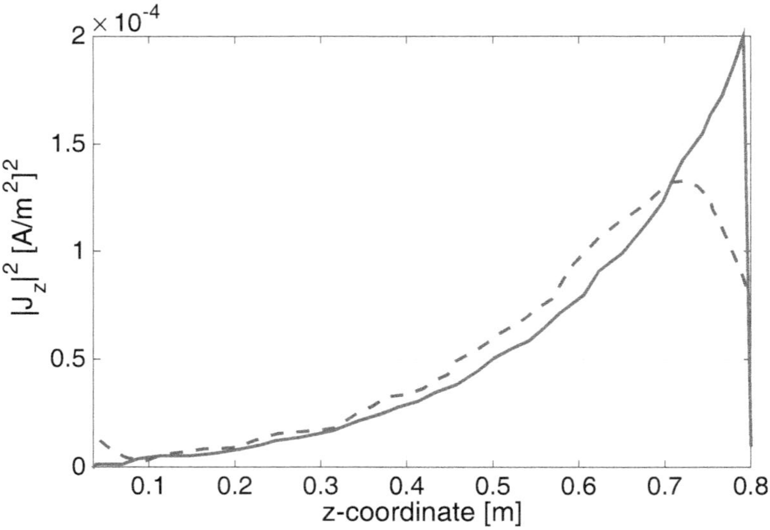

Figure 8.7 Square of the current density for the effective medium model (dashed) and the finite element solution (solid) as a function of position within the slab (which is positioned in $z \in (0, L)$). The structure is the same as in Fig. 8.6, illuminated at an angle of incidence $\theta = 40^\circ$ from the top. Note that the surface areas under the two curves (in this figure as well as Fig. 8.5) are the same because they are proportional to the Joule dissipation rates, which are seen to be equal from Fig. 8.6 (and Fig. 8.4) at the given angle of incidence.

The skin depth must be larger than the radius, due to the fact that the impedance used in defining κ (Eq. 8.39) is the static impedance, which differs from the quasi-static value by an imaginary inductive term $i\omega\mu/8\pi$ (see, for instance, Ref. (Ramo et al., 1994)). Requiring this term to be negligible is equivalent to requiring that $\delta^2/r^2 \gg 1$. Moreover, in the rescaling process, the skin depth/radius ratio is given by

$$\frac{\delta_\eta}{r_\eta} = \frac{\lambda}{\eta}\sqrt{\frac{1}{2\pi\kappa}}.$$

Since η approaches zero in the rescaling process, it is natural to expect the homogeneous model to be valid when the skin depth is large compared to the radius.

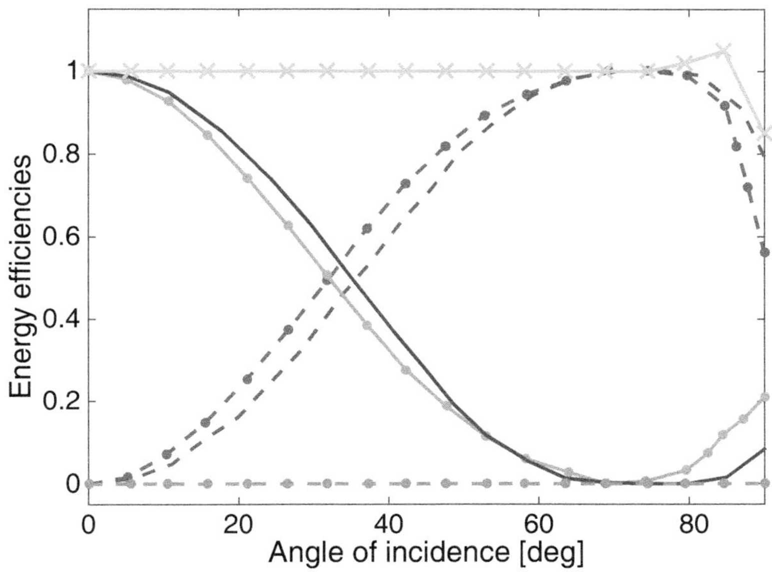

Figure 8.8 Transmission (solid), reflection (dot-dashed), and absorption (dashed) efficiency curves comparing the finite element solution (dot markers) and the effective medium (no markers) as a function of angle of incidence. The structure has a conductivity $\sigma = 1000\ (\Omega\text{m})^{-1}$, period $d = 0.01$ m, and dimensionless parameters $L/d = 50$, $\lambda/d = 8$, $r/d = 0.002$, and $\delta/d = 13$. The reflection remains low for angles of incidence of up to 80° even as the Joule absorption reaches almost 100% for $\theta > 60°$. Energy conservation is indicated by the × markers.

In addition, recall that the definition of γ in Eq. 8.38 fixes the capacitance of the wires to the value for thin, long wires. Consequently, the model is expected to hold for large L/d and for small r/d. To these, one should add the general requirement for all effective medium models: The wavelength must be large compared to the period.

Our study has made it possible to broadly determine the boundaries of the domain of applicability of the effective medium model. Roughly, one must have $\lambda/d \gtrapprox 7-12$, $\delta/r \gtrapprox 4-8$, $L/d \gtrapprox 20-30$, and $r/d \lessapprox 10$. The numerical exploration of the parameter space suggests that the skin depth/radius ratio is often the main limiting factor, particularly when considering highly conducting wires.

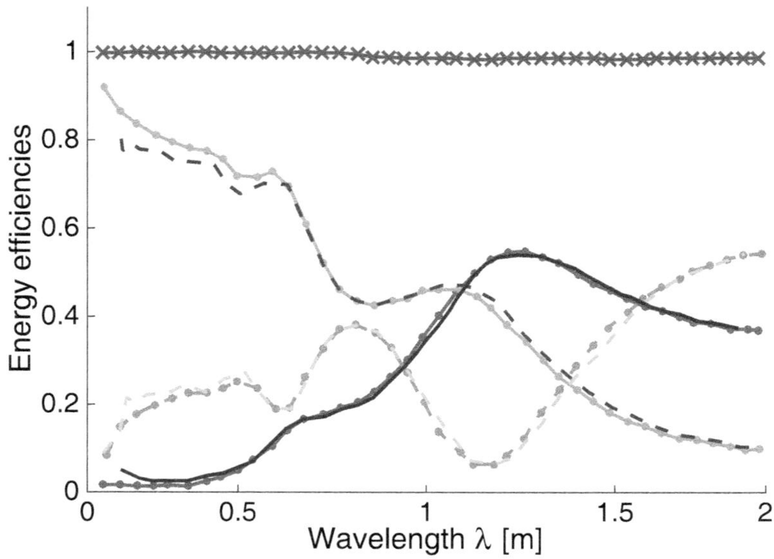

Figure 8.9 Transmission (solid), reflection (dot-dashed), and absorption (dashed) efficiency curves comparing the finite element solution (dot markers) and the effective medium (no markers) as a function of wavelength. Energy conservation for the finite element model is labeled with $\times$ markers. The structure has a conductivity $\sigma = 3000\ (\Omega\text{m})^{-1}$ (in the semiconductor domain), period $d = 0.01$ m, and dimensionless parameters $L/d = 60$, $r/d = 0.003$, and the angle of incidence is $\theta = 70^\circ$. δ/r runs approximately from 4 to 25 from left to right over the domain of the plot. The model fails around $\lambda \lessapprox 0.1\ \text{m} = 10d$.

The bed-of-nails structure is a medium exhibiting high absorption with low reflection. It requires a very low filling fraction of conducting material but exhibits near-perfect absorption over a wide range of angles of incidence, for sufficiently large thicknesses. The low filling fraction is useful because it allows the engineer to fill the space between the wires with materials satisfying other design constraints, such as mass density, or mechanical, chemical, or thermal robustness. The geometries studied here are transparent at normal incidence, but this aspect can easily be rectified by slanting the wires by about 20° with respect to the upper and lower boundaries. This design may, therefore, be used to obtain a near-perfect electromagnetic absorber for all angles of incidence in a very straightforward way, and with considerable freedom in the resulting mechanical, thermal, or chemical properties of the structure.

Appendix A

The Green's function for the following problem is derived.

$$\begin{aligned} p'' + \alpha^2 p &= \beta E_z \\ p'(-L/2) &= p'(L/2) = 0. \end{aligned} \tag{A.1}$$

The Green's function satisfies the equation

$$g'' + \alpha^2 g = \delta_{z_0},\ z_0 \in \left(-\frac{L}{2}, \frac{L}{2}\right) \tag{A.2}$$

and may be written as

$$g(z, z_0) = C u_1(z_<) u_2(z_>) \tag{A.3}$$

$$z_< = \min(z, z_0),\ z_> = \max(z, z_0)$$

Replacing form Eq. A.3 into Eq. A.2, one obtains finally:

$$g(z, z_0) = \frac{1}{\alpha \sin(\alpha L)} \cos(\alpha(z_< + L/2)) \cos(\alpha(z_> - L/2))$$

Appendix B

We now proceed to solve the homogeneous limit system Eq. 8.40. For convenience, we position it in $z \in (0, L)$. Since we are dealing with a system with translational invariance, a slab, we can split the problem into two independent polarization cases: TE, where the electric field is in the xy plane, and TM, where the magnetic field is in the xy plane. However, since we are considering thin wires (small volume fraction), the structure will be transparent to TE waves. We, therefore, only have to consider TM waves. We choose a coordinate system so that the plane of incidence is the xz plane, with angle of incidence θ, in which case our unknowns will be H_y and P_z. The translation invariance allows us to seek solutions to the form:

$$H_y = u(z)e^{i\alpha x},\ P_z = p(z)e^{i\alpha x}$$

with $\alpha = k_0 \sin\theta$. Inserting these into system 8.40, we obtain a system of equations for u and p:

$$\begin{cases} u''(z) + \left(k_0^2 - \alpha^2\right) u(z) & = \alpha\omega p(z) \\ p''(z) + \left(k_0^2 + \frac{2i\pi\gamma}{\kappa} - 2\pi\gamma\right) p(z) & = \frac{2\pi\alpha\gamma}{\omega} u(z), \ z \in [0, L] \end{cases} \tag{B.1}$$

with the boundary conditions $p' = 0$ at $z = 0$ and $z = L$, and u and u' continuous everywhere.

The transfer matrix T relates the field u and its derivative u' at the bottom and the top of the slab:

$$\begin{pmatrix} u(L) \\ u'(L) \end{pmatrix} = T \begin{pmatrix} u(0) \\ u'(0) \end{pmatrix}. \tag{B.2}$$

Once T is known, the reflection and transmission coefficients r and t can be obtained immediately from

$$r = e^{-2i\beta L}\frac{A+B}{A-B} \text{ and } t = \frac{2e^{-i\beta L}}{A-B} \tag{B.3}$$

$$A \equiv T_{11} - i\beta T_{12} \text{ and } B \equiv \frac{T_{21} - i\beta T_{22}}{i\beta}$$

where $\beta = k_0 \cos\theta = \sqrt{k_0^2 - \alpha^2}$.

We begin by integrating system B.1. Noting $\delta^2 = k_0^2 + \frac{2i\pi\gamma}{\kappa} - 2\pi\gamma$ for readability, we rewrite the system as

$$W''(z) = -MW(z) \tag{B.4}$$

where

$$W(z) = \begin{pmatrix} u(z) \\ p(z) \end{pmatrix}, \ M = \begin{pmatrix} \beta^2 & -\alpha\omega \\ -\frac{2\pi\alpha\gamma}{\omega} & \delta^2 \end{pmatrix}.$$

The matrix M can be diagonalized $M = QDQ^{-1}$ with $D = \text{diag}(K_u^2, K_p^2)$, so the system B.4 can be rewritten as $Q^{-1}W''(z) = -DQ^{-1}W(z)$. Since Q is constant and known, this can be integrated directly, and the general solution is then obtained as a sum of plane waves:

$$Q^{-1}W(z) = \begin{pmatrix} A_u^+ \exp(iK_u z) + A_u^- \exp(-iK_u z) \\ A_p^+ \exp(iK_p z) + A_p^- \exp(-iK_p z) \end{pmatrix} \tag{B.5}$$

The functions u and p are now expressed in terms of the elements of the matrix Q and the coefficients A_u^+, A_u^-, A_p^+, and A_p^-. By using the boundary conditions, we obtain p' as

$$\begin{aligned} p' = {} & iK_u Q_{21}(A_u^+ e^{iK_u z} - A_u^- e^{-iK_u z}) \\ & + iK_p Q_{22}(A_p^+ e^{iK_p z} - A_p^- e^{-iK_p z}). \end{aligned}$$

Setting this to zero at $z = 0, L$, we can obtain A_p in terms of A_u. Noting vectors

$$\underline{A_u} = \begin{pmatrix} A_u^+ \\ A_u^- \end{pmatrix}, \ \underline{A_p} = \begin{pmatrix} A_p^+ \\ A_p^- \end{pmatrix},$$

we introduce the matrix

$$C = -\frac{K_u Q_{21}}{K_p Q_{22}} \frac{1}{2i \sin(K_p L)} \times \begin{pmatrix} e^{iK_u L} - e^{-iK_p L} & e^{-iK_p L} - e^{-iK_u L} \\ e^{iK_u L} - e^{iK_p L} & e^{iK_p L} - e^{-iK_u L} \end{pmatrix}$$

so that

$$\underline{A_p} = C\underline{A_u}.$$

Equation B.5 can be rewritten as

$$W(z) = QE(z)\underline{A_u} \tag{B.6}$$

where $E(z)$ is defined as

$$E(z) = \begin{pmatrix} e^{iK_u z} & e^{-iK_u z} \\ C_{11}e^{iK_p z} + C_{21}e^{-iK_p z} & C_{12}e^{iK_p z} + C_{22}e^{-iK_p z} \end{pmatrix}.$$

With the help of Eq. B.6, we are in a position to construct the matrix $G(z)$ such that

$$\begin{pmatrix} u(z) \\ u'(z) \end{pmatrix} = G(z)\underline{A_u}.$$

By writing this equation at $z = 0$ and $z = L$, we obtain

$$\begin{pmatrix} u(z) \\ u'(z) \end{pmatrix} = G(L)G(0)^{-1} \begin{pmatrix} u(0) \\ u'(0) \end{pmatrix}.$$

Comparing with Eq. B.2, we obtain the result we seek,

$$T = G(L)G(0)^{-1},$$

leading to the reflection and transmission coefficients via Eqs. B.3.

References

Agha, Y. O., Zolla, F., Nicolet, A., and Guenneau, S. (2008). On the use of pml for the computation of leaky modes: An application to microstructured optical fibres, *COMPEL* **27**, 1, pp. 95–109.

Belov, P. A., Marques, R., Maslovski, S. I., Nefedov, I. S., Silveirinha, M., Simovski, C. R., and Tretyakov, S. A. (2003). Strong spatial dispersion in wire media in the very large wavelength limit, *Phys. Rev. B* **67**, 11, p. 113103.

Belov, P. A., Tretyakov, S. A., and Viitanen, A. J. (2002). Dispersion and reflection properties of artificial media formed by regular lattices of ideally conducting wires, *J. Electromagn. Waves Appl.* **16**, 8, pp. 1153–1170.

Bouchitté, G. and Felbacq, D. (2006). Homogenization of a wire photonic crystal: The case of small volume fraction, *SIAM J. Appl. Math.* **66**, 6, pp. 2061–2084.

Cabuz, A. I. (2007). Electromagnetic metamaterials: From photonic crystals to negative index composites, https://hal.archives-ouvertes.fr/tel-00161428.

Demésy, G., Zolla, F., Nicolet, A., Commandré, M., and Fossati, C. (2007). The finite element method as applied to the diffraction by an anisotropic grating, *Opt. Express* **15**, 26, pp. 18089–18102.

Demésy, G., Zolla, F., Nicolet, A., and Commandré, M. (2009). Versatile full-vectorial finite element model for crossed gratings, *Opt. Lett.* **34**, 14, pp. 2216–2218.

Dular, P., Nicolet, A., Genon, A., and Legros, W. (1995). A discrete sequence associated with mixed finite-elements and its gauge condition for vector potentials, *IEEE Trans. Mag.* **31**, 3, pp. 1356–1359.

Maslovski, S. I., Morgado, T. A., Silveirinha, M. G., Kaipa, C. S. R., and Yakovlev, A. B. (2010). Generalized additional boundary conditions for wire media, *New J. Phys.* **12**, 113047.

Nicolet, A., Guenneau, S., Geuzaine, C., and Zolla, F. (2004). Modelling of electromagnetic waves in periodic media with finite elements, *J. Comput. Appl. Math.* **168**, 1–2, pp. 321–329.

Pendry, J. B., Holden, A. J., Stewart, W. J., and Youngs, I. (1996). Extremely low frequency plasmons in metallic mesostructures, *Phys. Rev. Lett.* **76**, 25, pp. 4773–4776.

Pendry, J. B., Holden, A. J., Robbins, D. J., and Stewart, W. J. (1998). Low frequency plasmons in thin-wire structures, *J. Phys. Cond. Mat.* **10**, 22, p. 4785.

Ramo, S., Whinnery, J. R., and van Duzer, T. (1994). *Fields and Waves in Communication Electronics*, 3rd edn. (John Wiley and Sons).

Simovski, C. R. and Belov, P. A. (2004). Low-frequency spatial dispersion in wire media, *Phys. Rev. E* **70**, 4, p. 046616.

Zolla, F., Felbacq, D., and Bouchitté, G. (2006). Bloch vector dependence of the plasma frequency in metallic photonic crystals, *Phys. Rev. E* **74**, p. 056612.

Chapter 9

Resonant Problems

Didier Felbacq[a] **and Guy Bouchitté**[b]
[a]*Laboratory Charles Coulomb UMR CNRS-UM 5221, University of Montpellier, Place Bataillon, 34095 Montpellier Cedex 05, France*
[b]*Laboratory IMATH, University of Sud-Toulon-Var, BP 20132, 83957 La Garde Cedex, France*
didier.felbacq@umontpellier.fr

9.1 Introduction

In this chapter, the structures under study are made of a basic cell whose elements show a resonant behavior. The paradigmatic example is a dielectric rod with a sufficiently high index to have present a Mie resonance at a wavelength that is large with respect to the period. In order to understand the physical meaning of a Mie resonance, let us consider an electromagnetic cavity, that is, a medium surrounded by perfectly conducting wall. This cavity can support discrete electromagnetic modes at frequencies $\omega_1, \omega_2, \ldots$. When the wall is no longer perfectly conducting, the modes have a finite lifetime and the frequencies become complex with an imaginary part representing the exponential damping of the mode in time domain. This is what happens in a cylinder with a high permittivity: There are discrete modes, or resonances, characterized

Metamaterials Modeling and Design
Edited by Didier Felbacq and Guy Bouchitté

ISBN 978-981-4316-12-5 (Hardcover), 978-1-315-36500-8 (eBook)
www.panstanford.com

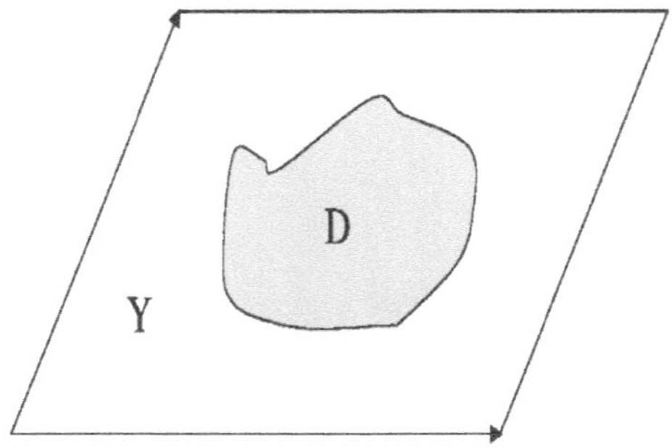

Figure 9.1 The basic cell of the dielectric metamaterial.

by complex frequencies that are poles of the scattering matrix. The existence of these resonances results in very interesting effective properties leading notably to an effective magnetic activity that can be used to design dielectric cloaks (Gaillot et al., 2008), Dirac cones (Huang et al., 2011), or zero-index metamaterials (Moitra et al., 2013).

A two-dimensional metamaterial made of a periodic collection of dielectric rods is considered (the basic cell Y) in Fig. 9.1. The dielectric rod has relative permittivity ε_i and cross-section D. It is embedded in a dielectric matrix ε_e.

9.2 $H_{||}$: A Two-Scale Approach

Our point is to show that near the resonances, the device behaves as if it had homogeneous electromagnetic parameters ε_h and μ_h. Of course, for this situation to be physically sound, the resonant wavelengths should be much larger than the period; otherwise, the medium could not be described by homogeneous parameters. That is why we request, as described by Pendry et al. (1999), that ε_i be much higher then ε_e. There is, therefore, a natural small parameter d/λ in the formulation of the scattering problem. The method that we employ consists of letting the small parameter tend to zero while rescaling the properties of the medium in order to keep the relevant physical phenomena, i.e., the resonances, unchanged. To do so, we choose a small number $\eta < 1$, and we proceed to the following operations:

- We multiply the radius of the rods and the period by η, while keeping constant the domain Ω where the rods are contained (the number N of rods is increased as $N \sim |\Omega|/\eta^2$).
- We divide the permittivity ε of the rods by η^2 (the optical diameter remains constant).

The wave is p-polarized, so the induction field reads $\mathbf{B}(\mathbf{x}) = u(\mathbf{x})\mathbf{e_z}$, but the vectorial form will prove useful for the analysis. In order to pass to the limit $\eta \to 0$, we use a two-scale expansion of $(\mathbf{E}, \mathbf{B})$:

$$\begin{aligned} \mathbf{B}_\eta(\mathbf{x}) &= \mathbf{B}_0(\mathbf{x}, \mathbf{x}/\eta) + \eta\mathbf{B}_1(\mathbf{x}, \mathbf{x}/\eta) + \dots \\ \mathbf{E}_\eta(\mathbf{x}) &= \mathbf{E}_0(\mathbf{x}, \mathbf{x}/\eta) + \eta\mathbf{E}_1(\mathbf{x}, \mathbf{x}/\eta) + \dots \end{aligned} \tag{B.1}$$

where the fields $\mathbf{E_0}$, $\mathbf{B_0}$ depend on both the real space variable $\mathbf{x}$ (the global variable) and on the Wigner–Seitz cell variable $\mathbf{y}$ (the local variable). The fields are periodic with respect to $\mathbf{y}$. Our point is to find the limit fields $\mathbf{E_0}$, $\mathbf{B_0}$. The local variable is, in fact, a hidden one: It is an internal degree of freedom. The true (observable) macroscopic fields $(\mathbf{E}_h, \mathbf{B}_h)$ are the averages of the microscopic fields $(\mathbf{E}_0, \mathbf{B}_0)$ over Y:

$$\mathbf{B}_h(\mathbf{x}) = \int_Y \mathbf{B}_0(\mathbf{x}, \mathbf{y})dy,\ \mathbf{E}_h(\mathbf{x}) = \int_Y \mathbf{E}_0(\mathbf{x}, \mathbf{y})d\mathbf{y}. \tag{B.2}$$

First, the behavior of the fields with respect to the local variables is investigated. This is the description of the microscopic behavior of the fields with respect to their internal degrees of freedom. Using the expansion (B.1) of the field, the $\nabla \times \cdot$ operator is transformed into:

$$\nabla \times \cdot \longrightarrow \nabla_x \times \cdot + \eta^{-1}\nabla_y \times \cdot$$

Plugging these expressions into Maxwell system and identifying the terms that correspond to the same power of η, we obtain the following system for the microscopic electric field:

$$\mathbf{curl}_y \times \mathbf{E}_0 = 0 \text{ on } Y\,,\ \mathbf{curl}_y \cdot \mathbf{E}_0 = 0 \text{ on } Y \setminus D \tag{B.3}$$

Besides: $\mathbf{E}_0 = 0$ on D and $\mathbf{E}_1 = 0$ on $Y \setminus D$. This system is of electrostatic type: $\mathbf{E}_0$ does not depend on the microscopic induction field, nor does it depend on the wavelength. As a matter of fact, on

$Y \setminus D$, $\mathbf{E}_0$ does not depend on the variable $\mathbf{y}$, as it can be deduced from system (B.3). Let us now turn to the magnetic field. The system satisfied by $\mathbf{B}_0$ is of an entirely different nature:

$$\begin{aligned} \mathbf{curl}_y\, \mathbf{B}_0 &= -i\omega\varepsilon_i \mathbf{E}_1 \quad \text{on } Y \\ \mathbf{curl}_y\, \mathbf{E}_1 &= i\omega \mathbf{B}_0 \quad\quad \text{on } D \end{aligned} \tag{B.4}$$

It is a microscopic Maxwell system that describes the microscopic behavior of the fields. The field $\mathbf{E}_1$ gives a first-order expansion of the field inside D: It replaces $\mathbf{E}_0$, which is null there. Using the fact that the fields are polarized, it comes: $\mathbf{B}_0(\mathbf{x}) = u_0(\mathbf{x})\mathbf{e}_z$. Plugging this expression into system (B.4) shows that the magnetic field is independent of $\mathbf{y}$ on $Y \setminus D$. Next, by combining the equation with system (B.4), one obtains:

$$\Delta_y u_0 + k^2 \varepsilon_i u_0 = 0 \text{ on } D\,,\ u_0 = C(x) \text{ on } Y \setminus D \tag{B.5}$$

By dividing u_0 by $C(x)$, we are led to considering the function $m(\mathbf{y})$ satisfying:

$$\Delta_y m + k^2 \varepsilon_i m = 0 \text{ on } D\,,\ m = 1 \text{ on } Y \setminus D \tag{B.6}$$

It holds $u_0(\mathbf{x}, \mathbf{y}) = C(x)m(\mathbf{y})$, and by averaging over Y, one deduces that the microscopic induction field is linked to the macroscopic one by: $u_0(\mathbf{x}, \mathbf{y}) = (m(\mathbf{y})/\mu_h)\, u_h(\mathbf{x})$ where μ_h, which shall be interpreted below as a relative permeability, is the mean value of m on Y: $\mu_h = \int_Y m(\mathbf{y}) d\mathbf{y}$. Having clarified the microscopic behavior of the fields, it remains to derive the equations that are satisfied by the macroscopic fields. The propagation equations read, for $\mathbf{y} \in Y \setminus D$:

$$\begin{aligned} \mathbf{curl}_x\, \mathbf{B}_0 + \nabla_y \times \mathbf{B}_1 &= -i\omega\varepsilon_0\varepsilon_e \mathbf{E}_0 \\ \mathbf{curl}_x\, \mathbf{E}_0 + \nabla_y \times \mathbf{E}_1 &= i\omega \mathbf{B}_0 \end{aligned} \tag{B.7}$$

The first equality is the Maxwell–Ampère equation with the extra-term $\nabla_y \times \mathbf{B}_1$, which is homogeneous to an electric displacement field. It represents the polarization due to the presence of the scatterers. More precisely, as u_0 does not depend on $\mathbf{y}$ on $Y \setminus D$, we obtain the following system satisfied by u_1:

$$\Delta_y u_1 = 0 \text{ on } Y \setminus D\,,\ \frac{\partial u_1}{\partial n} = -\mathbf{n} \cdot \nabla_x u_0 \text{ on } D, \tag{B.8}$$

where $\mathbf{n} = (n_1, n_2)$ is the normal to D. Let us introduce the auxiliary functions $w_1,\ w_2$, that satisfy $(j = 1, 2)$:

$$\Delta w_j = 0 \text{ on } Y \setminus D\,,\ \frac{\partial w_j}{\partial n} = -n_j \text{ on } \partial D \tag{B.9}$$

It then holds: $u_1 = w_1 \frac{\partial u_0}{\partial y_1} + w_1 \frac{\partial u_0}{\partial y_2}$. Therefore, there is a linear relation: $\nabla_y u_1 = \mathcal{P}(\mathbf{y}) \nabla_y u_0$ where

$$\mathcal{P}(\mathbf{y}) = \begin{pmatrix} 1 + \frac{\partial w_1}{\partial y_1} & \frac{\partial w_2}{\partial y_1} \\ \frac{\partial w_1}{\partial y_2} & 1 + \frac{\partial w_2}{\partial y_2} \end{pmatrix} \quad \text{(B.10)}$$

The average value of $\mathcal{P}$ over $Y \setminus D$ is denoted by $A_h = \int_{Y \setminus D} \mathcal{P}(y) dy$. It is the inverse of the effective permittivity tensor $\varepsilon_h (= A_h^{-1})$. The effective macroscopic equation can now be obtained by averaging system (B.7) on $Y \setminus D$:

$$\nabla \cdot (\varepsilon_h^{-1} \nabla (\mu_h^{-1} u_h)) + k^2 u_h = 0 \quad \text{(B.11)}$$

The macroscopic behavior of the system is characterized by an effective permittivity tensor ε_h and an effective permeability μ_h that depends on ω. This shows that the system exhibits an artificial magnetic activity. There are two huge differences between the effective permittivity and the effective permeability: The permittivity can be a matrix, so the medium can be anisotropic. However, the effective permeability is always a scalar; therefore, no anisotropic permeability can be obtained. Second, the permittivity is not frequency dependent; it is a static permittivity. However, the permeability depends on the frequency. Let us give a closer look at the system of equations that defines the effective permeability μ_h. The system (B.6) has a unique solution only if there is no function ψ such that ψ is null on $Y \setminus D$ and ψ satisfies the same Helmholtz equation on D. Otherwise, $m + \psi$ would still be a solution to (B.6). Following spectral theory (Kato, 1995), we denote $H = -\varepsilon_i^{-1} \Delta$ and we look for functions Φ satisfying the eigenvalue problem:

$$\Phi = 0 \text{ on } Y \setminus D, \; H\Phi = k^2 \Phi \text{ on } D. \quad \text{(B.12)}$$

We get a set of eigenvalues k_n^2 and a set of eigenfunctions $|\Phi_n\rangle$. The physical meaning of these eigenvalues can be understood by going back to the un-renormalized initial fiber, with permittivity ε_i. This fiber alone possesses resonant frequencies. They correspond to modes that are strongly localized inside the fiber. However, when there is a large number of fibers, these resonances are slightly shifted due to the coupling between the fibers, and these resonances are furthermore modified by the renormalization process. The eigenvalues of problem (B.12) are thus the renormalized Mie frequencies of the fiber.

For a given frequency k, we look for a solution m by expanding it on the basis $|\Phi_n\rangle$, by noting that $m - 1$ is null on $Y \setminus D$: $m(\mathbf{y}) = 1 + \sum_n m_n |\Phi_n\rangle$. The coefficients are obtained by inserting this expansion in (B.6). We get, after averaging, the effective permeability $\mu_h = \langle 1|m\rangle$ under the form:

$$\mu_h(k) = 1 + k^2 \sum_n \frac{|\langle \Phi_n|1\rangle|^2}{k_n^2 - k^2} \tag{B.13}$$

We have obtained a completely general expression for the effective permeability. It relies on the cavity modes of the fiber only. In the vicinity of a resonance k_n^2, we have: $\mu_h \sim 1 - k_n^2 |\langle \Phi_n|1\rangle|^2 (k^2 - k_n^2)^{-1}$ which shows, in complete generality, that the permeability exhibits anomalous dispersion near the resonances and becomes negative there. It should also be noted that only the eigenfunctions with nonzero mean value contribute. This is due to the fact that we have only kept the first-order terms in the expansions (B.1).

9.3 Numerical Results

9.3.1 Periodic Resonators

Let us give an explicit computation in the case of a square fiber. The eigenfunctions are $\Phi_{nm}(\mathbf{y}) = 2\sin(n\pi y_1)\sin(m\pi y_2)$ and the corresponding eigenvalues are $k_{nm}^2 = \pi^2(n^2 + m^2)$. The expansion of m on this basis leads to the following effective permeability:

$$\mu_h(k) = 1 + \frac{64a^2}{\pi^4} \sum_{(n,m)odd} \frac{k^2}{n^2 m^2 (\tilde{k}_{nm}^2 - k^2)} \tag{B.14}$$

where $\tilde{k}_{nm}^2 = k_{nm}^2/a^2\varepsilon_i$. Let us now turn to some numerical applications. First, we note that our analysis is supposed to work when there are Mie resonances at wavelengths large with respect to the period of the crystal. We choose $\varepsilon = 600 + 12i$ for our numerical computations, the point being to test the validity of the theory. This can be achieved by using ceramic rods in the gigahertz regime (Peng et al., 2007). Using a rigorous diffraction code for gratings

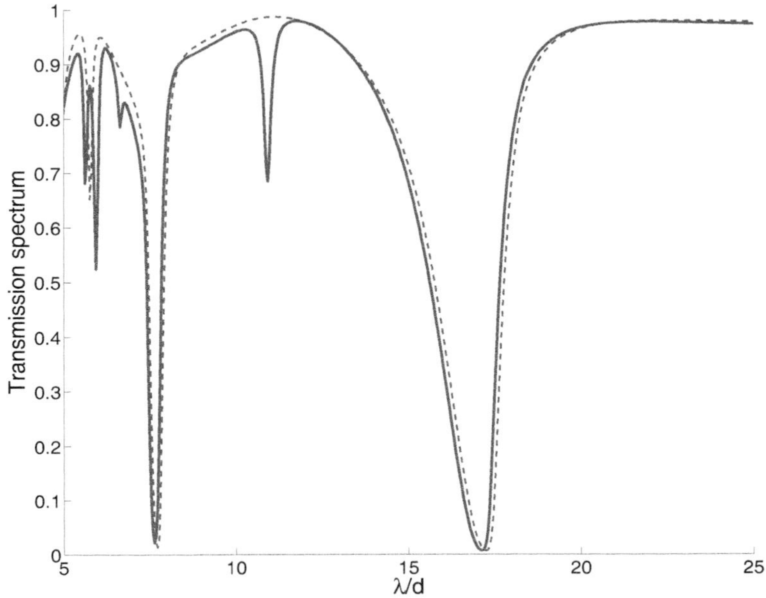

Figure 9.2 Modulus of the transmission for the metamaterial (solid line) and the homogenized material (dashed line).

(Nevière and Popov, 2003), we plot the transmission spectrum (dashed line in Fig. 9.2) for a stack of five diffraction gratings made of square rods. There is a band gap for λ/d between 8 and 12, due to a Mie resonance. In order to test our results, we plot the transmission spectrum of a homogeneous slab (solid line Fig. 9.2) with parameters $\varepsilon_h = 1.7$ (this value is obtained numerically from the solution to problem (B.9)) and μ_h given in (B.14). We see in Fig. 9.2 that both curves fit very well, indicating that although the wavelength is not that large, the dielectric metamaterial behaves as a homogeneous magnetic material. The discrepancy that is seen around $\lambda/d = 6.5$ is due to the presence of a Mie resonance of null mean value that is not taken into account in our theory. Only by expanding the fields to the second order could we incorporate this resonance in our global result.

9.4 $E_{||}$ Case: Green's Function Approach

In this section, the case of an electric field parallel to the rods is addressed. The standard periodic double-scale approach does not work properly here, because of the strong spatial dispersion, as explained in the introduction. A more down-to-earth method is used. It relies on characterizing the rods by electric and magnetic dipoles whose explicit expressions are derived in terms of the scattering matrix and then using a multiple scattering approach.

The electric field $\mathbf{E}^s$ scattered by an infinitely long dielectric rod of circular cross section, radius R, and permittivity ε in the far zone is given by (de Hulst, 1981):

$$\mathbf{E}^s(\mathbf{r}) = \sqrt{\frac{2}{\pi}} \frac{e^{ikr}}{\sqrt{kr}} e^{-i\frac{\pi}{4}} \left(b_0 + 2\sum_{n=1}^{+\infty} b_n \cos(n\theta) \right) \mathbf{u}_z \qquad \text{(B.15)}$$

where θ is the angle with respect to the direction of incidence, b_n is the nth-order Mie scattering coefficient of the circular rod, $\mathbf{k}$ ($|\mathbf{k}| = k = 2\pi/\lambda$) is the free space wavevector, and $\mathbf{u}_z$ is the unit vector defining the Oz-axis.

The scattered electric field can also be written at any point outside the rod in an integral form as (Felbacq et al., 1994):

$$\mathbf{E}^s(\mathbf{r}) = \frac{ik^2}{4} \int_C H_0^{(1)}(k|\mathbf{r} - \mathbf{r}'|)\,(\varepsilon - 1)\mathbf{E}(\mathbf{r}')d^2r' \qquad \text{(B.16)}$$

with $H_0^{(1)}$ the zeroth-order of the Hankel function of the first kind and C is the cross section of the rod. The different scattering orders of (B.15) can be found by developing the far-field expression of (B.16) into a series of multipoles. In the far zone ($k|\mathbf{r} - \mathbf{r}'| \gg 1$), $H_0^{(1)}$ can be described by its asymptotic form (Abramowitz and Stegun, 1965). The multipole expansion is introduced by writing $|\mathbf{r} - \mathbf{r}'| \simeq r - \mathbf{u_r} \cdot \mathbf{r}'$, where $\mathbf{r} = r\mathbf{u_r}$. This yields $\sqrt{k|\mathbf{r} - \mathbf{r}'|} = \sqrt{kr}$ and $e^{ik|\mathbf{r}-\mathbf{r}'|} = e^{ikr} \cdot e^{-ik\mathbf{u_r}\cdot\mathbf{r}'}$. The exponential $e^{-ik\mathbf{u_r}\cdot\mathbf{r}'}$ is then expanded in series as $e^{-ik\mathbf{u_r}\cdot\mathbf{r}'} = \sum_{n=0}^{\infty} \frac{(-ik\mathbf{u_r}\cdot\mathbf{r}')^n}{n!}$. Inserting these expressions in (B.16), we finally obtain the polarized multipole expansion of the electric field scattered by a circular rod in the far zone as:

$$\mathbf{E}^s(\mathbf{r}) = \sqrt{\frac{2}{\pi}} \frac{e^{ikr}}{\sqrt{kr}} e^{-i\frac{\pi}{4}} \sum_{n=0}^{\infty} \mathbf{f}_n(\mathbf{r}) \text{ with}$$

$$\mathbf{f}_n(\mathbf{r}) = \frac{ik^2}{4}\frac{(-ik)^n}{n!}\int_C (\mathbf{u}_\mathbf{r}\cdot\mathbf{r}')^n\,(\varepsilon-1)\mathbf{E}(\mathbf{r}')d^2r' \tag{B.17}$$

The successive terms of this expression can be identified with the classical dipole radiation fields at large distances. In particular, the zeroth ($n = 0$) and first ($n = 1$) components of (B.15) and (B.17) correspond to the electric and magnetic dipole radiations, respectively. By definition, the electric dipole moment per unit length is given by $\mathbf{p} = \int_C \mathbf{P}(\mathbf{r}')d^2r'$ with $\mathbf{P} = \varepsilon_0(\varepsilon - 1)\,\mathbf{E}$, the polarization per unit volume and ε_0 the free space permittivity. The magnetic dipole moment per unit length is given by $\mathbf{m} = \frac{1}{2}\int_C \mathbf{r}' \times \mathbf{J}(\mathbf{r}')d^2r'$, with $\mathbf{J} = \partial\mathbf{P}/\partial t$ the current density. For an incident plane wave propagating along the x direction, the electric and magnetic dipole moments per unit length can be written as a function of the scattering coefficients b_0 and b_1, respectively, as:

$$\begin{cases} \mathbf{p}/\varepsilon_0 = \dfrac{4b_0}{ik^2}\mathbf{u}_z \\ \mathbf{m}Z_0 = \dfrac{4b_1}{ik^2}\mathbf{u}_y \end{cases} \tag{B.18}$$

where $Z_0 = \sqrt{\mu_0/\varepsilon_0}$. These expressions can be used to represent the rods by point dipoles with moments $\mathbf{p}$ and $\mathbf{m}$, whenever the

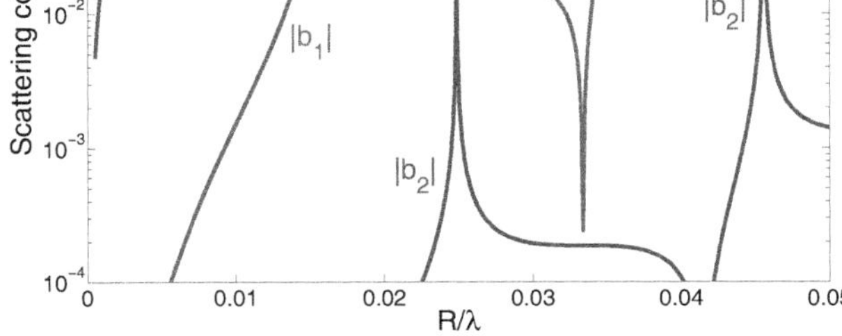

Figure 9.3 Complex magnitude of the b_0, b_1, and b_2 Mie scattering coefficients of a circular rod of radius R and permittivity $\varepsilon = 600$, in logarithmic scale versus the normalized frequency R/λ.

wavelength of light is much larger than the radius of the rods. Figure 9.3 sketches the complex amplitudes of the b_0, b_1, and b_2 coefficients of rods of permittivity $\varepsilon = 600$ as a function of the normalized frequency R/λ. We find that the b_0 and b_1 coefficients have much larger values than the b_2 coefficient (and consequently all coefficients with a higher order) over the whole frequency range of study. Therefore, these two coefficients play the main role in the optical properties of such rod-type structures.

Let us now consider an array of resonant dielectric rods. In the long-wavelength limit ($\lambda \gg a, R$), the corresponding array of dipoles can be described as an effective material with permittivity and permeability tensors $\bar{\bar{\varepsilon}}$ and $\bar{\bar{\mu}}$, respectively. In the case of planar propagation along the x-direction with the electric field along the z-axis, only ε_{zz} and μ_{yy} are required to define the effective index of the material $n_{\text{eff}} = \sqrt{\varepsilon_{zz}\mu_{yy}}$. Considering that the incident electric field amplitude has been normalized to unity and using the relations $|\mathbf{H}^i| = |\mathbf{E}^i|/Z_0$ and (B.18), the electric and magnetic polarizabilities per unit length of the dielectric rods are given by $\alpha^e_{zz} = p_z/\varepsilon_0 E^i_z = 4b_0/ik^2$ and $\alpha^m_{yy} = m_y/H^i_y = 4b_1/ik^2$, respectively. The permittivity ε_{zz} and permeability μ_{yy} can then be found from these expressions using an approach based on the quasi-periodic Green's function. The expression is then (Silveirinha, 2006):

$$\bar{\bar{\varepsilon}}_{zz} = 1 + \frac{\alpha^e_{zz}}{1 - C\alpha^e_{zz}}, \quad \bar{\bar{\mu}}_{zz} = 1 + \frac{\alpha^m_{zz}}{1 - C\alpha^m_{zz}} \tag{B.19}$$

where the interaction constant C is given by:

$$C = k^2\left(i/4 + \frac{1}{2\pi}\log\left(\frac{ka}{4\pi}\right) + \frac{\gamma}{2\pi} + \frac{1}{12} + \sum_{n\geq 1}\frac{1}{\pi|n|}\frac{1}{e^{\pi|n|}-1}\right).$$

To validate this approach, we compare the dispersion properties of a square array of high-permittivity rods calculated with the plane wave expansion method with those of the corresponding effective material using the dispersion relation $q_x = n_{eff}\omega/c$. The model is restricted to wave-vectors close to the Γ-point. The rods have a radius $R = 0.68a/3$, with a the lattice periodicity, and permittivity $\varepsilon = 600$, corresponding to the structure studied by Peng et al. (2007). Results are shown in Fig. 9.5. The dispersion curves (Fig. 9.5b) are in excellent agreement at wave-vectors close to

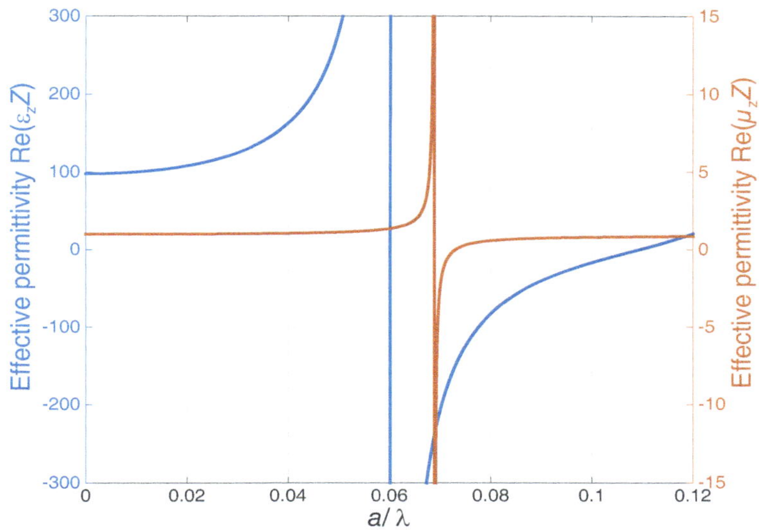

Figure 9.4 (a) Real parts of the effective permittivity ε_{zz} (blue solid line) and permeability μ_{yy} (red solid line) of a square array of rods ($R = 0.68a/3$, $\varepsilon = 600$) versus the reduced frequency a/λ.

the Γ-point. Although the multiple bands that appear for symmetry reasons at the resonant frequencies cannot be reproduced by the homogenization procedure, reliable conclusions on the origin of these bands can be given. In particular, the bands at a reduced frequency close to $a/\lambda = 0.07$ originate from the magnetic dipole resonance of the rods, taking place in the photonic band gap opened by the electric dipole resonance (see Fig. 9.5a). In this frequency range, both the permittivity and the permeability are negative, indicating a left-handed behavior in accordance with the interpretation of Peng et al. (2007). The right-handed bands are also well described by our model, which therefore makes it a reliable tool for understanding the correlation between the scattering properties of single rods and the dispersion properties of periodic rod-type structures.

Let us study the scalability of the metamaterial. The electric and magnetic dipole of dielectric rods are intrinsically related to their scattering coefficients. The scaling properties of rod-type metamaterials can thus be understood from the dependence of

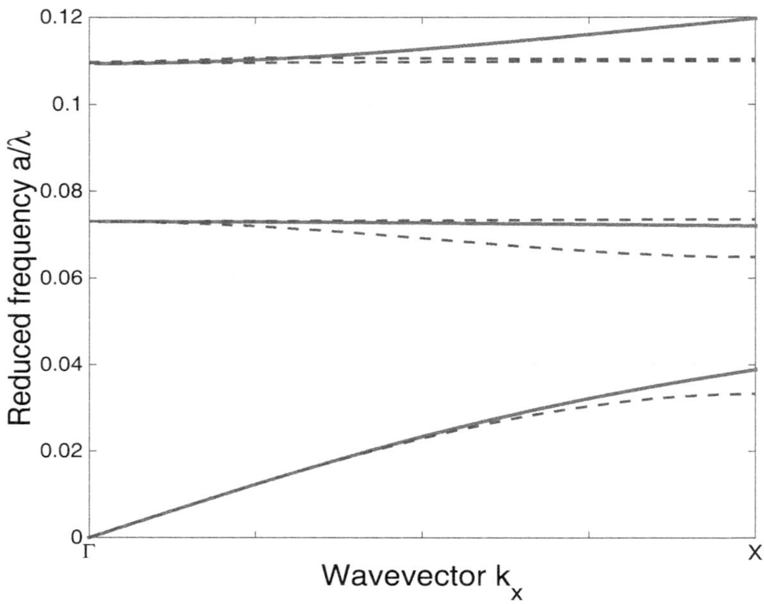

Figure 9.5 Dispersion curves of the PhC along the ΓX direction of the square array of rods (dashed lines) and that of the effective material (solid lines).

these coefficients with the permittivity ε of the rods and the free space wavelength λ. Figure 9.6 shows the evolution of the maxima of the modulus of the electric (b_0) and magnetic (b_1) coefficients of the rods with respect to their refractive index $n = \sqrt{\varepsilon}$ and to the wavelength-to-radius ratio λ/R. In the range of study, the wavelengths of resonance linearly depend on their refractive index. In particular, the magnetic dipole resonance observed in rods of permittivity $\varepsilon = 600$ ($n \simeq 24.5$) at reduced frequencies $a/\lambda \simeq 0.07$ ($\lambda/R \simeq 63$) is shifted to $a/\lambda \simeq 0.5$ ($\lambda/R \simeq 8.8$) in rods of permittivity $\varepsilon = 12$ ($n \simeq 3.5$). By calculating the complex amplitudes of the scattering coefficients of the rods, we can show that this permittivity is sufficiently high to insure the preponderance of the b_0 and b_1 coefficients over the higher-order ones in the frequency range of interest. The left-handed behavior is then expected to hold with a permittivity $\varepsilon = 12$.

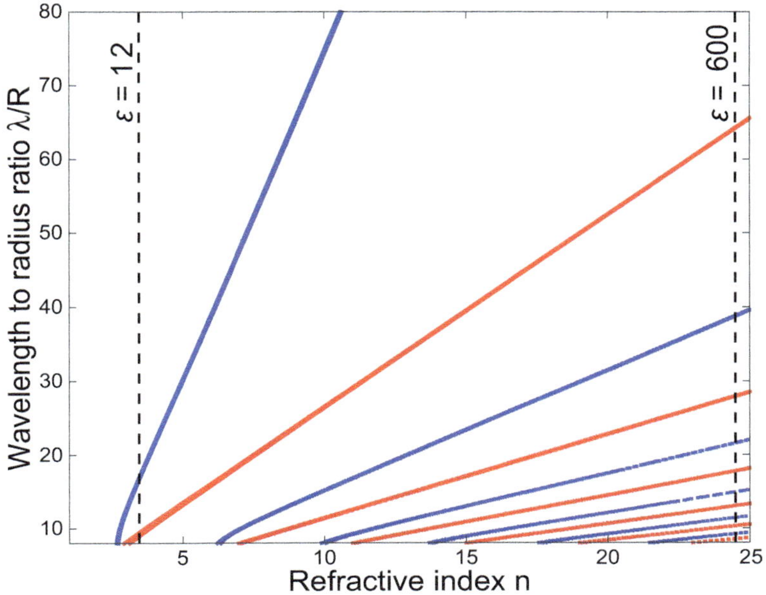

Figure 9.6 First maxima of the complex magnitude of the b_0 (blue solid lines) and b_1 (red solid lines) Mie scattering coefficients of dielectric rods with radius R as a function of their refractive index $n = \sqrt{\varepsilon}$ and the wavelength-to-radius ratio λ/R.

This is verified by comparing the photonic band structures and second-band isofrequency curves of the two metamaterials. Results are shown in Fig. 9.9(a)–(d). The dispersion properties of both structures exhibit very similar features. The bands located in the left-handed frequency range at reduced frequencies $a/\lambda \simeq 0.07$ are pushed up to $a/\lambda \simeq 0.5$, as expected.

This confirms the above demonstration that this left-handed behavior results from an overlap of the electric and magnetic dipole resonances. Second, the isofrequency curves of both structures exhibit a strong spatial dispersion. This shows that having a large wavelength-to-period ratio does not necessarily insure an isotropic response (Belov et al., 2003; Cabuz et al., 2008). The magnetic dipole term is, in fact, responsible for the first-order spatial dispersion (Landau et al., 1984). Since non-local effects are very sensitive to the

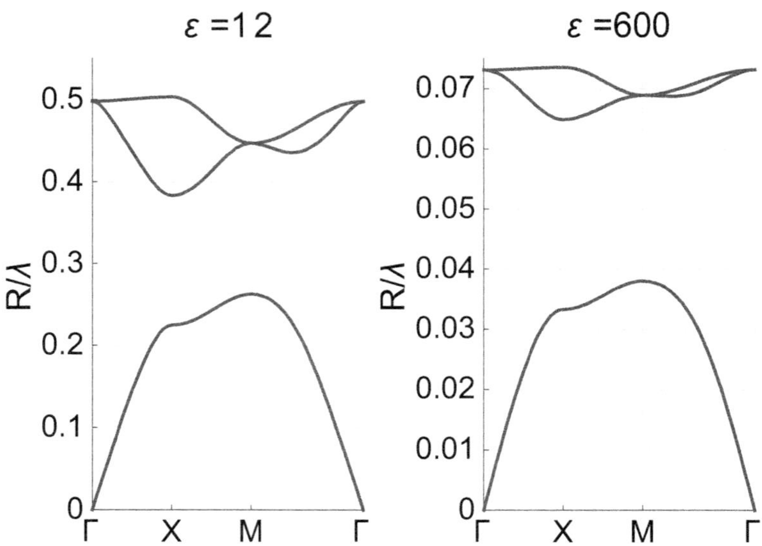

Figure 9.7 Band structure for the dielectric rod metamaterials (left: $\varepsilon = 12$, right: $\varepsilon = 600$).

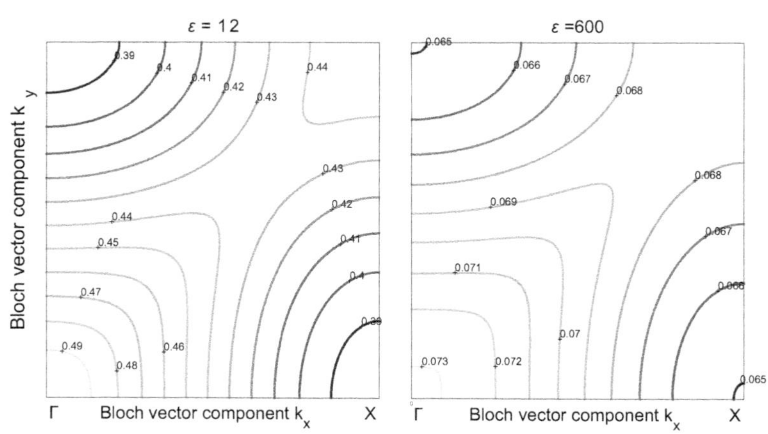

Figure 9.8 Isofrequency curves for the second band of the dielectric rod metamaterials.

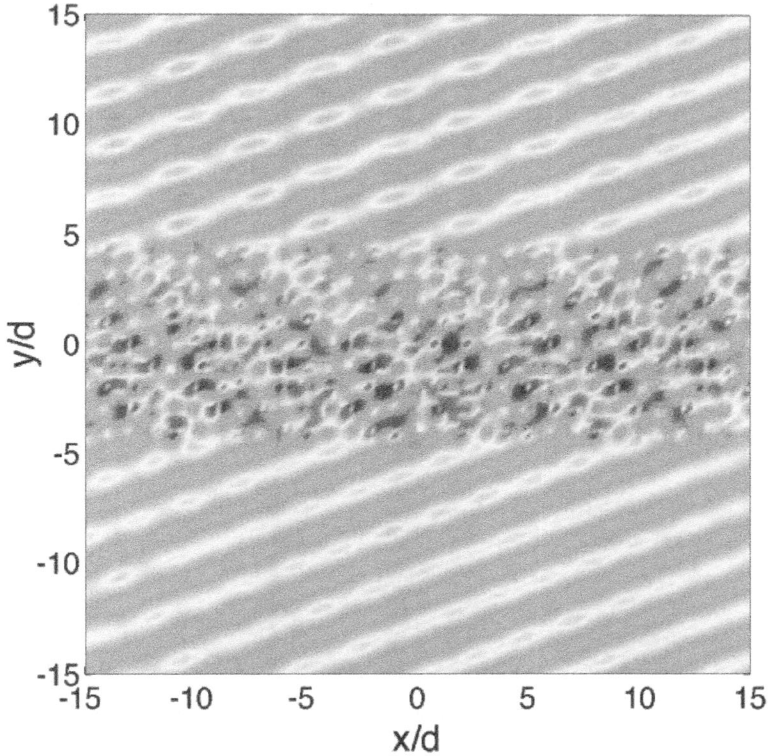

Figure 9.9 Steady-state amplitude of the electric field of a $E_{||}$-polarized plane wave at a reduced frequency $a/\lambda = 0.45$ incident at an angle of 20° on the low-permittivity metamaterial.

symmetry of the structure, the coupling between the rods plays an important role in both the spectral and spatial responses.

To prove the concept of a left-handed behavior, a full-wave calculation of an s-polarized plane wave at the reduced frequency $a/\lambda = 0.45$ incident at an angle of 20° on the lower-permittivity metamaterial is performed using a multiple scattering approach. The steady-state amplitude of the electric field is shown in Fig. 9.9(e). The phase of the propagating field in the metamaterial is opposite to that of the field in free space. This effect, characteristic of a left-handed behavior, can be tuned to the telecommunications wavelengths ($\lambda \simeq 1.55$ μm) by using silicon rods ($\varepsilon \simeq 12$) of radius

$R \simeq 160$ nm and a lattice of periodicity $a \simeq 700$ nm. Silicon material could, therefore, open the route toward the realization of all-dielectric metamaterials operating at optical frequencies. It is worth noting that very similar structures have been fabricated and characterized some years ago (Xu et al., 2001). The underlying origin of their dispersion properties actually relies on the collective response of the resonant rods, which defined them as true metamaterials on the same level as metallic metamaterials, with the advantages that dielectric structures are scale-invariant and exhibit no intrinsic loss.

References

Abramowitz, M. and Stegun, I. (1965). *Handbook of Mathematical Functions* (Dover Publication Inc.).

Belov, P. A., Marques, R., Maslovski, S. I., Nefedov, I. S., Silveirinha, M., Simovski, C. R., and Tretyakov, S. A. (2003). Strong spatial dispersion in wire media in the very large wavelength limit, *Phys. Rev. B* **67**, 11, p. 113103.

Cabuz, A. I., Felbacq, D., and Cassagne, D. (2008). Spatial dispersion in negative-index composite metamaterials, *Phys. Rev. A* **77**, 1, p. 013807.

de Hulst, H. C. V. (1981). *Light Scattering by Small Particles* (Dover Publication Inc., New York).

Felbacq, D., Tayeb, G., and Maystre, D. (1994). Scattering by a random set of parallel cylinders, *J. Opt. Soc. Am. A* **11**, p. 2526.

Gaillot, D. P., Croënne, C., and Lippens, D. (2008). An all-dielectric route for terahertz cloaking, *Opt. Express* **16**, pp. 3986–3992.

Huang, X., Lai, Y., Hang, Z. H., Zheng, H., and Chan, C. T. (2011). Dirac cones induced by accidental degeneracy in photonic crystals and zero-refractive-index materials, *Nat. Mat.* **10**, pp. 582–586.

Kato, T. (1995). *Perturbation Theory for Linear Operators* (Springer-Verlag, Berlin).

Landau, L., Lifschitz, E., and Pitaevski, L. (1984). *Electrodynamics of Continuous Media* (Pergamon Press Ltd., Oxford).

Moitra, P., Yang, Y., Anderson, Z., Kravchenko, I. I., Briggs, D. P., and Valentine, J. (2013). Realization of an all-dielectric zero-index optical metamaterial, *Nat. Photon* **7**, pp. 791–795.

Nevière, M. and Popov, E. (2003). *Light Propagation in Periodic Media: Differential Theory and Design*, Optical Engineering (Marcel Dekker Inc., New York).

Pendry, J. B., Holden, A. J., Robbins, D. J., and Stewart, W. J. (1999). Magnetism from conductors, and enhanced non-linear phenomena, *IEEE Trans. Microw. Theory Tech.* **47**, p. 2075.

Peng, L., Ran, L., Chen, H., Zhang, H., Kong, J. A., and Grzegorczyk, T. M. (2007). Experimental observation of left-handed behavior in an array of standard dielectric resonators, *Phys. Rev. Lett.* **98**, p. 157403.

Silveirinha, M. G. (2006). Nonlocal homogenization model for a periodic array of ϵ-negative rods, *Phys. Rev. E* **73**, p. 046612.

Xu, Y., Sun, H.-B., Ye, J.-Y., Matsuo, S., and Misawa, H. (2001). Fabrication and direct transmission measurement of high-aspect-ratio two-dimensional silicon-based photonic crystal chips, *J. Opt. Soc. Am. B* **18**, pp. 1084–1091.

Section IV

Mathematical Annex

Appendix A
Mathematical Annex

- *Numbers* used in the book are natural integers $0, 1, 2, \cdots \in \mathbb{N}$ or signed integers $\cdots, -1, 0, 1, \cdots \in \mathbb{Z}$ mainly as indices. Physical quantities take their values in (sets of) real numbers $\mathbb{R}$ or complex numbers $\mathbb{C}$. Real and complex numbers are called *scalars*. A complex number z can be viewed as a pair (a, b) of real numbers denoted by $z = a + ib$ where $i^2 = -1$. The real number $a = \Re\mathrm{e}(z)$ is called the real part, whereas the real number $b = \Im\mathrm{m}(z)$ is called the imaginary part. The *complex conjugate* of z is $\bar{z} = a - ib$ and its *modulus* or absolute value is $|z| = \sqrt{z\bar{z}} = \sqrt{a^2 + b^2}$. The exponential representation $z = |z|e^{i\theta(z)}$ involves the modulus and the *argument* $\theta(z)$. The function $\theta(z)$ is usually chosen as being a continuous function on an open set of $\mathbb{C}$, called a continuous determination of the argument. Two continuous determinations on the same set differ by an integer multiple of 2π and the principal determination is defined by $\theta(z) : \mathbb{C}\backslash\{\Re\mathrm{e}(z) \leq 0\} \rightarrow]-\pi, +\pi[$ and given by $\theta = 2\arctan\left(\dfrac{b}{a + \sqrt{a^2 + b^2}}\right)$.
- *Vector space*: Given scalars $\mathbb{R}$ or $\mathbb{C}$, a vector space V is a set of elements, called vectors, with two operations: the addition of two vectors and the product of a vector by a scalar, both defined axiomatically in a way quite obvious for the common intuition and giving a new vector as the result. Given a set of vectors $\mathbf{v}^i$ of a vector space V and a set of scalars a_i, the combination of products by a scalar and vector additions is a new vector called a *linear combination* and denoted $a_i\mathbf{v}^i$.

This expression uses the *Einstein summation convention* on repeated indices: $a_i\mathbf{v}^i$ denotes $\sum_i a_i\mathbf{v}^i$. If things are written with care, it always involves an upper and a lower index.
The null scalar 0 and the null vector $\mathbf{0}$ are defined so that $0\mathbf{v} = \mathbf{0}$ and $\mathbf{0} + \mathbf{v} = \mathbf{v}$ for any vector. Given vectors are said to be *linearly independent* if the only linear combination that gives $\mathbf{0}$ is $\sum_i 0\mathbf{v}^i$, else they are said to be linearly dependent. The maximum number of vectors that you can find in a set of linearly independent vectors of V is called the *dimension* $\dim(V)$ of the vector space. A maximum set of linearly independent vectors is called a *basis* and any vector can be expressed as a linear combination of the basis vectors.
The canonical examples of vector spaces are the sets of n-tuples of scalars: $(b^1, \cdots, b^n)$ or $(c^1, \cdots, c^n)$ with $b^1, \cdots, b^n, c^1, \cdots, c^n \in \mathbb{C}$. Explicitly, the addition is given by $(b^1, \cdots, b^n) + (c^1, \cdots, c^n) = (b^1 + c^1, \cdots, b^n + c^n)$ and the product by a scalar is given by $a(b^1, \cdots, b^n) = (ab^1, \cdots, ab^n)$ with $a \in \mathbb{C}$.
Given a set of vectors $\mathbf{v}^1, \cdots, \mathbf{v}^p$, they *span* a vector space $V = \text{span}(\mathbf{v}^1, \cdots, \mathbf{v}^p)$, which is the set of all the linear combinations of those vectors.

- *Multi-linear forms*: a p-multi-linear form is a map $f : \{\mathbf{v}^1, \mathbf{v}^2, \cdots, \mathbf{v}^p\} \in V \to f(\mathbf{v}^1, \mathbf{v}^2, \cdots, \mathbf{v}^p) \in \mathbb{C}$ from p vectors to a scalar (complex numbers $\mathbb{C}$ are taken as the default set of scalars) such that $f(\mathbf{v}^1, \cdots, a_i\mathbf{v}^i, \cdots, \mathbf{v}^p) = a_i f(\mathbf{v}^1, \cdots, \mathbf{v}^i, \cdots, \mathbf{v}^p)$, i.e., the map is linear in all entries. Particular cases are the *linear forms* that map linearly a single vector to a scalar and *bilinear forms* that map two vectors to a scalar.
  Note that the set of p-multilinear forms for a given p is obviously a vector space itself since a linear combination of m p-forms $f_1, \cdots, f_m$ is defined as a new m-multilinear form by $(a^i f_i)(\mathbf{v}^1, \cdots, \mathbf{v}^p) = a^i f_i(\mathbf{v}^1, \cdots, \mathbf{v}^p)$. In particular, the set of linear forms on a vector space V is called the *algebraic dual vector* space and is denoted by V^*. In the finite dimension case, the dual space V^* is *isomorphic* to V, i.e., there is a one-to-one correspondence between the spaces preserving the various operations. For instance, if a vector $\mathbf{v}$ of V is a set of n

numbers $(b^1, \cdots, b^n)$, a general linear form f is also given by a set of n numbers $(c_1, \cdots, c_n)$ so that the action of the linear form on the vector is $f(\mathbf{v}) = b^1 c_1 + \cdots + b^n c_n = b^i c_i$.
Note the lower indices for the coefficients of forms.
To stress on the duality, the action of f can be written as a *duality product* $f(\mathbf{v}) =< f, \mathbf{v} >$.
In the case of the canonical example, a basis is given by vectors $\mathbf{e}_i = (0, \cdots, 1, \cdots, 0)$ where 1 in the ith position is the only nonzero component.
Given a basis $\mathbf{e}_i$ of a vector space V, the set of vectors $\mathbf{e}^i$ of V^* such that $< \mathbf{e}_i, \mathbf{e}^j >= \delta_i^j$, where δ_j^i is the *Kronecker delta symbol* equal to 1 when $i = j$ and else equal to 0, is a basis of V^*, called the *dual basis* and $< b^i \mathbf{e}_i, c_j \mathbf{e}^j >= b^i c_i$.

- Functional analysis and function spaces: Some sets of functions are vector spaces of particular interest, called function spaces or functional vector spaces. The main characteristic of such spaces is that they are usually infinite dimensional and not isomorphic to their dual spaces. For instance, the set of square integrable functions on the $[0, 2\pi]$ interval is a functional vector space. A basis is the infinite set of trigonometric functions $\{1, \cos(x), \sin(x), \sin(2x), \cos(2x), \cdots\}$.
- Important examples of function spaces: Let Ω be a bounded open set in $\mathbb{R}^n$ or $\mathbb{R}^n$ itself. We denote by $C^m(\Omega)$ the set of functions that are continuous in Ω together with all their first m derivatives.
In a multi-variable context, the following notation is introduced: $\alpha = (\alpha_1 \cdots \alpha_n) \in \mathbb{N}^n$ is a multi-index with

$$|\alpha| = \sum_i \alpha_i \text{ and } D^\alpha = (-i)^{|\alpha|} \frac{\partial^{|\alpha|}}{\partial^{\alpha_1} x^1 \ldots \partial^{\alpha_n} x^n}.$$

The first m derivatives of a function f are the $D^\alpha f$ with $|\alpha| \leq m$.
The subset consisting of the functions that have a compact support (i.e., which vanish outside a compact subset of Ω) is denoted by $C_0^m(\Omega)$. The corresponding space of infinitely differentiable functions (also called smooth functions) is denoted by $C^\infty(\Omega)$ and $C_0^\infty(\Omega)$.

- *Tensor product*: The tensor product is a fundamental operation of linear algebra, which transforms multilinear forms to linear ones. Given two vector spaces V and W, there exists a unique vector space (up to an isomorphism) denoted by $V \otimes W$ and such that for any vector space U, the space of linear maps from $V \otimes W$ to U is isomorphic to the space of bilinear maps from $V \times W$ to U where $V \times W$ is the Cartesian product, i.e., the set of pairs $(v \in V, w \in W)$. The vector space $V \otimes W$ is called the tensor product.
For example, in the case of functional spaces, the tensor product of two functions $f(x)$ and $g(x)$ of a single variable is the two-variable function given by $f \otimes g(x, y) = f(x)g(y)$. As another example, consider a vector space V of finite dimension n and its dual space V^*. The tensor product $V^* \otimes V^*$ associates to a pair of linear forms on V, i.e., an element of $V^* \times V^*$ given by the coefficients a_i and b_i (a single index), a bilinear form on V given by the coefficients $(a \otimes b)_{i,j} = a_i b_j$ (two indices) such that the action on two vectors given by the coefficients v^i and w^i is the scalar $(a \otimes b)_{i,j} v^i w^j = a_i b_j v^i w^j$. Tensor product is an important tool to manipulate multilinear maps as vectors spaces. For instance, if V is a vector space and V^* its dual, p-multilinear maps form the $\otimes^p V^*$ vector space. A basis of this space may be built by taking the tensor products of p basis vectors of V^* and the dimension of this space is $\dim(\otimes^p V^*) = \dim(V)^p$.
- Direct sum: A subspace U of V is a subset of V such that any linear combination of elements of U is in U, which is, therefore, itself a vector space. Suppose that a vector space V has two subspaces U and W such that any vector $\mathbf{v} \in V$ can be written in a unique way as the sum $\mathbf{u} = \mathbf{v}+\mathbf{w}$ where $\mathbf{u} \in U$ and $\mathbf{w} \in W$. The subspace V is, in this case, the *direct sum* of the vector spaces U and W, which is denoted by $V = U \oplus W$. One has $\dim(V) = \dim(U) + \dim(W)$.
- Scalar product and norm: New operations have often to be added to the bare structure of a vector space.
The *scalar product* is a form $(\mathbf{v}, \mathbf{w}) \to \mathbb{C}$ with the following properties:

- Linear in the first variable, i.e.,
  $(a\mathbf{u} + b\mathbf{v}, \mathbf{w}) = a(\mathbf{u}, \mathbf{w}) + b(\mathbf{v}, \mathbf{w})$.
- $(\mathbf{v}, \mathbf{w}) = \overline{(\mathbf{w}, \mathbf{v})}$
- positive, i.e., $(\mathbf{v}, \mathbf{v}) \geq 0$ and the equality arises only when $\mathbf{v} = \mathbf{0}$.

In the case of a real vector space, this form is simply bilinear, but in the complex case, it is *sesqui-linear*, i.e., semi-linear in the second argument:$(\mathbf{v}, a\mathbf{w}) = \bar{a}(\mathbf{v}, \mathbf{w})$. A vector space with a scalar product is called a *pre-Hilbert space*.
The *norm* $\|\mathbf{v}\|$ is the (positive) square root of the scalar product of a vector $\mathbf{v}$ with itself: $\|\mathbf{v}\|^2 = (\mathbf{v}, \mathbf{v})$.
In the case of the canonical example, the canonical scalar product is given by

$$((b^1, \cdots, b^n), (c^1, \cdots, c^n)) = b^1\overline{c^1} + \cdots + b^n\overline{c^n}.$$

Two norms $\|.\|_1$ and $\|.\|_2$ are *equivalent* if there exist two real strictly positive constants c and C such that $c\|\mathbf{v}\|_1 \leq \|\mathbf{v}\|_2 \leq C\|\mathbf{v}\|_1$.
A vector that has a norm equal to 1 is called a unit vector or a normalized vector. Two vectors are *orthogonal* if their scalar product is equal to 0.
If V is a vector space with a norm $\|.\|_V$, this defines a norm on the dual space V^* by $\|f\|_{V^*} = \frac{\sup_{\mathbf{v}\in V} f(\mathbf{v})}{\|\mathbf{v}\|_V} = \sup_{\|\mathbf{v}\|_V=1} f(\mathbf{v})$ for any linear form $f \in V^*$.
We use the name of the concerned vector space as a subscript to indicate the vector product or the norm when this is necessary, but it will be omitted if there is no ambiguity.

- The norm determines a *topology*, i.e., a system of subsets called open sets, on the vector space (which is, therefore, called a topological vector space), which is necessary to introduce the concepts of convergence and limit. Two equivalent norms determine the same topology.
  The distance between two vectors is defined by $\rho(\mathbf{u}, \mathbf{v}) = \|\mathbf{u} - \mathbf{v}\|$ and the (open) ball of center $\mathbf{u}$ and radius $\epsilon \in \mathbb{R}$ is the set of vectors $\mathbf{v} \in B(\mathbf{u}, \epsilon)$ so that $\rho(\mathbf{u}, \mathbf{v}) < \epsilon$. A set Ω is called *open* if for any $\mathbf{u} \in \Omega$, there exists an ϵ such that the ball $B(\mathbf{u}, \epsilon)$ is contained in Ω. A set E is a *closed* if its complement (i.e., elements that are not in E) is open. The *closure* $\overline{\Omega}$ of a

set is the smallest closed set that contains it and the *interior* $Int(\Omega)$ is the largest open set contained in it.[a] The *boundary* $\partial\Omega$ is the set of elements that are in the closure and not in the interior.
A vector space with a norm that is complete (i.e., every *Cauchy sequence* of the space $\{\mathbf{u}_n\}$, i.e., such that $\rho(\mathbf{u}_n, \mathbf{u}_m) \to 0$ as $n, m \to +\infty$, has its limit in the space) is called a *Banach space*.

- Quotient: If W is a vector subspace of the vector space V, equivalence classes in V can be defined in the following way: Two elements of V are equivalent if they only differ by a vector of W. The set of equivalence classes is itself a vector space called the *quotient space* V/W and one has $V = W \oplus (V/W)$. If a norm $\|.\|_V$ is defined on the vector space V, a norm on the quotient space may be defined by $\|\mathbf{u}\|_{V/W} = \inf_{\mathbf{v}\in\mathbf{u}} \|\mathbf{v}\|_V$ for any equivalence class $\mathbf{u}$ of V/W.
- It is usual to build integration upon *Lebesgue measure theory*, which is quite technical and mathematically demanding. We just recall here that a fundamental concept is the one of *null measure set*. For all practical purpose, null measure sets on $\mathbb{R}$ are finite and denumerable sets of points (although there are non-denumerable null measure sets), e.g., the set of rational numbers, and null measure sets on $\mathbb{R}^n$ are denumerable sets of subsets of $\mathbb{R}^p$ with $p < n$. Two functions are equal *almost everywhere (a.e.)* if they only differ on a null measure set.
For instance, $\mathcal{L}^p(\Omega)$ is the set of functions f such that $\int_\Omega |f(\mathbf{x})|^p d\mathbf{x} < \infty$ where $d\mathbf{x}$ denotes the Lebesgue measure on Ω. The sets of functions that are equal a.e. are equivalent classes. The set $\mathcal{N}(\Omega)$ of functions equal to zero a.e. is a subspace of $\mathcal{L}^p(\Omega)$ and one defines $L^p(\Omega) = \mathcal{L}^p(\Omega)/\mathcal{N}(\Omega)$. L^p spaces are Banach spaces with the norm

$$\|f\|_{L^p} = (\int_\Omega |f(\mathbf{x})|^p d\mathbf{x})^{1/p}.$$

$L^\infty(\Omega)$ is the Banach space of functions bounded a.e. on Ω, which has been given the norm $\|f\|_{L^\infty} = \operatorname{ess\,sup}_\Omega |f(\mathbf{x})|$

[a]Note that if Ω is an open set, we have $\mathrm{Int}(\Omega) = \Omega$.

where "ess" stands for "essentially," i.e., the norm is the smallest value M for which $|f(\mathbf{x})| \leq M$ a.e.

If Ω is the whole $\mathbb{R}^n$, the integral may be defined only locally, i.e., on any compact subset. $L^p_{\mathrm{loc}}(\mathbb{R}^n)$ is the space of functions f such that $f\varphi \in L^p(\mathbb{R}^n)$ for any $\varphi \in C_0^\infty(\mathbb{R}^n)$.

One of the main results of the integration theory is the *Lebesgue dominated convergence theorem*: If $\{f_n\}$ is a sequence of functions in $L^1(\Omega)$ and if there exists a function $g \in L^1(\Omega)$ so that $|f_n(\mathbf{x})| \leq g(\mathbf{x})$ a.e. Then, if this sequence is convergent a.e. to a function f, this function is also integrable and $\int_\Omega f_n(\mathbf{x})d\mathbf{x} \to \int_\Omega f(\mathbf{x})d\mathbf{x}$. This result is sometimes called the *Lebesgue bounded convergence theorem* if one considers a constant $M > 0$ instead of the function g.

As an easy corollary, series and integral can be swapped with confidence provided the following holds: Let us consider that for every integer N, the previous sequence $f_n \in L^1(\Omega)$ is such that $|\sum_{n=1}^N f_n(\mathbf{x})| \leq g(\mathbf{x})$ a.e. Then, if $\sum_{n=1}^\infty f_n(\mathbf{x})$ is convergent a.e. to a sum function $f(\mathbf{x})$, this function is also integrable, the series $\sum_{n=1}^\infty \int_\Omega f_n(\mathbf{x})d\mathbf{x}$ is convergent, and $\int_\Omega f(\mathbf{x})d\mathbf{x} = \int_\Omega \sum_{n=1}^\infty f_n(\mathbf{x})d\mathbf{x} = \sum_{n=1}^\infty \int_\Omega f_n(\mathbf{x})d\mathbf{x}$.

- *Hilbert spaces*: Important examples of infinite dimensional vector spaces are the Hilbert spaces. A Hilbert space is a complete space with a scalar product (i.e., a pre-Hilbert Banach space). For instance, given a domain Ω of $\mathbb{R}^n$, the space of square integrable functions $L^2(\Omega)$ (introduced in the previous section) with the scalar product

$$(f, g) = \int_\Omega f(\mathbf{x})\overline{g(\mathbf{x})}d\mathbf{x} \text{ is a Hilbert space.}$$

One of the important properties of the Hilbert spaces is the *Riesz representation theorem*: Given any linear form f on a Hilbert space H, there exists one and only one vector $\mathbf{u}$ such that for all vectors $\mathbf{v} \in H$, one has $f(\mathbf{v}) = (\mathbf{u}, \mathbf{v})$. Hilbert spaces are, therefore, *reflexive spaces*, i.e., they are isomorphic to their dual spaces. A common abuse of notation is to identify a Hilbert space with its dual and to write $(\mathbf{u}, \mathbf{v}) = <\mathbf{u}, \mathbf{v}>$, where $\mathbf{v}$ denotes both the element of the Hilbert space in the scalar product and its corresponding linear form by the Riez theorem in the duality product.

Via the scalar product, the notion of orthogonality is available in Hilbert spaces. Given a Hilbert space H and V a Hilbert subspace, i.e. a subspace which is also a Hilbert space, the set $V^{\perp}$ of elements of H orthogonal to all the elements of V is also a Hilbert subspace of H and one has $H = V \oplus V^{\perp}$.

- Operators and functionals: Various operations may be defined on vector spaces. Mappings from a vector space onto a vector space are usually called *operators*, while mappings from a vector space onto scalars are called *functionals*. Given two Banach spaces V and W, $\mathcal{L}(V, W)$ is the set of *linear operators* from V to W, i.e., operators that preserve linear combinations. If $L \in \mathcal{L}(V, W)$, for all $\mathbf{v}^1, \mathbf{v}^2 \in V$ and $a, b \in \mathbb{C}$, we have $L(a\mathbf{v}^1 + b\mathbf{v}^2) = aL(\mathbf{v}^1) + bL(\mathbf{v}^2)$. The set $\mathcal{L}(V, V)$ is written $\mathcal{L}(V)$. The space V is called the *domain* $\mathrm{dom}(L)$ of the operator L, and the vector space spanned by the elements of W, which can be obtained by the action of the operator on an element of V, is called the *range, image*, or *codomain* $\mathrm{cod}(L)$ of L. The *kernel* or *nullspace* $\ker(L)$ is the subspace $\ker(L) = \{\mathbf{v} \in V = \mathrm{dom}(L), L(\mathbf{v}) = \mathbf{0} \in W\}$.

  A linear operator L is a *bounded* if there is a constant C such that $\|L(\mathbf{v})\|_W \leq C\|\mathbf{v}\|_V$ for all $\mathbf{v} \in V$. A linear operator is continuous (i.e., if $\mathbf{v}^n \to \mathbf{v}$, then $L(\mathbf{v}^n) \to L(\mathbf{v})$) if and only if it is bounded.

  The subset $\mathcal{B}(V, W)$ of bounded linear operators is a normed linear space with the norm $\|L\|_{\mathcal{B}} = \sup_{\|\mathbf{v}\|_V = 1} \|L(\mathbf{v})\|_W$.

  A bounded operator is *compact* if the image of every bounded sequence contains a convergent sub-sequence. In particular, the identity map I such that $I(\mathbf{v}) = \mathbf{v}$ is not compact on an infinite dimensional space.

  If the image of a linear operator has a finite dimension, the operator is a *finite rank operator*. All the finite rank operators are compact.

  Linear operators in $\mathcal{L}(V, W)$ can be combined linearly by just taking the linear combination of their action on the resulting vectors, which gives a structure of vector space to $\mathcal{L}(V, W)$

itself. The *composite* of two operators L, M in $\mathcal{L}(V)$ is given by their successive applications: $L \circ M : \mathbf{v} \to L(M(\mathbf{v}))$. This can be viewed as a product that makes $\mathcal{L}(V)$ an algebra.
In functional spaces, important linear operators are obtained by combinations of multiplications by functions and partial derivatives. Parenthesis are often omitted when expressing the action of an operator and one writes, for instance, $L\mathbf{v}$ instead of $L(\mathbf{v})$.

- A *matrix* is a linear operator A between finite dimensional vector spaces, which can be represented by a rectangular array of scalars $[a_{ij}]$ (called the elements of the matrix) so that $(A\mathbf{v})_i = \sum_j a_{ij}v_j$. Note that in the context of matrix algebra, one is often not very careful about upper and lower indices. The multiplication by a scalar and the addition of matrices are obvious from the vector space structure of linear operators. The composite of two matrices is given by their successive applications: $C\mathbf{v} = AB\mathbf{v}$ and the corresponding array for C is given by $[c_{ij}] = [\sum_k a_{ik}b_{kj}]$ called the *matrix product*, which is not commutative. If the domain and image vector spaces have the same dimension, the matrix is a *square matrix*. A square matrix I such that $AI = IA = A$ for all the square matrices A of the same dimensions is a *unit matrix*. The *trace* $Tr(A)$ of a square matrix $A = [a_{ij}]$ is the sum of its diagonal elements: $Tr(A) = \sum_i a_{ii}$. If the columns of elements of a square matrix A are considered a set of n vectors of dimension n (they are, in fact, the $A\varepsilon_i$ vectors), a scalar can be built from the matrix via a n-linear form. One considers the totally skew-symmetric n-linear form,[b] which gives 1. when it is fed with the columns of the unit matrix, i.e., the ε_i. The scalar resulting from the action of this form on the columns of the matrix A is called the *determinant* $det(A)$. If the determinant of a matrix A is not equal to zero, the matrix is *regular* and there exists a matrix A^{-1} called the *inverse matrix* of A such that $A^{-1}A = AA^{-1} = I$ and else the matrix is *singular*.

[b]See the section below on n-covectors for further details.

A vector **v** can be considered a single column matrix and the action of a matrix on such a vector is merely a matrix product.

- *Fourier transformation*: The Fourier transformation $\mathcal{F}$ is an extremely important linear operator, which can be defined on $L^2(\mathbb{R}^n)$.
  Given a function $f \in L^2(\mathbb{R}^n) : \mathbf{x} \in \mathbb{R}^n \to \mathbb{R}$ and if the duality product on $\mathbb{R}^n$ is denoted by
  $$< \mathbf{k}, \mathbf{x} >= k_i x^i,$$
  the Fourier transform $\mathcal{F}[f] = \hat{f}$ is given by
  $$\hat{f}(\mathbf{k}) = \frac{1}{(2\pi)^{n/2}} \int_{\mathbb{R}^n} e^{-i<\mathbf{k},\mathbf{x}>} f(\mathbf{x}) d\mathbf{x}.$$
  It is not obvious that the previous integrals exist for all functions in $L^2(\mathbb{R}^n)$. Technically, one has to start with a space where the existence is obvious, $L^1(\mathbb{R}^n)$ or $C_0^\infty(\mathbb{R}^n)$, and then one extends the operator to $L^2(\mathbb{R}^n)$.
  The inverse transform is given by
  $$f(\mathbf{x}) = \frac{1}{(2\pi)^{n/2}} \int_{\mathbb{R}^n} e^{i<\mathbf{k},\mathbf{x}>} \hat{f}(\mathbf{k}) d\mathbf{k}.$$
  One of the fundamental properties of the Fourier transformation is that it makes differentiation algebraic:
  $$\widehat{D^\alpha u}(\mathbf{k}) = \mathbf{k}^\alpha \widehat{u}(\mathbf{k}) \text{ and } \widehat{\mathbf{x}^\alpha u}(\mathbf{k}) = D^\alpha \widehat{u}(\mathbf{k})$$
  where $\mathbf{x}^\alpha = (x^1)^{\alpha_1} \cdots (x^n)^{\alpha_n}$ and $\mathbf{k}^\alpha = k_1^{\alpha_1} \cdots k_n^{\alpha_n}$.
  A fundamental property is the *Parseval–Plancherel theorem*: The Fourier transform is an isometry in the L^2 norm, i.e., it conserves the scalar product and the norm:
  $$(f, g)_{L^2} = (\mathcal{F}[f], \mathcal{F}[g])_{L^2}.$$
- *Convolution*: Another useful operation is the convolution. Given two functions $f, g \in L^2(\mathbb{R}^n)$, the convolution product is defined by
  $$f \star g(\mathbf{x}) = \int_{\mathbb{R}^n} f(\mathbf{y}) g(\mathbf{x}-\mathbf{y}) d\mathbf{y} =< f(\mathbf{y}), g(\mathbf{x}-\mathbf{y}) >_{\mathbf{y}}$$
  where the subscript **y** indicates that the integration involved in the duality product is performed along this variable.
  Surprisingly, this operation is commutative and associative: $f \star g = g \star f$ and $(f \star g) \star h = f \star (g \star h) = f \star g \star h$.

The fundamental result relating the Fourier transform and the convolution is the *Faltung theorem*: If f and g are two functions having Fourier transform and such that the convolution exists and is integrable: $\mathcal{F}[f \star g] = \mathcal{F}[f]\mathcal{F}[g]$.

- *Distribution*: The duality in infinite dimensional function space leads to a fundamental tool of mathematical physics: distributions or generalized functions. The space of test functions $\mathcal{D}$ is defined as the set of functions infinitely differentiable on $\mathbb{R}^n$ and with a bounded support, i.e., $C_0^\infty(\mathbb{R}^n)$. Those functions are gentle enough to be integrated on their whole domain and differentiated everywhere as many times you like. The *dual topological space* $\mathcal{D}'$ of continuous (this requirement makes the difference with the algebraic dual), linear forms on $\mathcal{D}$ is the space of distributions. A fundamental example is the Dirac delta distribution (often improperly called Dirac delta function in physics) δ, which associates to a test function $\phi \in \mathcal{D}$ its value at 0: $\delta(\phi) =< \delta, \phi >= \phi(0)$. The space of distributions is larger than the space of test functions and in a sense, contains it. On one hand, with any test function ϕ is associated a distribution D_ϕ such that the action of this distribution on another test function χ is given by $< D_\phi, \chi >= \int_{\mathbb{R}^n} \phi(\mathbf{x})\chi(\mathbf{x})\, d\mathbf{x}$, and on the other hand, there is no test function associated with the Dirac distribution.
Another important distribution is $\text{vp}\{1/\|\mathbf{x}\|\}$ (where vp stands for the *Cauchy principal value*) defined by

$$< \text{vp}\{1/\|\mathbf{x}\|\}, \varphi >= \lim_{\epsilon \to 0} \int_{\mathbb{R}^n - \{\|\mathbf{x}\| < \epsilon\}} \varphi(\mathbf{x})/\|\mathbf{x}\|\, d\mathbf{x}.$$

Nevertheless, there is a common and useful abuse of notation in physics, which writes the action of the Dirac delta distribution and other singular distributions (those which do not correspond to a test function) using the integral symbol $< \delta, \phi >= \int_{\mathbb{R}^n} \delta(\mathbf{x})\phi(\mathbf{x})d\mathbf{x} = \phi(0)$. In the same spirit, one writes $< \delta_{\mathbf{y}}, \phi >= \int_{\mathbb{R}^n} \delta(\mathbf{x} - \mathbf{y})\phi(\mathbf{x})d\mathbf{x} = \phi(\mathbf{y})$.
We will often use the duality product notation to write the integral of the product of two functions f, g on their common domain of definition Ω: $< f, g >= \int_\Omega fg\, d\mathbf{x}$ in the real case and $< f, g >= \int_\Omega f\overline{g}\, d\mathbf{x}$ in the complex case.

Note that this corresponds also to the scalar product on L^2: $(f, g)_{L^2} =< f, g >$.[c]

- Operations such as derivation, Fourier transformation, and convolution can be applied to distributions. As far as the Fourier transformation is concerned, not all the distributions have a transformation and one has to reduce the space of distributions by increasing the space of test functions (see here how the duality plays). The set of rapidly decreasing functions, $\mathcal{S}$, is introduced, i.e., functions that decrease more rapidly to zero than any power of $\|\mathbf{x}\|$ when $\|\mathbf{x}\| \to \pm\infty$. The set of tempered distribution, $\mathcal{S}'$, is the topological dual space of $\mathcal{S}$, i.e., the set of continuous linear functionals on $\mathcal{S}$ and any tempered distribution admit a Fourier transform. Operators on distributions are described by giving the resulting distribution, but this one is itself described through its action on test functions φ. Therefore, one has the following definitions,[d] given only in the case of test functions and distributions on $\mathbb{R}$ for the sake of simplicity:
  - Derivative of a distribution: $< \frac{dD}{dx}, \varphi >= - < D, \frac{d\varphi}{dx} >$. One often introduces the notation $\{dD(x)/dx\}$ to mean that the derivation is taken in the sense of the functions and not of the distributions. For instance, if a function $f : \mathbb{R} \to \mathbb{R}$ is C^∞ except in a single point $x = a$ where there is a discontinuity

$$\lim_{\epsilon \to 0^+} f(a + \epsilon) - \lim_{\epsilon \to 0^-} f(a + \epsilon) = disc_f(a)$$

    (also denoted $[f]_a$), one has $\frac{df}{dx} = \{\frac{df}{dx}\} + disc_f(a)\ \delta(x - a)$.
  - Multiplication of a distribution by a function:

$$< fD, \varphi >=< D, f\varphi > .$$

    Yes, this seemingly trivial operation needs a definition! You can always multiply a distribution by a test function, but you cannot, for instance, multiply a function with a discontinuity at the origin with the Dirac distribution. An

[c] See the remark above on the abuse of notation due to the reflexive nature of the Hilbert spaces.

[d] Which are chosen to match the definitions for functions when the distribution can be associated with a function.

important negative property of distributions is that the product of two distributions does not always exist.

- Fourier transform of a tempered distribution:

$$< \mathcal{F}[D], \varphi >=< D, \mathcal{F}[\varphi] > .$$

For instance, we have $\mathcal{F}[\delta(x)] = \chi_{\mathbb{R}}(k)$ (where χ_Ω is the characteristic function of the set Ω: $\chi_\Omega(x) = 1$ if $x \in \Omega$ and $\chi_\Omega(x) = 0$ if $x \notin \Omega$) and $\mathcal{F}[\text{sgn(x)}] = \frac{i}{\pi}\,\text{pv}\left(\frac{1}{k}\right)$ where $\text{sgn}(x) = \frac{x}{|x|}$.

- Convolution of two distributions:

$$< D(x) \star E(x), \varphi(x) >=< D \otimes E(x, y), \varphi(x + y) >,$$

where the right-hand duality product takes place in $\mathbb{R}^2$ with x and y variables.

Another interesting example of tempered distribution is the *Dirac comb* $Ш(x) = \sum_{n\in\mathbb{Z}} \delta(x - n)$, which is its own Fourier transform $\mathcal{F}[Ш(x)] = Ш(k) = \sum_{n\in\mathbb{Z}} \delta(k - n)$. Introducing this distribution in the Parseval–Plancherel theorem (extended to the duality product between $\mathcal{S}$ and $\mathcal{S}'$) together with a function $\varphi \in \mathcal{S}$ gives the *Poisson summation formula* $\sum_{n\in\mathbb{Z}} \varphi(n) = \sum_{n\in\mathbb{Z}} \widehat{\varphi}(n)$, a useful trick for convergence acceleration.

- *Sobolev spaces*: These are the Banach spaces defined for integers m, p by $W^{m,p}(\Omega) = \{u : u \in L^p(\Omega),$
$D^\alpha u \in L^p(\Omega)$ for $|\alpha| \leq m\}$.
The spaces $W^{m,2}(\Omega) = H^m(\Omega)$ are of particular interest because they are Hilbert spaces with the scalar product defined by $(u, v)_{H^m} = \sum_{|\alpha|\leq m}(D^\alpha u, D^\alpha v)_{L^2}$ and the corresponding norm $\|u\|^2_{H^m} = (u, u)_{H^m}$.
It can be shown that $u \in H^m(\mathbb{R}^n)$ if and only if

$$(1 + k^2)^{m/2}\widehat{u}(\mathbf{k}) \in L^2(\mathbb{R}^n)$$

and that the norm

$$\|u\|'_{H^m} = \left(\int_{\mathbb{R}^n} (1 + k^2)^{m/2}\widehat{u}(\mathbf{k})\, d\mathbf{k}\right)^{1/2}$$

with $k^2 = \sum_i (k_i)^2$ is equivalent to the first defined norm.
The advantage of this definition is that it can be extended to any real value s and, therefore, allows the definition of

negative and/or fractional index spaces H^s. Note that in the case of negative indices, the elements of H^s are not all in $L^2 = H^0$ and these spaces are rather distribution spaces than genuine function spaces.
The *Rellich–Kondrachov theorem* states that we have the embeddings $H^s(\mathbb{R}^n) \subset H^t(\mathbb{R}^n)$ if $s > t$ and that the corresponding inclusion maps are compact .
The *Sobolev lemma* states that

$$H^s(\mathbb{R}^n) \subset C^k(\mathbb{R}^n) \text{ if } s > k + n/2.$$

$H^{-s}(\mathbb{R}^n)$ is the topological dual space of $H^s(\mathbb{R}^n)$ according to the classical duality pairing corresponding to the L^2 scalar product.
Given Ω, the space $H_0^s(\Omega)$ is the closure of $C_0^\infty(\Omega) = \mathcal{D}(\Omega)$ in $H^s(\Omega)$ (i.e., every element in $H_0^s(\Omega)$ is the limit according to the norm of $H^s(\Omega)$ of a sequence of functions in $C_0^\infty(\Omega)$).
If $\partial\Omega$ is "regular enough," $H_0^s(\Omega)$ is the subspace of elements of $H^s(\Omega)$ equal to zero on $\partial\Omega$. In the case $\Omega = \mathbb{R}^n$, $H_0^s(\Omega)$ is the same as $H^s(\Omega)$, but it is a strict subset in the other cases.
$H^{-s}(\Omega)$ is the topological dual of $H_0^s(\Omega)$.
One has the following situation: $H_0^s(\Omega) \subset L^2(\Omega) \subset H^{-s}(\Omega)$ called a *Gelfand triplet* or a *rigged Hilbert space*. The space L^2 is called the *pivot space* and its scalar product provides the duality pairing. $H^{-s}(\Omega)$ can be, without contradiction, "larger" than $H_0^s(\Omega)$ (it is a distribution space) and "equal" since it is isomorphic according to the Riesz theorem.
Another fundamental example of rigged Hilbert space is $\mathcal{D}(\Omega) \subset L^2(\Omega) \subset \mathcal{D}'(\Omega)$ where the "rigging" spaces are not Hilbert ones.

- *Trace theorems* answer the question of the regularity and restriction of functions (and distributions) on the boundary of a domain or another lower dimensional sub-domain. Consider Ω an open domain of $\mathbb{R}^n$ and a piecewise smooth hyper-surface Γ of co-dimension 1 (i.e., of dimension $n-1$) contained in $\overline{\Omega}$ (and in particular which may coincide with the boundary $\partial\Omega$). The *restriction operator* γ is introduced by $\gamma : C^\infty(\overline{\Omega}) \to C^\infty(\Gamma)$, $\gamma(\mathbf{u}) = \mathbf{u}|_\Gamma$. This operator can be extended to a continuous operator on some Sobolev spaces:

If the hyper-surface Γ is either compact or a portion of a hyper-plane, for $s > 1/2$ the operator γ can be extended to a continuous operator $\gamma : H^s(\Omega) \to H^{s-1/2}(\Gamma)$.
Considering the case $s = 1$ as an example, an alternative definition of the space $H^{1/2}(\Gamma)$ can be defined as the quotient space $H^1(\Omega)/H_0^1(\Omega)$ with the quotient norm $\|\mathbf{u}\|_{H^{1/2}(\Gamma)} = \inf_{\{\mathbf{v}\in H^1(\Omega),\gamma(\mathbf{v})=\mathbf{u}\}} \|\mathbf{v}\|_{H^1(\Omega)}$. Accordingly, one has $H^1(\Omega) = H_0^1(\Omega) \oplus H^{1/2}(\Gamma)$ and the trace inequality $\|\mathbf{u}|_\Gamma\|_{H^{1/2}(\Gamma)} \leq C\|\mathbf{u}\|_{H^1(\Omega)}$.

- Given a linear operator $L \in \mathcal{L}(V, V)$, the *adjoint operator* is $L^* \in \mathcal{L}(V^*, V^*)$ such that

$$< L\mathbf{u}, \mathbf{v} > = < \mathbf{u}, L^*\mathbf{v} >, \forall \mathbf{u} \in V, \forall \mathbf{v} \in V^*.$$

In the functional case, given a differential operator L: $C^\infty(\Omega) \to C^\infty(\Omega)$ of order m defined by $L = \sum_{|\alpha|\leq m} a_\alpha(x) D^\alpha$, the *formal adjoint* L^* is such that

$$< L\varphi, \chi > = < \varphi, L^*\chi >, \forall\varphi, \forall\chi \in C_0^\infty(\Omega).$$

Integration by parts shows that $L^*\varphi = \sum_{|\alpha|\leq m} (-1)^{|\alpha|} D^\alpha(a_\alpha\varphi)$.
For real matrices, the adjoint of a matrix $A = [a_{ij}]$ is called the *transposed matrix* $A^T = [a_{ji}]$ where lines are written as columns. For complex matrices, the adjoint is the complex conjugate transposed or Hermitian transposed $A^H = [\overline{a_{ji}}]$. A matrix A is *symmetric* if $A = A^T$ and it is (symmetric) *definite positive* if $(\mathbf{v}, A\mathbf{v}) > 0, \ \forall\mathbf{v} \neq \mathbf{0}$. A matrix A is *Hermitian* if $A = A^H$. The transposed $\mathbf{v}^T$ of a column vector $\mathbf{v}$ is a row vector, i.e., a single row matrix. The matrix product $\mathbf{v}^T\mathbf{w}$ is the canonical scalar product of real finite dimensional vectors. The matrix product $\mathbf{v}\mathbf{w}^T = [v_i w_j]$ resulting in a general matrix is called the *dyadic product* of the two vectors.
An operator L is *self-adjoint* if $L = L^*$. In the case of unbounded operators, physicists often confuse self-adjoint operators with merely Hermitian (in the complex case) or symmetric (in the real case) operators or even with formally self-adjoint operators and so we do in this book. An *extension* B of an operator A of domain $\text{dom}(A)$ is an operator with a domain $\text{dom}(B) \supset \text{dom}(A)$ such that $B|_{\text{dom}(A)} = A$

and this situation is denoted by $B \supset A$. An operator is self-adjoint if $A^* = A$, but the most common situation is $A^* \supset A$, which corresponds to symmetric and Hermitian operators. We will disregard this distinction except in the following simple example. Consider A_0 an operator acting on $H_0^1([a, b])$ (i.e., the set of square integrable functions φ defined on the interval $[a, b]$, with a square integrable derivative, and verifying the boundary conditions $\varphi(a) = \varphi(b) = 0$), defined by $A_0\varphi = i\frac{d\varphi}{dx}$. This operator is Hermitian:

$$< A_0\varphi, \chi >= \int_a^b i\frac{d\varphi}{dx}\overline{\chi}dx = \int_a^b \overline{i\frac{d\chi}{dx}}dx + i[\varphi\overline{\chi}]_a^b =< \varphi, A_0^*\chi >$$

with $A_0^*\chi = \frac{d\chi}{dx}$ for any function $\chi \in H^1([a, b])$, which is obtained because of the boundary conditions on φ. There are no boundary conditions needed on $\chi \in H^1([a, b]) \supset H_0^1([a, b])$ and obviously $A_0^* \supset A_0$. Given $\theta \in [0, 2\pi[$, the operator $A_\theta\varphi = i\frac{d\varphi}{dx}$ with the domain $\text{dom}(A_\theta) = H_\theta^1([a, b]) = \{\varphi \in H^1([a, b]), \varphi(a) = e^{i\theta}\varphi(b)\}$ is now considered, the boundary term becomes

$$i(\varphi(b)\overline{\chi(b)} - \varphi(a)\overline{\chi(a)}) = i\varphi(b)(\overline{\chi(b)} - \overline{e^{-i\theta}\chi(a)}).$$

This term vanishes if and only if $\chi(a) = e^{i\theta}\chi(b)$, i.e., $\chi \in H_\theta^1([a, b])$ and $A_\theta^* = A_\theta$ is a self-adjoint operator. This is, in fact, a one-parameter family of self-adjoint operators with $A_0 \supset A_\theta \supset A_0^*$ but with $A_\theta \neq A_{\theta'}$ for $\theta \neq \theta'$.
Choosing $\chi \in C_0^\infty$ in the duality product avoids any question about the boundary condition and leads to the careless concept of formally self-adjoint operator A such that

$$< A\varphi, \chi >=< \varphi, A\chi > \text{ for any } \chi \in C_0^\infty.$$

One says that $Lf = g$ *weakly* for $f, g \in L^1_{\text{loc}}$ if

$$< g, \varphi >=< f, L^*\varphi > , \forall\varphi \in C_0^\infty(\Omega).$$

- The *Green's function* of an operator L with constant coefficients is a distribution $G(\mathbf{x} - \mathbf{y}) \in \mathcal{D}' \otimes \mathcal{D}'$ such that $L_\mathbf{x}^*G(\mathbf{x} - \mathbf{y}) = \delta(\mathbf{x} - \mathbf{y})$ where the fact that the operator acts on the $\mathbf{x}$ variables is emphasized. One also has to request that $G|_{\partial\Omega} = 0$ or to consider the free space Green's function, but we stay on a rather formal level here and let such considerations on side. The Green's function is usually used to (formally) invert the differential operator. Consider once again $Lf = g$. By means of the fact that the Dirac distribution is the neutral element of the convolution, one has $f(\mathbf{x}) = f \star \delta =< f(\mathbf{y}), \delta(\mathbf{x}-\mathbf{y}) >_{\mathbf{y}} = < f(\mathbf{y}), L^{*}G(\mathbf{x}-\mathbf{y}) >_{\mathbf{y}} =< Lf(\mathbf{y}), G(\mathbf{x}-\mathbf{y}) >_{\mathbf{y}} =< g(\mathbf{y}), G(\mathbf{x}-\mathbf{y}) >_{\mathbf{y}} = g \star G$. The formula $f = g \star G$ is only valid in free space, i.e., if the domain is $\mathbb{R}^n$. In the case of a bounded domain, boundary conditions have to be taken into account as it will be explained later.

- Up to now, the geometrical domains were open sets of $\mathbb{R}^n$. A serious treatment of the geometrical framework of physics requires the concept of *manifold*. A *manifold* M is a set of points that is locally homeomorphic to $\mathbb{R}^n$ in the sense that any neighborhood of a point can be continuously mapped on an open set of $\mathbb{R}^n$ (n is the same for all the points of M and is called the *dimension* of the manifold). With such a mapping, the points in the neighborhood can be distinguished by an ordered set of n real numbers called the *(local) coordinates*. Nevertheless, we cannot hope to be always able to find a set of coordinates that covers the whole manifold at once. Therefore, it is allowed to cover the manifold with several overlapping open sets each endowed with a particular coordinate system. The regularity of the manifold is given by the regularity of the so-called *transition functions*: Considering two coordinate systems defined on overlapping opens sets U and $\tilde{U}$ by the mappings $\varphi : U \rightarrow \mathbb{R}^n$ and $\tilde{\varphi} : \tilde{U} \rightarrow \mathbb{R}^n$, the invertibility of the continuous coordinate mappings induces a transition mapping $\tilde{\varphi}\varphi^{-1} : \varphi(U \cap \tilde{U}) \subset \mathbb{R}^n \rightarrow \tilde{\varphi}(U \cap \tilde{U}) \subset \mathbb{R}^n$. The regularity of the manifold is the one of the transition functions: A differentiable C^m manifold is such that all the transition functions are C^m.

  The $\mathbb{R}^n$ are trivial examples of manifolds, and in fact the only manifold used here is $\mathbb{R}^3$. In this case, why bother about manifolds? Because $\mathbb{R}^n$ can also be considered a vector space, for instance. On the one hand, in "vector space $\mathbb{R}^n$," two elements can be added but not in "manifold $\mathbb{R}^n$" and on the other hand, changing all the n-tuples $\{x^1, \cdots, x^n\}$ in "manifold $\mathbb{R}^n$" to n-tuples $\{y^1 = (x^1)^3, \cdots, y^n = (x^n)^3\}$ is a valid global change of coordinates leaving the manifold unchanged but makes no sense in "vector space $\mathbb{R}^n$"!

- The position of a point in a manifold M of dimension n is given by an ordered set of n numbers $(x^1, \cdots, x^n)$ called the coordinates. Each coordinate can also be viewed as a function on the manifold.

  A *curve* γ is an application from an interval of $\mathbb{R}$ on the manifold M: $\mathbf{r}(t) = (x^1(t), \cdots, x^n(t))$ where t is the parameter. If $f(x^1, \cdots, x^n)$ is a scalar function on the space, the composition of this function with the curve gives a function from $\mathbb{R}$ to $\mathbb{R}$: $f(x^1(t), \cdots, x^n(t))$. The derivation with respect to t of this function gives, applying the chain rule: $\frac{d}{dt} f(x^1(t), \cdots, x^n(t)) = \frac{\partial f}{\partial x^1}\frac{dx^1}{dt} + \cdots + \frac{\partial f}{\partial x^n}\frac{dx^n}{dt}$. This expression can be viewed as the duality product $< df, \mathbf{v}_\gamma >$ of the covector $df = \frac{\partial f}{\partial x^1} dx^1 + \cdots + \frac{\partial f}{\partial x^n} dx^n = \frac{\partial f}{\partial x^i} dx^i$ with the vector $\mathbf{v}_\gamma = \frac{dx^1}{dt}\frac{\partial .}{\partial x^1} + \cdots + \frac{dx^n}{dt}\frac{\partial .}{\partial x^n} = \frac{dx^i}{dt}\frac{\partial .}{\partial x^i}$. The vector $\mathbf{v}_\gamma$ is the tangent to the curve given in the form of a first-order linear differential operator and the covector df is the *differential* of the function. One takes the general definitions:

  - A *vector* (in the geometric sense) at a point of the manifold is a first-order linear differential operator on the functions on the manifold. In a coordinate system $(x^1, \cdots, x^n)$, a vector $\mathbf{v}$ at a point of coordinates $(p^1, \cdots, p^n)$ is represented by a set of n numbers $(v^1, \cdots, v^n)$, the *components* of the vector, so that the result of the action of this vector on a function $f(x^1, \cdots, x^n)$ is the scalar: $\mathbf{v}(f) = v^1 \frac{\partial f}{\partial x^1} + \cdots + v^n \frac{\partial f}{\partial x^n}|_{(x^1=p^1, \cdots, x^n=p^n)}$.
  - In a coordinate system, a basis for the vectors are the partial derivatives with respect to the coordinates, so that a vector can be written $\mathbf{v} = v^1 \frac{\partial}{\partial x^1} + \cdots + v^n \frac{\partial}{\partial x^n}$.
  - A *covector* at a point of the space is a linear form on the vectors at this point.
  - In a coordinate system, a basis for the covectors is the differential of the coordinates so that a covector can be written $\alpha = \alpha_1 dx^1 + \cdots + \alpha_n dx^n$. They form a basis dual

to the partial derivatives: $< dx^i, \frac{\partial.}{\partial x^j} >= \delta^i_j$ and, therefore, $< \mathbf{v}, \alpha >= \alpha_i v^i$.

- A *(co)vector field* is a set of (co)vectors so that with each point of the manifold is associated a (co)vector. In a coordinate system, it is represented by a set of n functions on the coordinates. The result of the action of a covector field on a vector field is a scalar function on the coordinates.
- A covector field is also called a *1-form*.
- The differential of a scalar function is a 1-form.
- In older terminology, vectors were called *contravariant vectors* and covectors were called *covariant vectors*.
- Vector fields and 1-forms have both n components, i.e., they can be represented (at least locally) by sets of n functions of the coordinates. There is a strong temptation to say that the sets of vector fields and 1-forms are n-dimensional vector spaces, but this forgets the fact that the components are functions that are themselves elements of infinite dimensional functional spaces.

- Geometrical tensor spaces are generated by tensor products of vector and covectors. For example, $A = A^i_{\ jk} dx^i \otimes \frac{\partial.}{\partial x^j} \otimes \frac{\partial.}{\partial x^k}$ is a rank 3 tensor, once contravariant and twice covariant and the $A^i_{\ jk}$ are its n^3 components in the x^i coordinate system. A tensor field is a set of tensors so that with each point of the manifold is associated a tensor. A rank k ($k \in \mathbb{N}$) tensor field has n^k components.
- Skew-symmetric tensors play a fundamental role in differential geometry. A *k-covector* ω is a totally skew-symmetric tensor of rank k ($k \in \mathbb{N}$), i.e., it is a k-linear form on vectors such that the swapping of two vectors changes the sign of the resulting scalar: $\omega(\cdots, \mathbf{v}^i, \cdots, \mathbf{v}^j, \cdots) = -\omega(\cdots, \mathbf{v}^j, \cdots, \mathbf{v}^i, \cdots)$. Note that k-covectors are identically zero if $k > n$.

  Given $\{1, \cdots, n\}$, the set of integers from 1 to n, a *permutation* is a bijection $\sigma \in \mathcal{P}_n : \{1, \cdots, n\} \rightarrow \{1, \cdots, n\}$. A *transposition* is a permutation that swaps two elements and leaves the others at the same place. Any permutation can

be decomposed in a finite sequence of transpositions. The *signature* $\varepsilon(\sigma)$ of a permutation is the number equal to 1 if the number of transposition is even and to -1 if it is odd. It does not depend, of course, on the particular set of transpositions used to describe the permutation. The *Levi–Civita symbol* $\varepsilon_{i_1 i_2 \cdots i_n}$ is equal to $\varepsilon(\sigma)$ if the indices are a permutation σ of $\{1, \cdots, n\}$: $\{i_1 = \sigma(1), \cdots, i_n = \sigma(n)\}$ and equal to 0 if some indices are repeated. The general properties of k-covectors is $\omega(\mathbf{v}^1, \cdots, \mathbf{v}^k) = \varepsilon(\sigma)\omega(\mathbf{v}^{\sigma(1)}, \cdots, \mathbf{v}^{\sigma(k)})$.
The dimension of the vector space formed by the k-covectors on a vector space of dimension n is $\binom{n}{k} = \frac{n!}{k!(n-k)!}$.
A *k-form* is a field of k-covectors so that it is a map from the sets of k vector fields on M to the scalar functions on M. Note that k-forms are usually called *differential forms* or *exterior forms*. The vector space of k-forms on a manifold M is denoted by $\bigwedge^k(M)$.
The *exterior product* of $\alpha \in \bigwedge^k(M)$ and $\beta \in \bigwedge^j(M)$ is $\alpha \wedge \beta \in \bigwedge^{k+j}(M)$ such that

$$\alpha \wedge \beta(\mathbf{v}^1, \cdots, \mathbf{v}^{k+j}) =$$
$$\frac{1}{k!j!}\sum_{\sigma \in \mathcal{P}_{k+j}} \varepsilon(\sigma)\alpha(\mathbf{v}^{\sigma(1)}, \cdots, \mathbf{v}^{\sigma(k)})\beta(\mathbf{v}^{\sigma(k+1)}, \cdots, \mathbf{v}^{\sigma(k+j)}).$$

The main properties of the exterior product of forms are:

$\alpha \wedge \beta = (-1)^{jk}\beta \wedge \alpha, \quad \forall \alpha \in \bigwedge^k(M), \beta \in \bigwedge^j(M)$.
$\alpha \wedge (\beta + \gamma) = \alpha \wedge \beta + \alpha \wedge \gamma$, for all forms α, β, γ.
$(\alpha \wedge \beta) \wedge \gamma = \alpha \wedge (\beta \wedge \gamma) = \alpha \wedge \beta \wedge \gamma$, for all forms α, β, γ.

The exterior products of coordinate differentials form a basis for the forms so that any k-form α can be written as:

$$\alpha = \alpha(x^1, \cdots, x^n)_{i_1 \cdots i_k} dx^{i_1} \wedge \cdots \wedge dx^{i_k}.$$

- n-forms have a single component and can be written in the form $f(x^1, \cdots, x^n)dx^1 \wedge \cdots \wedge dx^n$. If f is everywhere different from zero on the manifold, the n-form is called a *volume form*. Given an n-form ω, n vector fields $\mathbf{v}^i$ and a matrix A with constant elements acting on the vector fields as a linear operator, we have $\omega(A\mathbf{v}^1, \cdots, A\mathbf{v}^n) = det(A)\omega(\mathbf{v}^1, \cdots, \mathbf{v}^n)$, which gives the geometrical meaning of the determinant.

- Another fundamental operation is the *exterior derivative* of a form defined by:

$$d\alpha = d\alpha_{i_1\cdots i_k} \wedge dx^{i_1} \wedge \cdots \wedge dx^{i_k} = \frac{\partial \alpha_{i_1\cdots i_k}}{\partial x^i} dx^i \wedge dx^{i_1} \wedge \cdots \wedge dx^{i_k}.$$

In the right hand, member $d\alpha_{i_1\cdots i_k}$ denotes the differential of the component $\alpha_{i_1\cdots i_k}$ (these coefficients are functions of the coordinates) and the implicit summation on repeated indices is used. From this definition, it is clear that

$$d : \bigwedge^{k}(M) \to \bigwedge^{k+1}(M).$$

The exterior derivative of a function is its differential; the exterior derivative of an n-form is zero. The main properties of the exterior derivative of forms are:

d is linear.
$d(\alpha \wedge \beta) = (d\alpha) \wedge \beta + (-1)^k \alpha \wedge d\beta, \quad \forall \alpha \in \bigwedge^k(M)$
(*Leibnitz rule*).
$dd\alpha = 0$, for all forms α.

- Given an open set Σ of dimension $p \leq n$ in a manifold M of dimension n, sufficiently regular, and, for the simplicity of the presentation, which can be covered by a single local coordinate system $\{x^1, \cdots, x^n\}$ so that Σ can be parameterized by $\Sigma : \{x^1(\xi^1, \cdots, \xi^p), \cdots, x^n(\xi^1, \cdots, \xi^p), \forall \{\xi^1, \cdots, \xi^p\} \in \Omega \subset \mathbb{R}^p\}$. Depending on the context, Σ can be, in fact, a *submanifold* (i.e., a manifold of dimension p together with a regular map of this manifold into M), a hyper-surface or a chain that corresponds to different objects from a formal point of view.[e] The *integral* $\int_\Sigma \alpha$ of a p-form α on Σ is defined by:

$$\int_\Sigma \alpha = \int \cdots \int_\Omega \alpha_{i_1\cdots i_p} \det\left(\frac{\partial(x^{i_1}, \cdots, x^{i_p})}{\partial(\xi^1, \cdots, \xi^p)}\right) d\xi^1 \cdots d\xi^p,$$

where the indices $i_1, \cdots, i_p$ run between 1 and n (without repetition), $\det\left(\frac{\partial(x^{i_1}, \cdots, x^{i_p})}{\partial(\xi^1, \cdots, \xi^p)}\right)$ are the Jacobians (i.e., the determinants of the matrices whose elements are the partial derivatives of the x^i with respect to the ξ^j) and

[e] The case $p = 1$ corresponds to a curve.

$d\xi^1 \cdots d\xi^p$ the Lebesgue measure on $\mathbb{R}^p$. This definition is, in fact, independent of the system of coordinates and of the parametrization used in the definition. The extension to the case where several local coordinate systems are necessary is purely technical and is based on the use of a partition of the unity.

The integration of a form is a linear operation. In order to emphasize the duality between the p-dimensional Σ and the p-forms, the abstract notation $< \Sigma, \alpha >= \int_\Sigma \alpha$ can be used. The main property is the *Stokes theorem*. Given a $(p-1)$-form α and an open set Σ of dimension p such that its boundary $\partial\Sigma$ (of dimension $p-1$) is regular enough, we have:

$$\int_\Sigma d\alpha = \int_{\partial\Sigma} \alpha.$$

- The marriage of differential forms and distribution theory leads to the concept of *de Rham current*. First, a test p-form on a smooth manifold M (of dimension n) is a p-form the coefficients of which are $C_0^\infty(M)$ functions, i.e., infinitely differentiable and with a compact support. A p-current C is a continuous linear form on the test $(n-p)$-forms φ. The associated duality product is denoted by $< C, \varphi >$. The set of currents is, therefore, the topological dual of the space of test forms.

  With a p-form α is associated a p-current C_α, denoted as α by abuse of notation, such that $< C_\alpha, \varphi >= \int_M \alpha \wedge \varphi$, for all test $(n-p)$-forms φ.

  The lower dimensional submanifolds or hyper-surfaces give currents similar to the singular distributions[f] with an $(n-p)$-dimensional submanifold Σ is associated a p-current C_Σ, denoted as Σ by abuse of notation, such that

  $$< C_\Sigma, \varphi >= \int_\Sigma \varphi, \text{ for all test } (n-p) \text{ -forms } \varphi.$$

  The product of a p-current C by a C^∞ q-form α (not necessarily with a bounded support) is a $(p+q)$-current

[f]A first technical step is to consider formal linear combinations of those geometrical objects in order to give them a vector space structure.

defined by

$$< C \wedge \alpha, \varphi > = < C, \alpha \wedge \varphi >,$$

for all test $(n-p-q)$-form φ, and $C \wedge \alpha = (-1)^{pq}\alpha \wedge C$. For instance, with a pair (Σ, α), where Σ is an $(n-p)$-dimensional submanifold and α is a $C^\infty(\Sigma)$ q-form (which needs only be defined on the support of Σ), is associated a $(p+q)$-current $C_{\Sigma\wedge\alpha}$, denoted as $\Sigma \wedge \alpha$ by abuse of notation, such that $< C_{\Sigma\wedge\alpha}, \varphi > = \int_\Sigma \alpha \wedge \varphi$, for all test $(n-p-q)$-forms φ.

The exterior derivative of a p-current C is defined by $< dC, \varphi > = (-1)^{p-1} < C, d\varphi >$, for all test $(n-p-1)$-forms φ. If the current is associated with a form, the definition coincides with the former definition of the exterior derivative. In the case of a p-current associated with an $(n-p)$-dimensional manifold Σ, the previous definition via the Stokes theorem gives:

$$< d\Sigma, \varphi > = (-1)^{p-1} < \Sigma, d\varphi > = (-1)^{p-1} < \partial\Sigma, \varphi >$$

hence $d\Sigma = (-1)^{p-1}\partial\Sigma$. For the p-current C and the C^∞ q-form α:

$$d(C \wedge \alpha) = dC \wedge \alpha + (-1)^p C \wedge d\alpha.$$

If Σ is an $(n-1)$-dimensional submanifold (and its associated 1-current) and if ω is a p-form discontinuous on Σ, i.e., the components in any coordinate system are differentiable in the complement of Σ in M except across Σ where they suffer a jump $[\omega]_\Sigma$, then[g] $d\omega = \{d\omega\} + \Sigma \wedge [\omega]_\Sigma$.

- All those geometrical notions, the exterior product, the exterior derivative, and the integration of a form, do not rely on the definition of a scalar product or a norm and are, therefore, purely topological and differential but not metric.
- **Riemmanian spaces**: A scalar product on the tangent vectors of a manifold can be defined as a rank 2 totally covariant symmetric tensor (field) $\mathbf{g}$ called the *metric*. In a coordinate system, this tensor can be written as $\mathbf{g} =$

[g]See the derivation of a discontinuous function in the section on the distributions above for the notation $\{d\omega\}$.

$g_{ij}dx^i \otimes dx^j$ where $g_{ij} = g_{ji}$ (the n^2 coefficients form a positive definite matrix with $n(n-1)/2$ independent values) and the scalar product of two vectors can be written as $(\mathbf{v}, \mathbf{w}) = g_{ij}v^i w^j$.

If the coefficients g_{ij} are considered to form a matrix, the coefficients of the inverse matrix are denoted by g^{ij} (with $g^{ik}g_{kj} = \delta^i_j$) and define a rank 2 totally contravariant symmetric tensor $g^{ij}\frac{\partial}{\partial x^i} \otimes \frac{\partial}{\partial x^j}$.

In the context of differential forms, the metric is mostly involved in the *Hodge star operator* $* : \bigwedge^p(M) \to \bigwedge^{(n-p)}(M)$, which maps p-forms on $(n-p)$-forms. The p- and $(n-p)$-covectors have the same number of components and the map is linear, one-one, and $** = (-1)^{p(n-p)}$. For any p-form α expressed in an arbitrary (covector) basis $\{\varepsilon^{i_1}, \cdots, , \varepsilon^{i_n}\}$ by $\alpha_{j_1\cdots j_p}\varepsilon^{i_1} \wedge \cdots \wedge \varepsilon^{i_p}$, the action of the Hodge operator $*$ is given by

$$*(\alpha_{j_1\cdots j_p}\varepsilon^{i_1} \wedge \cdots \wedge \varepsilon^{i_p}) = \frac{1}{(n-p)!}\varepsilon_{i_1\cdots i_n}|\det[g_{ij}]|^{1/2}\alpha_{j_1\cdots j_p}g^{i_1 j_1}\cdots g^{i_p j_p}\varepsilon^{i_{p+1}} \wedge \cdots \wedge \varepsilon^{i_n}.$$

The Hodge operator allows the definition of a scalar product on the vector spaces $\bigwedge^p(M)$ of p-forms, which makes them Hilbert spaces[h] by setting

$$(\alpha, \beta) = \int_M \alpha \wedge *\beta.$$

The *coderivative* $\delta = (-1)^{n(p+1)+1} * d*$ is the formal adjoint of the exterior derivative[i] since one has $(d\alpha, \beta) = (\alpha, \delta\beta)$ for forms with $C_0^\infty(M)$ components. One has also $\delta\delta = 0$.

The *Laplace–Beltrami operator* Δ (*Laplacian* for short) is defined by $\Delta = (d+\delta)(d+\delta) = \delta d + d\delta$ and is self-adjoint.

- Of course, many of the previous operations are simpler and indeed even trivial in the *three-dimensional Euclidean space* $\mathbb{E}^3$ that is $\mathbb{R}^3$. This is considered a manifold and is equipped with a special metric such that there exist global coordinates

[h] The situation is, in fact, more subtle since it depends on the regularity of the components of the forms as functions of the coordinates, but we consider here that they are all in $L^2(M)$.

[i] Which could have then been denoted by d^*.

called *Cartesian coordinates* $\{x^1 = x, x^2 = y, x^3 = z\}$ where the metric has the form $\mathbf{g} = dx \otimes dx + dy \otimes dy + dz \otimes dz$. In these coordinates, the Hodge operator has the following action:
$*dx = dy \wedge dz, *dy = dz \wedge dx, *dz = dx \wedge dy,$
$*(dx \wedge dy) = dz, *(dz \wedge dx) = dy, *(dy \wedge dz) = dx,$
$*1 = dx \wedge dy \wedge dz, *(dx \wedge dy \wedge dz) = 1.$

- There are, in fact, several mathematical structures that can be considered "natural" descriptions of our three-dimensional perception of "space." $\mathbb{R}^3$ as a bare manifold obviously lacks structures, but defining Cartesian coordinates is too arbitrary since it involves, for instance, the choice of a distinguished point, the origin O of coordinates $\{0, 0, 0\}$. Another candidate is the "vector space $\mathbb{R}^3$," which we note $\mathbb{V}^3$ to make the distinction with the manifold. The addition of two points is now a valid but meaningless operation. A sounder choice is to consider $\mathbb{A}^3$, the affine space associated with $\mathbb{V}^3$ (loosely speaking obtained by forgetting the origin). The physical points are elements of $\mathbb{A}^3$; they cannot be added, but their differences are vectors of $\mathbb{V}^3$ and the elements of $\mathbb{V}^3$ operate on points of $\mathbb{A}^3$ as displacements. The introduction of the Euclidean distance gives the affine Euclidean space, which we still call $\mathbb{E}^3$ and where displacements (and not points) have a length. The distance between two points is then defined. Of course, in practice, an origin point and three mutually orthogonal unit vectors are chosen such that they define a particular Cartesian coordinate system.
- The peculiarities of $\mathbb{E}^3$ allow a simpler setting called *vector analysis*, which takes advantage of the Cartesian coordinates but forget almost all the geometric relevance! We do not want to advocate for giving up vector analysis, but we just want it to be considered as a computational trick rather than a genuine geometrical framework. In $\mathbb{E}^3$, 0-forms are scalar functions $v(x, y, z)$ and 3-forms are pseudo-scalars or *densities* $\rho(x, y, z)dx \wedge dy \wedge dz = \rho *1$. Both fields have only one single component so that they are merged into the single concept of *scalar field* using the Hodge operator. Similarly, the

1-forms $\alpha_x dx + \alpha_y dy + \alpha_z dz$ are the *field intensities* and the 2-forms $\beta_x dy \wedge dz + \beta_y dz \wedge dx + \beta_z dx \wedge dy$ are the *flux densities* and both have three components so that they are merged in the concept of *vector field*. In this case, those vector fields are considered *proxies* for 1-forms and 2-forms.

A vector field is written as $\mathbf{v} = v_x \varepsilon^x + v_y \varepsilon^y + v_y \varepsilon^y$ where the unit vectors ε are as well unit 1-forms as unit 2-forms associated with Cartesian coordinates. This is safe as long as only Cartesian coordinates are used. The scalar product of two vectors is called the *dot product* and is defined by $\mathbf{v} \cdot \mathbf{w} = v_x w_x + v_y w_y + v_z w_z$ and the associated norm is, of course, $|\mathbf{v}|^2 = \mathbf{v} \cdot \mathbf{v}$. Dot product can be traced back in differential geometry as the scalar product of two 1-forms or of two 2-forms but also to the metric free exterior product of a 1-form and a 2-form. This is usually enough to cloud the geometrical meaning of vector analysis computations! The *cross product* of two vectors is defined by

$$\mathbf{v} \times \mathbf{w} = (v_y w_z - v_z w_y)\varepsilon^x + (v_z w_x - v_x w_z)\varepsilon^y + (v_x w_y - v_y w_x)\varepsilon^z.$$

It comes mostly from the exterior product of two 1-forms.[j]

The exterior derivative gives rise to several operators.

The exterior derivative of a 0-form $df = \frac{\partial f}{\partial x}dx + \frac{\partial f}{\partial y}dy + \frac{\partial f}{\partial z}dz$ corresponds to the *gradient* of a scalar field:

$$\mathbf{grad}\ f = \frac{\partial f}{\partial x}\varepsilon^x + \frac{\partial f}{\partial y}\varepsilon^y + \frac{\partial f}{\partial z}\varepsilon^z.$$

For a 1-form $\mathbf{v} = v_x dx + v_y dy + v_z dz$, the exterior derivative $d\mathbf{v} = (\frac{\partial v_z}{\partial y} - \frac{\partial v_y}{\partial z})dy \wedge dz + (\frac{\partial v_x}{\partial z} - \frac{\partial v_z}{\partial x})dz \wedge dx + (\frac{\partial v_y}{\partial x} - \frac{\partial v_x}{\partial y})dx \wedge dy$ corresponds to the *curl* of a vector field:

$$\mathbf{curl}\,\mathbf{v} = \left(\frac{\partial v_z}{\partial y} - \frac{\partial v_y}{\partial z}\right)\varepsilon^x + \left(\frac{\partial v_x}{\partial z} - \frac{\partial v_z}{\partial x}\right)\varepsilon^y + \left(\frac{\partial v_y}{\partial x} - \frac{\partial v_x}{\partial y}\right)\varepsilon^z.$$

For a 2-form $\mathbf{w} = w_x dy \wedge dz + w_y dz \wedge dx + w_z dx \wedge dy$, the exterior derivative $d\mathbf{w} = (\frac{\partial w_x}{\partial w} + \frac{\partial w_y}{\partial y} + \frac{\partial w_z}{\partial z})dx \wedge dy \wedge dz$ corresponds to the *divergence* of a vector field:

$$\text{div}\ \mathbf{w} = \frac{\partial w_x}{\partial w} + \frac{\partial w_y}{\partial y} + \frac{\partial w_z}{\partial z}.$$

[j]There are many other features in differential geometry not introduced here, such as the Lie derivative and the inner product. In electromagnetism, the cross product of the velocity together with the magnetic flux density in the Lorentz force is, in fact, the inner product of a vector with a 2-form.

Alternative notations for these operations use the nabla operator ∇: $\mathbf{grad}\ f = \nabla f$, $\mathbf{curl}\,\mathbf{v} = \nabla \times \mathbf{v}$, and $\text{div}\ \mathbf{v} = \nabla \cdot \mathbf{v}$. The Leibnitz rule is expressed in the classical formulae:

$$\begin{aligned}
\mathbf{grad}(fg) &= f\ \mathbf{grad}\, g + g\ \ \mathbf{grad}\ f. \\
\mathbf{curl}(f\mathbf{v}) &= f\ \mathbf{curl}\,\mathbf{v} - \mathbf{v} \times \mathbf{grad}\ f. \\
\text{div}(f\mathbf{v}) &= f\ \text{div}\,\mathbf{v} + \mathbf{v} \cdot \mathbf{grad}\ f. \\
\text{div}(\mathbf{v} \times \mathbf{w}) &= \mathbf{w} \cdot \mathbf{curl}\,\mathbf{v} - \mathbf{v} \cdot \mathbf{curl}\,\mathbf{w}.
\end{aligned}$$

and of course $dd = 0$ is nothing else than $\mathbf{curl}\ \mathbf{grad} = \mathbf{0}$ and $\text{div}\ \mathbf{curl} = 0$.

The Laplacian of a scalar field is

$$\Delta f = \text{div}\ \ \mathbf{grad}\ f = \frac{\partial^2 f}{\partial x^2} + \frac{\partial^2 f}{\partial y^2} + \frac{\partial^2 f}{\partial z^2}.$$

The Laplacian of a vector field is

$$\Delta \mathbf{v} = \mathbf{grad}\ \ \text{div}\,\mathbf{v} - \mathbf{curl}\,\mathbf{curl}\,\mathbf{v} = (\Delta v_x)\varepsilon^x + (\Delta v_y)\varepsilon^y + (\Delta v_z)\varepsilon^z.$$

Note that Laplacians are self-adjoint operators.

In vector analysis, a (rank 2) tensor (field) is a linear operator transforming a vector (field) into a vector (field). In a particular system of coordinates, it is usually given in the form of an array of 9 coefficients, hence the common confusion with a square matrix. A "vector analysis" tensor is denoted by $\underline{\underline{\alpha}}$ and its action on a vector $\mathbf{v}$ is simply denoted by $\underline{\underline{\alpha}}\mathbf{v}$.

The Stokes theorem corresponds to various integral equalities:

If Γ is a curve with initial point $\mathbf{a}$ and end point $\mathbf{b}$, for any scalar field whose gradient exists:

$$\int_\Gamma \mathbf{grad}\ f \cdot d\mathbf{l} = f(\mathbf{b}) - f(\mathbf{a})$$

where $d\mathbf{l}$ is the line element.

If Σ is a surface with boundary $\partial\Sigma$ (with an orientation inherited from Σ), for any vector field $\mathbf{v}$ whose curl exists:

$$\iint_\Sigma \mathbf{curl}\,\mathbf{v} \cdot \mathbf{n}\, ds = \int_{\partial\Sigma} \mathbf{v} \cdot d\mathbf{l}$$

where ds is the surface element and $\mathbf{n}$ is the normal vector on Σ.

If V is a volume with boundary ∂V, for any vector field $\mathbf{v}$ whose divergence exists:

$$\iiint_V \operatorname{div} \mathbf{v}\, d\mathbf{v} = \iint_{\partial V} \mathbf{v} \cdot \mathbf{n}\, ds$$

where $d\mathbf{v}$ is the volume element and $\mathbf{n}$ is the outer normal vector on ∂V.

Often, multiple integrals are simply denoted by a single $\int$ just as in differential geometry.

The simplest way to give natural definitions of all the integrals involved here above is to go back to the definition of the integration of differential forms. *Moreover, the traditional notations of vector analysis rely on metric concepts such as the normal vector (involving both orthogonality and unit length) and the scalar product while the definition of these integrals do not require any metric.*

Given a unit vector ε, the *directional derivative* $\frac{\partial \cdot}{\partial \varepsilon}$ is defined by $\frac{\partial f}{\partial \varepsilon} = \mathbf{grad}\ f \cdot \varepsilon$. Another fundamental integral identity is the *Green's formula*:

$$\iiint_V (f\Delta g - g\Delta f) d\mathbf{v} = \iint_{\partial V} \left(f\frac{\partial g}{\partial \mathbf{n}} - g\frac{\partial f}{\partial \mathbf{n}} \right) ds\,.$$

- Using the Leibnitz rule to integrate by parts, it is easy to find the formal adjoint to the vector analysis operators:

$$\mathbf{grad}^* = -\operatorname{div},\ \mathbf{curl}^* = \mathbf{curl},\ \text{and}\ \operatorname{div}^* = -\,\mathbf{grad}\,.$$

- The identity $d\omega = \{d\omega\} + \Sigma \wedge [\omega]_\Sigma$ for de Rham currents leads to the following identities in vector analysis in the case of a function f and a vector field $\mathbf{v}$ differentiable in the complement of a surface Σ but undergoing the jumps $[f]_\Sigma$ and $[\mathbf{v}]_\Sigma$ across Σ, respectively (the direction chosen to cross the surface is given by $\mathbf{n}$, the normal vector to the surface Σ):

$$\begin{aligned} \mathbf{grad}\, f &= \{\mathbf{grad}\, f\} + \mathbf{n}[f]_\Sigma \delta_\Sigma. \\ \mathbf{curl}\, \mathbf{v} &= \{\mathbf{curl}\, \mathbf{v}\} + \mathbf{n} \times [\mathbf{v}]_\Sigma \delta_\Sigma. \\ \operatorname{div}\, \mathbf{v} &= \{\operatorname{div}\, \mathbf{v}\} + \mathbf{n} \cdot [\mathbf{v}]_\Sigma \delta_\Sigma. \end{aligned}$$

where it is necessary to denote explicitly the singular distribution δ_Σ associated with the surface Σ (by definition $\int_{\mathbb{R}} \delta_\Sigma \varphi(\mathbf{x}) d\mathbf{x} = \varphi|_\Sigma$). Note that metric concepts creep again in a place where they are not necessary.

- $L^2(\mathbb{R}^3, \mathbb{C}^3) = [L^2(\mathbb{R}^3)]^3 = \mathbb{L}^2(\mathbb{R}^3)$ denotes the space of square integrable functions on $\mathbb{R}^3$ with values in $\mathbb{C}^3$ where the scalar product is defined by $(\mathbf{v}, \mathbf{w}) = \int_{\mathbb{R}^3} \mathbf{v}(\mathbf{x}) \cdot \overline{\mathbf{w}(\mathbf{x})} d\mathbf{x}$.
- The Green's function G of three-dimensional scalar Laplacian (in free space) is such that $\Delta_{\mathbf{p}} G(\mathbf{p} - \mathbf{q}) = \delta(\mathbf{p} - \mathbf{q})$ where $\mathbf{p}, \mathbf{q} \in \mathbb{E}^3$ and $\Delta_{\mathbf{p}}$ indicates that the derivatives in the Laplacian are taken with respect to the coordinates of $\mathbf{p}$ (but this is seldom explicitly indicated when there is no ambiguity). The expression of this Green's function is $G(\mathbf{p} - \mathbf{q}) = \frac{-1}{4\pi|\mathbf{p}-\mathbf{q}|}$. The displacement vector $\mathbf{p} - \mathbf{q}$ is traditionally called $\mathbf{r}$ and its norm r and one can be written $G = \frac{-1}{4\pi r}$.
  In the two-dimensional scalar Laplacian case, $G = \frac{1}{2\pi} \ln(r)$.
  For the scalar Helmholtz equation, one has $(\Delta + k^2)H = \delta$ and $H = -\frac{e^{ikr}}{4\pi r}$.
  In the two-dimensional scalar Helmholtz equation case, $H = \frac{1}{4i} H_0^{(1)}(kr)$ where $H_0^{(1)}$ denotes a Hankel function.
  The Green's function is physically interpreted as the field (or potential) generated by a (monopolar) point source. The introduction of Green's functions in Green's formula gives identities, which are the basis of the *boundary element method*.
- The differential geometry on $\mathbb{R}^2$ provides a nice framework to introduce *complex analysis* on $\mathbb{C}$. Given two Cartesian coordinates x and y and the Euclidean metric $\mathbf{g} = dx \otimes dx + dy \otimes dy$, the associated Hodge star operator has the following action on 1-forms: $*dx = dy$ and $*dy = -dx$. This action is a $\pi/2$ rotation counterclockwise equivalent to the action of a multiplication by i in the complex plane. The knowledge of this Hodge star on the plane does not determine the Euclidean geometry but only the conformal geometry, i.e., the metric up to a scalar factor so that only angles but not lengths are relevant. In the complex plane, the complex variable $z = x + iy$ and its complex conjugate $\overline{z} = x - iy$ play the role of independent variables. Their differentials $dz = dx + idy$ and $d\overline{z} = dx - idy$ are a basis for the 1-forms. The dual basis for the vectors are the differential

operators $\frac{\partial}{\partial z} = \frac{1}{2}(\frac{\partial}{\partial x} - i\frac{\partial}{\partial y})$ and $\frac{\partial}{\partial \bar{z}} = \frac{1}{2}(\frac{\partial}{\partial x} + i\frac{\partial}{\partial y})$. Note that $\frac{\partial}{\partial \bar{z}}\frac{\partial}{\partial z} = \frac{\partial}{\partial z}\frac{\partial}{\partial \bar{z}} = \frac{1}{4}\Delta$ and $dz \wedge d\bar{z} = -2i(dx \wedge dy)$.

A complex function of the complex variable $f : \mathbb{C} \to \mathbb{C}$ is associated with a pair (p, q) of real functions on $\mathbb{R}^2$ so that $f(z, \bar{z}) = p(x, y) + iq(x, y)$. Those complex functions are usually denoted by $f(z)$, which is fully justified in the case of holomorphic functions as explained below.

Given two complex functions f and g, the exterior derivative of the 0-form f is

$$df = \frac{\partial f}{\partial z}dz + \frac{\partial f}{\partial \bar{z}}d\bar{z} = \left(\frac{\partial p}{\partial x} + i\frac{\partial q}{\partial x}\right)dx + \left(\frac{\partial p}{\partial y} + i\frac{\partial q}{\partial y}\right)dy$$

and the exterior derivative of the 1-form $f\,dz + g\,d\bar{z}$ is

$$d(f\,dz + g\,d\bar{z}) = \left(\frac{\partial g}{\partial z} - \frac{\partial f}{\partial \bar{z}}\right)dz \wedge d\bar{z}.$$

Let γ be a closed curve in $\mathbb{C}$ (and, therefore, also in $\mathbb{R}^2$), i.e., a map $\gamma : t \in [a, b] \subset \mathbb{R} \to z(t) \in \mathbb{C}$ such that $\gamma(a) = \gamma(b)$, and such that there are no other multiple points. The line integral of the complex 1-form $f dz$ associated with f (note that it is not a general 1-form since there is no term in $d\bar{z}$) is $\int_\gamma f dz = \int_\gamma (p\,dx - q\,dy) + i\int_\gamma (p\,dy + q\,dx)$. The exterior derivative is $d(f dz) = -\frac{\partial f}{\partial \bar{z}}\,dz \wedge d\bar{z}$.

If D is a bounded domain of $\mathbb{C}$ (and, therefore, also of $\mathbb{R}^2$), the domain integral of a complex 2-form $f\,dz \wedge d\bar{z}$ is

$$\int_D f\,dz \wedge d\bar{z} = -2\int_D p\,dx \wedge dy + 2i\int_D q\,dx \wedge dy.$$

In this case, the Stokes theorem becomes $\int_D d(f dz) = \int_{\partial D} f dz$. As for a 1-form $f\,d\bar{z}$, one has

$$\int_D d(f d\bar{z}) = \int_{\partial D} f d\bar{z}$$

where $\int_\gamma f d\bar{z} = \int_\gamma (p\,dx + q\,dy) + i\int_\gamma (q\,dx - p\,dy)$.

A function $f(z)$ is *holomorphic* or *analytic* on a topologically simple domain (contractible to a point) $D \subset \mathbb{C}$ if one of the following five equivalent conditions are satisfied:

(1) If at each point of D, f is $\mathbb{C}$-differentiable, i.e., if the limit $f'(z) = \lim_{|h|\to 0}\frac{f(z+h)-f(z)}{h}$ exists and is independent of the direction of $h \in \mathbb{C}$ (Fréchet derivative). In this case $f'(z) = \frac{\partial f(z)}{\partial z} = \frac{1}{2}(\frac{\partial p}{\partial x} + \frac{\partial q}{\partial y}) - \frac{i}{2}(\frac{\partial p}{\partial y} - \frac{\partial q}{\partial x})$.

(2) If the 1-form $f dz$ is closed, i.e., its exterior derivative vanishes, on D: $d(f\ dz) = -\frac{\partial f}{\partial \bar{z}}\ dz = 0$ at each point of D, i.e., if the *Cauchy–Riemann conditions* $\frac{\partial p}{\partial x} - \frac{\partial q}{\partial y} = 0$ and $\frac{\partial p}{\partial y} + \frac{\partial q}{\partial x} = 0$ are verified. This condition means that $f(z)$ behaves as if it were independent of $\bar{z}$. The operator $\frac{\partial}{\partial \bar{z}}$ is called the Cauchy–Riemann operator, and since it is a factor of the Laplacian, a holomorphic function f is also a harmonic function, i.e., $\Delta f = 0$.

(3) If $df = g\ dz$, i.e., if the exterior derivative of f has no $d\bar{z}$ component.

(4) If the line integrals $\int_\gamma f\ dz = 0$ for any closed curve γ contained together with its interior in D. As any closed curve can be defined as a boundary $\partial\Omega$ with $\Omega \subset D$, by the Stokes theorem $\int_{\partial\Omega} f dz = \int_\Omega d(f dz) = 0$.

(5) If f can be represented by an infinite power series $f(z) = \sum_{n=0}^{\infty} a_n (z - z_0)^n$ in a neighborhood of any point of z_0 of D.

In the case of the function $1/z$ on a domain Ω, including the origin $z = 0$, this function is holomorphic everywhere on Ω except at the origin where there is a singularity called a *pole* of order 1. Such a function, which is holomorphic except for a discrete set of points, is called *meromorphic*. The fact that the domain in which the function is holomorphic is no more topologically simple has important consequences: Consider for instance a disc $D(r, 0)$ of radius r and center $z = 0$. Using the polar representation $z = re^{i\theta}$, $dz = ire^{i\theta} d\theta$ on $\partial D(r, 0)$ and one has, for any $n \in \mathbb{Z}$,

$$\int_{\partial D(r,0)} z^n dz = \int_0^{2\pi} e^{in\theta} r^n ire^{i\theta} d\theta = ir^{n+1} \int_0^{2\pi} e^{i(n+1)\theta} d\theta = 2i\pi$$

if $n = -1$ and 0 if $n \neq -1$. As $1/z$ is holomorphic on any domain not including the origin, its domain integral is null and this fact can be used to easily prove that $\int_{\partial\Omega} \frac{1}{z} dz = 2i\pi$ for any Ω, including the pole $z = 0$. A holomorphic function f on Ω, including the origin $z = 0$, can be written as an infinite power series $f(z) = \sum_{n=0}^{\infty} a_n z^n$ where the z^n are all holomorphic functions and where $a_0 = f(0)$. Therefore, $\int_{\partial\Omega} \frac{f(z)}{z} dz = \int_{\partial\Omega} \sum_{n=0}^{\infty} a_n z^{n-1} dz = 2i\pi f(0)$. It is easy to shift the pole at any ζ inside a domain Ω where f is holomorphic to obtain the *Cauchy's integral formula* or *Cauchy theorem*:

$$f(\zeta) = \frac{1}{2i\pi} \int_{\partial\Omega} \frac{f(z)dz}{z-\zeta}.$$

If the function f is meromorphic, it can be represented by a *Laurent series* $f(z) = \sum_{n=-p}^{\infty} a_n (z-z_0)^n$ (involving negative powers of z) in a neighborhood D of a pole z_0 of order p. The coefficient a_{-1} denoted by $R(z_0; f)$ is called the *residue* and we have: $R(z_0; f) = \lim_{z\to z_0}(z-z_0)f(z) = \frac{1}{2i\pi}\int_\gamma f(z)dz$ for a curve γ such that it is contained in D, and that z_0 is the only pole contained in its interior.

Given two complex functions f and g on D, a duality product $< f, g >= \int_D f\overline{g}\, dz \wedge d\overline{z}$ can be introduced so that $-\frac{\partial}{\partial \overline{z}}$ is the formal adjoint of $\frac{\partial}{\partial z}$ since

$$< \frac{\partial f}{\partial z}, g >= - < f, \frac{\partial g}{\partial \overline{z}} > + \int_{\partial D} f\overline{g}\, d\overline{z}.$$

The Green's function of $\frac{\partial}{\partial z}$ is given by $\frac{\partial}{\partial \overline{z}}\frac{1}{\pi z} = \delta$, where δ is the Dirac distribution on $\mathbb{R}^2$, and it can be justified by $\int_{\partial D} f \frac{1}{2i\pi z} dz = f(0) = \int_D f\delta \frac{dz\wedge d\overline{z}}{-2i} = \int_D f \frac{\partial g}{\partial \overline{z}} \frac{dz\wedge d\overline{z}}{-2i} = \frac{1}{2i}\int_D d(fg\ dz) = \frac{1}{2i}\int_{\partial D} fg\ dz$. The theory of hyperfunctions is a distribution theory on $\mathbb{C}$ where test functions are holomorphic.

- *Landau notation*: Given x_0 and two functions f and g defined in a neighborhood of $x_0 \in \overline{\mathbb{R}}$ (i.e., real numbers, including the cases $x_0 \to \pm\infty$), $f = O(g)$ if and only if $\frac{f(x)}{g(x)} = O(1)$, which means that $|\frac{f(x)}{g(x)}|$ stays bounded in a neighborhood of x_0. Moreover, $f = o(g)$ if and only if $\frac{f(x)}{g(x)} = o(1)$, which means that $\lim_{x\to x_0} \frac{f(x)}{g(x)} = 0$. It is legitimate to write, say, $2x = O(x) = o(x^2)$ for $x \to \infty$ with the understanding that we are using the equality sign in an unsymmetrical (and informal) way, in that we do not have, for example, $o(x^2) = O(x)$.

  Landau notation is a practical tool to give asymptotic behaviors, e.g., for $\nu \in \mathbb{R}$,

$$J_\nu(x) = \sqrt{\frac{2}{\pi x}} \cos\left(x - (2\nu+1)\frac{\pi}{4}\right) + O\left(\frac{1}{x\sqrt{x}}\right) \text{ as } x \to +\infty.$$

Landau notation is also handy in computer science, e.g., in describing the efficiency of an algorithm. It is common to say that an algorithm requires $O(n^3)$ steps, e.g., without needing to specify exactly what is a step; for if $f = O(n^3)$, then $f = O(An^3)$ for any positive constant A.

- Perfect notations probably do not exist since absolutely rigorous ones should be intractable! It is difficult to avoid ambiguities and collisions. For instance, using $\star$ for the convolution, $*$ for the Hodge star operator, and the upper index * for algebraic duals may not be appreciated by presbyopic readers. Adopting nonambiguous but nonstandard notations may be a solution to avoid collisions. The danger is to lose the reader in a cumbersome formal deciphering game. Therefore, we prefer to leave some ambiguities, which may be removed by understanding.

  Moreover, we leave on side with regret some important issues that are not explicitly used here, such as the orientation of manifolds and twisted forms, the Hodge orthogonal decomposition theorem for forms that generalizes the Helmholtz decomposition theorem for vector fields.

  There is, of course, a huge number of books on the mathematical tools for physics, but almost everything in this appendix can be found in a more detailed and rigorous version in the formidable Ref. (Choquet-Bruhat et al., 1982). The classical reference for functional analysis is Ref. (Yosida, 1980), but a more readable book for the physicist interested in the functional analysis for partial differential equations is Ref. (Folland, 1995) and a concise introduction to basic analysis and functional analysis is Ref. (Friedman, 1982). One of the best reference for distribution theory, including de Rham currents, is still Ref. (Schwartz, 1966). There are many good books on geometrical methods in physics, such as the very pedagogical Ref. (Bamberg and Sternberg, 1991) or the quite comprehensive Ref. (Nakahara, 1990) but a very concise presentation aimed at electromagnetism is Ref. (Bossavit, 1991).

References

Bamberg, P. and Sternberg, S. (1991). *A Course in Mathematics for Students of Physics: vol. 1 and vol. 2* (Cambridge University Press, Cambridge).

Bossavit, A. (1991). *Notions de géométrie différentielle pour l'étude des courants de Foucault et des méthodes numériques en électromagnétisme*, Vol. Méthodes numériques en électromagnétisme (A. Bossavit, C. Emson, I. Mayergoyz) (Eyrolles, Paris), see also the English translation httpee also the English translation http://www.lgep.supelec.fr/mse/perso/ab/DGSNME.pdf.

Choquet-Bruhat, Y., Witt-Morette, C. D., and Dillard-Bleick, M. (1982). *Analysis, Manifolds and Physics*, revised edn. (Elsevier, Amsterdam).

Folland, G. B. (1995). *Introduction to Partial Differential Equations* (Princeton University Press, Princeton).

Friedman, A. (1982). *Foundation of Modern Analysis* (Dover, New York).

Nakahara, M. (1990). *Geometry, Topology and Physics* (IOP, Bristol).

Schwartz, L. (1966). *Théorie des Distributions*, 3rd edn. (Hermann, Paris).

Yosida, K. (1980). *Functional Analysis*, 6th edn. (Springer, Berlin).

Index